Series in Biomedical Engineering

Series Editor: Joachim H. Nagel

Series in Biomedical Engineering

Editor-in-Chief

Prof. Dr. Joachim H. Nagel
Institute of Biomedical Engineering
University of Stuttgart
Seidenstrasse 36
70174 Stuttgart
Germany
E-mail: jn@bmt.uni-stuttgart.de

The International Federation for Medical and Biological Engineering, IFMBE, is a federation of national and transnational organizations representing internationally the interests of medical and biological engineering and sciences. The IFMBE is a non-profit organization fostering the creation, dissemination and application of medical and biological engineering knowledge and the management of technology for improved health and quality of life. Its activities include participation in the formulation of public policy and the dissemination of information through publications and forums. Within the field of medical, clinical, and biological engineering, IFMBE's aims are to encourage research and the application of knowledge, and to disseminate information and promote collaboration. The objectives of the IFMBE are scientific, technological, literary, and educational.

The IFMBE is a WHO accredited NGO covering the full range of biomedical and clinical engineering, healthcare, healthcare technology and management. It is representing through its 58 member societies some 120.000 professionals involved in the various issues of improved health and healthcare delivery.

Previous Editions:

Spaan, J. et al: BIOMED, Biopacemaking, 2007, ISBN 978-3-540-72109-3

Sekihara, K., Nagarajan, S.: BIOMED, Adaptive Spatial Filters for Electromagnetic Brain Imaging, 2008, ISBN 978-3-540-79369-4

Madhavan, G., Oakley, B.; Kun, L.(Eds.): BIOMED, Career Development in Bioengineering and Biotechnology, 2009, ISBN 978-0-387-76494-8

Boris Rubinsky (Ed.)

Irreversible Electroporation

Springer

Prof. Boris Rubinsky
University of California, Berkeley
Dept. Mechanical Engineering
6124 Etcheverry Hall
Berkeley, CA 94720-1740
USA
E-mail: rubinsky@me.berkeley.edu

and

Hebrew University of Jerusalem
"Center for Bioengineering in the
Service of Humanity and Society"
Givat Ram Campus
78 Ross Bldg
Jerusalem, Israel
E-mail: rubinsky@cs.huji.ac.il

ISBN 978-3-642-26199-2 e-ISBN 978-3-642-05420-4

DOI 10.1007/978-3-642-05420-4

Series in Biomedical Engineering ISSN 1864-5763

Softcover reprint of the hardcover 1st edition 2009

Typeset & Cover Design: Scientific Publishing Services Pvt. Ltd., Chennai, India.

Printed in acid-free paper

9 8 7 6 5 4 3 2 1

springer.com

IFMBE

The International Federation for Medical and Biological Engineering (IFMBE) was established in 1959 to provide medical and biological engineering with a vehicle for international collaboration in research and practice of the profession. The Federation has a long history of encouraging and promoting international cooperation and collaboration in the use of science and engineering for improving health and quality of life.

The IFMBE is an organization with membership of national and transnational societies and an International Academy. At present there are 53 national members and 5 transnational members representing a total membership in excess of 120 000 professionals worldwide. An observer category is provided to groups or organizations considering formal affiliation. Personal membership is possible for individuals living in countries without a member society. The IFMBE International Academy includes individuals who have been recognized for their outstanding contributions to biomedical engineering.

Objectives

The objectives of the International Federation for Medical and Biological Engineering are scientific, technological, literary, and educational.Within the field of medical, clinical and biological engineering its aims are to encourage research and the application of knowledge, to disseminate information and promote collaboration.

In pursuit of these aims the Federation engages in the following activities: sponsorship of national and international meetings, publication of official journals, cooperation with other societies and organizations, appointment of commissions on special problems, awarding of prizes and distinctions, establishment of professional standards and ethics within the field as well as other activities which in the opinion of the General Assembly or the Administrative Council would further the cause of medical, clinical or biological engineering. It promotes the formation of regional, national, international or specialized societies, groups or boards, the coordination of bibliographic or informational services and the improvement of standards in terminology, equipment, methods and safety practices, and the delivery of health care.

The Federation works to promote improved communication and understanding in the world community of engineering, medicine and biology.

Activities

Publications of the IFMBE include: the journal Medical and Biological Engineering and Computing, the electronic magazine IFMBE News, and the Book Series on Biomedical Engineering. In cooperation with its international and regional conferences, IFMBE also publishes the IFMBE Proceedings Series. All publications of the IFMBE are published by Springer Verlag.

Every three years the IFMBE hosts a World Congress on Medical Physics and Biomedical Engineering in cooperation with the IOMP and the IUPESM. In addition, annual, milestone and regional conferences are organized in different regions of the world, such as Asia Pacific, Europe, the Nordic-Baltic and Mediterranean regions, Africa and Latin America.

The administrative council of the IFMBE meets once a year and is the steering body for the IFMBE. The council is subject to the rulings of the General Assembly, which meets every three years.

Information on the activities of the IFMBE are found on its web site at: http://www.ifmbe.org.

Preface

Non-thermal irreversible electroporation is a new minimally invasive surgical procedure with unique molecular selectivity attributes – in fact it may be considered the first clinical molecular surgery procedure. Non-thermal irreversible electroporation is a molecular selective mode of cell ablation that employs brief electrical fields to produce nanoscale defects in the cell membrane, which can lead to cell death, without an effect on any of the other tissue molecules. The electrical fields can be produced through contact by insertion of electrode needles around the undesirable tissue and non-invasively by electromagnetic induction. This new addition to the medical armamentarium requires the active involvement and is of interest to clinical physicians, medical researchers, mechanical engineers, chemical engineers, electrical engineers, instrumentation designers, medical companies and many other fields and disciplines that were never exposed in their training to irreversible electroporation or to a similar concept. This edited book is designed to be a comprehensive introduction to the field of irreversible electroporation to those that were not exposed or trained in the field before and can also serve as a reference manual. Irreversible electroporation is broad and interdisciplinary. Therefore, we have made an attempt to cover every one of the various aspects of the field from an introductory basic level to state of the art. Readers will find in this book enough information to begin applying their specific training to the field of irreversible electroporation and also to understand at a basic level aspects of this interdisciplinary field in which they are not trained.

The book is divided into three parts. The first part is a compilation of background knowledge required in order to work in the field of irreversible electroporation. It consists of chapters one to four, which provide a basic introduction to all fundamental aspects of irreversible electroporation, with particular emphasize on the cellular and molecular aspects. The second part contains recently gained knowledge on non-thermal irreversible. It is made of chapters five to nine and deals with the macroscopic aspect of non-thermal irreversible electroporation. The section is emphasizing aspects that make this field unique and new, in particular the effect of irreversible electroporation on tissues and treatment planning. The third part of the book is practical engineering design and the most recent clinical applications. Chapters ten to twelve provide an insight into the engineering technology and clinical applications. Chapter thirteen is stand-alone and was added to provide a different perspective on irreversible electroporation. In the past the only practical application of irreversible application was in the food industry. This chapter is self-contained and provides a broad review of the use of pulsed electrical fields in the food industry and their significance. Although different from the medical applications, the knowledge gained in the food industry may find some use in the medical applications.

The first part of the book begins with a chapter by Antoni Ivorra and Boris Rubinsky. The chapter is an updated historical review of work in the field of irreversible electroporation; a biophysical phenomenon that may have been already recorded in the 1754 book of J.A. Nollet "*Recherches sur les causes particulieres des phénoménes électriques*." The review is comprehensive and contains discussions of over 100 key references.

Chapter two by Antoni Ivorra, is entitled "Tissue electroporation as a bioelectric phenomenon: basic concepts". It begins with a very basic introduction to electrical engineering and concludes with the latest studies on the bioelectronics of electroporation. The basic introduction part of the chapter is for readers with minimal training in electrical engineering. The chapter builds on these basic concepts and continues with an introduction to bioelectronics, which may be of benefit also to electrical engineers that were not exposed to bioelectronics before. This leads to the most up-to-date findings in the bioelectronics of irreversible electroporation. The chapter was written by a leading expert in the field of bioelectronics in such a way as to introduce a reader with no background in electrical engineering to the field while also providing relevant new information to experts in the field.

Most of the fundamental experimental studies on electroporation were done on cells. The third chapter by Avigail Ben-Or and Boris Rubinsky was written to introduce the reader to the fundamental biophysical aspects of electroporation that have emerged from experiments conducted with cells. Since, up to date, a uniform comprehensive theory that explains the mechanism(s) that underlies the electroporation phenomena has not yet been proposed, it is interesting to look into the experimental facts that such a theory needs to encompass. From this chapter, readers with no previous training in electroporation should gain a basic understanding of the biophysical phenomena which occur during electroporation at the single cell level.

A major challenge in the field of irreversible electroporation is to develop a fundamental understanding of the biophysical phenomena that occur at the sub cellular level and to develop mathematical models to analyze the process. The existence of multiple competing processes involving mechanical, electrical chemical phenomena makes it difficult to pinpoint the exact mechanism of irreversible electroporation in cells subject to electroporation pulses. Wanda Krassowska Neu and John C. Neu are leading biophysicists and mathematicians in the field of electroporation. In the fourth chapter of the book they have undertaken a unique review of the fundamental molecular scale processes, which may be responsible for cell death by irreversible electroporation and present their mathematical models. First, they review the theory of irreversible electroporation in planar membranes and in cells under voltage-clamp conditions. Then they introduce mathematical modeling studies of the processes taking place during irreversible electroporation and evaluate their potential role in cell death. Finally, based on the current understanding of cell electroporation, they summarize the potential mechanisms of cell death during irreversible electroporation. This chapter can benefit researchers and practitioners in the field by providing them with an understanding of the biophysics of the phenomenon and a glimpse into the mathematics of analysis.

Researchers in biophysics, mechanical, electrical, thermal and chemical engineers and in applied mathematics who have not worked in the field of electroporation before will find in this chapter an excellent and well documented starting point for entering the field.

Chapter five deals with a most fundamental aspect of the field of non-thermal irreversible electroporation and is co-authored by Rafael Davalos, Paulo A. Garcia and Jon Edd. Rafael Davalos and Jon Edd played pivotal roles in developing the field of non-thermal irreversible electroporation. In one of our casual conversations, Rafael Davalos, then a graduate student in my laboratory at Berkeley, asked me "why shall we not use irreversible electroporation to ablate tissue?" In retrospect, when reading Chapter 1, it will become evident that over the years others have also asked this question, some in a documented form and others in an anecdotal way. One focus of the three decades of research in my laboratory was minimally invasive surgery and tissue ablation and we have been working on almost every technology of tissue ablation and medical imaging of tissue ablation in existence. It took me several months to come up with an answer to the question of Rafael Davalos. In this time I was thinking; is there a need for another method of minimal invasive tissue ablation among the many that already exist? I then remembered the work of R.C. Lee, who from the mid eighties was showing in his studies on thermal damage due to electrical Joule heating that what might be irreversible electroporation, also occurs together with the thermal damage. It is important to understand that electrical fields in tissue produce a number of phenomena simultaneously. One of them is irreversible electroporation. Simultaneously, electrical fields also produce Joule heating through dissipation of electrical energy. In fact Joule heating has been used for tissue ablation for decades. Unlike irreversible electroporation that selectively affects only the cell membrane in a treated volume, thermal ablation affects every molecule in the treated volume. Therefore, it occurred to me that if one could find electrical fields that produce only irreversible electroporation but without Joule heating thermal damage than a new and unique tissue ablation would exist – one that selectively affects only one type of molecule in a treated volume. A molecular specific tissue ablation method did not exist before and it would have great advantages as will become obvious in Chapter six. However, at the time this idea occurred to me it was not obvious that electrical fields that produced only irreversible electroporation without thermal damage exist or that a substantial volume of tissue can be treated with such electrical fields to make non-thermal irreversible electroporation a viable method for tissue ablation. The issue is that both the electrical fields required for thermal ablation and irreversible electroporation are high and their effect is simultaneous, as found by R.C. Lee. This is when I came back to Rafael Davalos, several months later, and suggested that we try to determine through a mathematical model involving the electrical field equation and the bioheat equation if such electrical fields exist. In the first model of non-thermal irreversible electroporation Rafael Davalos has shown that these fields exist. While limited in the volume of tissue they affect, this volume is compatible with many of the needs of minimally invasive tissue ablation surgical procedures. Jon Edd, also a graduate student in my laboratory at the time, later confirmed experimentally the existence of non-thermal irreversible electroporation and the accuracy of the mathematical models of non-thermal irreversible electroporation. Liana Horowitz and Narayan Raju provided important support to those experiments. It is obvious that to use non-thermal irreversible electroporation as a precise molecular selective minimally invasive tissue ablation method treatment specific electrical fields have to be applied. Chapter five introduces the mathematical models that can be used to design non-thermal irreversible protocols. The chapter should give an insight to the users of the technique as well as introduce scientists who work in mathematical modeling of thermal and electrical phenomena to a field of mathematical modeling to which they could contribute.

Chapter six by Boris Rubinsky is a summary of all the experimental studies on non-thermal irreversible electroporation. This is the result of extensive work done with Gary Onik and Paul Mikus with valuable support from William Hamilton: while I took a leave of absence without pay from the UC Berkeley to develop the field. The most important aspect of this chapter is the mounting evidence on the value of the molecular selective nature of non-thermal irreversible electroporation (NTIRE). With NTIRE it is possible to ablate undesirable tissues around and within blood vessels, ducts and other sensitive structures in the body without affecting their structural and mechanical integrity. This opens the door to types of surgery that were not possible in the past. This chapter should provide insight to medical practitioners on the phenomena that occur in tissue during NTIRE for designing surgical procedures and to researchers in life sciences who will pursue experimental research on NTIRE.

As discussed in chapter five, mathematical modeling is an important part of designing appropriate NTIRE protocols. Chapter seven by Nataša Pavšelj and Damijan Miklavčič, a leading researcher in the field of electroporation engineering and computation, presents an introduction to various numerical tools that can be used for modeling NTIRE. In chapter eight by Anže Županič and Damijan Miklavčič, there is a thorough introduction to treatment planning for NTIRE. Chapters five to eight give an insight into the special aspects of NTIRE, what is possible to accomplish medically with NTIRE and how to design a treatment protocol for NTIRE.

Chapter nine by Lluis Mir, the pioneer of the technique of electrochemotherapy with reversible electroporation, brings here a discussion of other tissue treatment techniques that employ pulsed electrical fields. The chapter discusses electrochemotherapy and also the use of nano-second pulses for molecular treatment in the interior of the cell. There is no doubt that the ability of selective electrical fields to affect various types of molecules in the body non-invasively with induction or minimally invasive with conduction will have an effect on modern surgery. Because these techniques are complementary it is very likely that they will be used in combination in the future. The chapter was brought here to make readers who are entering the field through NTIRE aware of the existence of other pulsed electrical fields technologies.

Chapter ten by Gary Onik and Boris Rubinsky brings results from the first clinical use of NTIRE. The studies were performed in the prostate to test the safety of the procedure. They confirmed the unique attributes of this molecular selective procedure. Kenneth Thomson, is currently the physician with the broadest clinical experience in NTIRE. He has treated liver, kidney, lung and lymph node cancers. Both, the papers of Gary Onik and Ken Thomson will be valuable to clinical physicians interested in getting a first hand insight into the procedure and learn from their clinical experience.

Chapter 12 by the group of Ruggero Cadossi, is a thorough description of the engineering principles and design concepts of the irreversible electroporation devices. This chapter will be valuable for design engineers as well as companies in medical device design. For medical practitioners, this chapter will be useful in understanding how the devices that they use function.

Chapter 13 by Alex Golberg, Judith Fischer and Boris Rubinsky is different from the other chapters in the book and it is meant as stand-alone. Reading chapter one it should become evident that most of the earlier work on irreversible electroporation was done for applications related to treatment of food and sterilization of food. Obviously the use of irreversible electroporation for the food industry is different from the use in medicine. The difference in needs is evident from the first sections of this chapter. The application of irreversible electroporation in the food industry is also different. However, we thought that the large body of work on irreversible electroporation for the food industry may provide some additional insight to practitioners and scientist in the medical use of NTIRE. Chapter 13 was therefore brought here for completeness.

I am certain that I write on behalf of all the contributors to this work to state our hope that this edited book will contribute towards further advances in the science and applications of the emerging field of non-thermal irreversible electroporation.

Berkeley California
Jerusalem Israel
September 2009

Boris Rubinsky

Contents

Historical Review of Irreversible Electroporation in Medicine

Antoni Ivorra* and Boris Rubinsky

Department of Mechanical Engineering, University of California, Berkeley, CA 94720 USA
* Corresponding author E-mail: antoni.ivorra@gmail.com

Abstract. The objective of this chapter is to present a historical review of the field of irreversible electroporation (IRE) in the context of its medical applications. Although relevant scientific observations were made since the 18th century, the electroporation phenomenon was not identified as an increase of membrane permeability until mid 20th century. After that, multiple applications of reversible electroporation emerged in vitro (DNA electrotransfer) and in vivo (electrogenetherapy and electrochemotherapy). Irreversible electroporation was tested commercially in the 60s as a bactericidal method for liquids and foods but its use in the context of medical applications was not studied until the early 2000s as an ablative method. The cell destruction mechanism of IRE is not based on thermal damage and this fact provides to IRE an important advantage over other physical ablation methods: the extracellular scaffolding, including the vessels, is preserved. Several surgical applications are now under study or even under clinical trial: ablation of hepatocarcinomas, ablation of prostate tumors, treatment of atrial fibrillation and treatment of vascular occurrences such as restenosis and atherosclerotic processes.

Introduction

Electroporation is the phenomenon in which cell membrane permeability to ions and macromolecules is increased by exposing the cell to short high electric field pulses. Such increase in permeability is, presumably, related to the formation of nano-scale defects or pores in the cell membrane; from which the term electro-"poration" stems. Under some conditions (e.g. extremely large field magnitude), membrane permeabilization is permanent and the process leads to cell lysis. It is in this sense of permanent permeabilization that most authors define irreversible electroporation (IRE). However, it must be noted that temporary permeabilization can also cause a severe disruption of the cell homeostasis that can finally result in cell death, either necrotic or apoptotic. Therefore, in a broader sense, IRE could be defined as the permanent or temporal membrane electroporation process that causes cells to die.

The primary use of irreversible electroporation is to induce the death of undesirable cells without causing excessive heating. Recently, irreversible electroporation has begun to be employed as tool in the surgeon armamentarium for minimally invasive ablation of undesirable tissue. The goal of this chapter is to present a historical review of the field of irreversible electroporation from the first

B. Rubinsky (Ed.): Irreversible Electroporation, BIOMED, pp. 1–21.
springerlink.com

reports of the phenomenon in the 18th century to modern applications in minimally invasive surgery. We intentionally omit to mention membrane electroporation theories [1] and models; the present book contains the most recent study of models for the electroporation phenomenon.

The 18th and 19th Centuries

Probably the first scientific description of a phenomenon suggestive of irreversible electroporation in tissue can be found in the 1754 book of J.A. Nollet "*Recherches sur les causes particulieres des phénoménes électriques.*" [2]. In experiments with electrical fields Nollet noticed the formation of red spots on the skin of humans and animals in areas where electrical sparks were applied. This same phenomenon has been studied recently by J.P. Reilly [3] and, according to him, the red spots may be the consequence of stratum corneum degradation due to thermal damage. However, while thermal Joule heating effects cannot be ruled out, it is also quite probable that those red spots were in fact caused by damage to the capillaries by irreversible electroporation; two facts would support this hypothesis: 1) erythemas are common in skin electroporation [4, 5] and 2) it does not seem likely that the electrical generators at the time, (static electricity generators having been invented by Otto von Guericke in 1663), should be able to cause significant heating.

The interest in the science of electricity in biological materials emerged throughout the 18th century. As a matter of fact, discoveries in electricity and physiology were interlinked at those dates. In a series of experiments started in 1780, Luigi Galvani discovered that when a dead frog was placed on an iron grating and a bronze hook touched the spinal cord the frog's muscle twitched. His explanation to the phenomenon was based on what he called "animal electricity". However it was Alessandro Volta who found the correct explanation: the presence of two different metals in the same electrolyte (frog's body fluids) had created a DC current that had stimulated the frog's muscles. That lead to the invention of the voltaic pile (he replaced the frog's fluids by brine-soaked paper), which was the first device able to produce steady electric current and that became a basic element for later discoveries in electromagnetism.

In 1802, J.W. Ritter, who did a significant number of experiments and discoveries in electrophysiology, observed what was later called *Ritter's opening tetanus*: a contraction that occasionally occurs when a strong current passing through a stretch of muscle-nerve preparation is interrupted [6]. This phenomenon was not understood at the time but, as it will be discussed later, it was hypothesized in the mid 20th century that it was due to the "breakdown" of the cell membrane and such hypothesis leaded to the discovery of the electroporation phenomenon as we know it nowadays.

What may be perhaps the first work focusing on an irreversible electroporation phenomenon can be found in the 1898 study by G.W. Fuller entitled "Report on the investigations into the purification of the Ohio river water at Louisville Kentucky." [7]. In his report, an experiment is mentioned in which multiple high voltage discharges have some bactericidal effect on a water sample. Temperature was found not to increase significantly because of the treatment. Electrolytic effects cannot be completely ruled out (overall the pulses are applied for several minutes). However, on the basis of the way in which irreversible electroporation is currently used for sterilization of fluids, Fuller's reported bactericidal effect is most likely due to irreversible electroporation.

Towards the end of the 19th century, therapeutic uses of electricity are frequently reported in the medical literature. Most applications are based on thermal effects or on electrochemical phenomena. Use of short strong pulses is not cited. However, a book by A.D. Rockwell [8] reports experiments performed during the late 1800s in which "Under the discharges of the Leyden jar the red corpuscles (of the blood, i.e. red blood cells) change their shape and lose their color". This is probably a description of hemolysis induced by irreversible electroporation [9]. Leyden jars were the original capacitors and they were used to accumulate the charge generated by the electrostatic generators so that it was possible to produce high-voltage and high-current short pulses as it is required for electroporation of cells in suspension.

As a conclusion for this brief review on irreversible electroporation during the 18th and 19th centuries it can be said that some effects that we now know to be the consequence of IRE had been already observed before the beginning of the 20th century. Nevertheless, it seems that no explanation based on an increase of cell membrane permeability or based on membrane rupture was proposed.

First Half of the 20th Century

Observations on the effects of electrical fields on tissue made in the early 20th century can, in hindsight, be related to electroporation. In a 1913 set of lectures, A.J. Jex-Blake reviews knowledge on the lethal effects of human made electricity and lighting [10]. He notes that burns observed in industrial accidents with electricity are related to thermal effects whereas electrical injuries from lighting do not seem to be always from thermal origin. At the present it is accepted that some of the injurious effects of lighting are caused by irreversible electroporation [11].

An extraordinary non-lethal effect of lighting on humans is the emergence of red Lichtenberg figures on the skin that disappear in a few days. Quite probably these figures have the same origin than Nollet's red spots ("It is thought that they represent red blood cells extravasated into the superficial layers of the skin from capillaries secondary to the dielectric breakdown of the skin and subsequent massive electron shower" [12]).

Jex-Blake also cites works of the XVIII century in which animals were recovered from heart failure induced by electrical current (fibrillation term not coined at the time) by means of another electrical shock. In the context of electroporation this is relevant because now it is thought that electroporation plays a significant role in defibrillation [13].

In the 1930s the thermal effect of electrical fields on biological materials was well established, (e.g. [14, 15]). The 1936 report of G.M. McKinley [16] is relevant to this review in the sense that, from his own observations and from experiments performed during the 20s and 30s by other researchers, McKinley concludes that damage caused to living tissues by high frequency fields (10 to 100 MHz) cannot be only from thermal origin, particularly in the case of nervous tissue. He even proposes that this special "agent" associated with the electrical field can be used as a minimally invasive ablative method that will be selective to some specific tissues. However, in his study using chick embryos, there is not enough methodological data to conclude that electroporation was being performed. In fact his description of inductive current effects suggests that this may be a thermal effect. Nevertheless, the paper proposes that electrical fields produce other mechanisms of damage to biological cells in addition to thermal. Therefore, it is quite likely that attempts to understand the nature of these mechanisms were made in the ensuing years.

As indicated earlier, in 1951 A.L. Hodgkin [17] proposed that the Ritter's opening tetanus phenomenon is associated with " ...(the breakdown)... of the insulating properties of the membrane ... under the influence of the abnormally high potential difference". It is obvious that this explanation describes a phenomenon akin to what is now referred to as irreversible electroporation. The wording is suggestive of the concept of breakdown of a cell membrane viewed as a dielectric layer. As a matter of fact, the notion that the cell membrane could be modeled as a thin dielectric layer had appeared quite earlier. For instance, in 1925, H. Fricke [18] was able to hypothesize a reasonable value for the membrane thickness (30 nm instead of the actual 7 nm) by analyzing the passive electrical properties of red blood cells under the assumption that cell membrane acts electrically as a thin dielectric layer. Once the cell membrane was viewed as a thin dielectric it was reasonable to expect that some sort of dielectric rupture phenomenon could exist in the case of living cells as it happens in most dielectrics. A common breakdown mechanism in dielectrics is the *avalanche breakdown*: when the dielectric is subjected to a sufficiently high electric field, some bound electrons are freed, accelerated and then those electrons can liberate additional electrons during collisions in a process that leads to a dramatic conductivity increase and, in some cases, to permanent physical damage of the dielectric material. Now it is accepted that electroporation is not due to dielectric rupture by electron avalanche [19] but the idea that membrane breakdown could be caused by excessive transmembrane voltage surely helped to understand some experimental observations that are related to electroporation.

These highlights from research in the first half of the 20th century suggest that during this period further observations on the effects of electrical fields on biological materials were made that were consistent with the phenomenon of irreversible electroporation. The concept that the cell membrane is a dielectric and that it can irreversible breakdown under the application of an electrical field seems to have become accepted. Perhaps the central characteristics of the findings relevant to irreversible electroporation during this period is the realization that while electricity can induce damage to biological materials through thermal effects there is also another mechanism associated with electricity that induces damage and which is not thermal. In addition evidence seems to be building that electrical fields can produce irreversible damage to the cell membrane.

1950 to 1970

Considering that research on nerve stimulation with electricity was done since Galvani, it should not be surprising that the first systematic work on what is now known as irreversible and reversible electroporation was done on nerves. In 1956, B. Frankenhaeuser and L. Widén publish a study that attempts to explain the phenomenon of *anode break excitation* [20]. The anode break excitation phenomenon is described as the finding that that there is a change in the normal nerve conductivity behavior when electrical pulses are applied on nerve nodes with (in their study) amplitudes that are up to ten times the normal threshold and the pulse duration increased from less than 1 ms to more than 100 ms. Frankenhaeuser and Widén state that the actual phenomenon is known since, at least, 1898 [21]. In an explanation to the phenomenon, which is indicative of reversible and irreversible electroporation, they write "It is, therefore, concluded that the strong (electrical) shock damaged to a large extent the nodes... and that these nodes were more or less inactivated". In addition they write "It may be concluded that the effect caused by the strong shock is to a fair extent reversible." It should be emphasized that several prior studies, including that of Hodgkin [17], are brought in support of their conclusion.

R. Stämpfli produced, between 1950 and 1960, in collaboration with A. F. Huxley and others, a series of studies, describing irreversible and reversible electroporation on a frog nerve membrane. In a paper entitled "*Membrane potential of a Ranvier Node measured after electrical destruction of its membrane*" Stämpfli and Willi [22] write: "We had confirmed the observation of Frankenhaeuser and Widén [20], showing that anode break excitation in myelinated nerve can be elicited by strong positive pulses. We were able to show that such pulses produce a breakdown of membrane resistance and potential, if they increase the membrane potential by 70 to 110 mV, which corresponds to a voltage gradient of approximately half a million V/cm across the membrane. If only one short pulse is given the membrane recovers immediately after the breakdown like an electrolytic condenser. If very strong positive pulses of the order of 10 V are applied, the membrane is destroyed irreversibly." In a further study with a single insulated Ranvier node of a frog nerve fiber, entitled "*Reversible electrical breakdown of the excitable membrane of a Ranvier node*", Stämpfli reports that 5 second pulses which induce membrane potentials of about 120 mV to 140 mV (corresponding to voltage gradients across the membrane of roughly half a million V/cm) can cause the breakdown of the membrane resistance. Under certain conditions this breakdown is irreversible whereas in others it is reversible [23]. He describes the phenomenon as akin to the breakdown of the dielectric field of a capacitor. The observed reversibility is attributed to the fact that in his particular device the electrical field is stopped at the onset of the breakdown of the membrane and therefore the membrane can recover.

Throughout the history of irreversible electroporation the field has advanced with research carried out in parallel in the area of biomedicine and in the area of food processing technology. In food technology the irreversible electroporation topic is referred to as *Pulsed Electric Field* processing or *Electroplasmolysis* in reference to the lysis of cell membranes in tissue, for extracting their contents, or the bactericidal effect in treatment of fluids. The bactericidal non-thermal effects of electrical fields, reported first in 1898 [7], remained an area of research in the food industry through the first and second half of the 20th century and until today [24]. However, apparently during this period it was not obvious to researchers whether electric fields had a bactericidal effect beyond those expected from thermal or electrochemical causes. For instance, A.J.H. Sale and W.A. Hamilton, who produced seminal papers in electroporation in 1967 (see next paragraph), cite a review published in 1949 that finds accounts both for and against such non-thermal effects of electric fields [25] and the same author of the report, H. Burton, publishes a letter in *Nature* in 1950 [26] in which he opposes the conclusions from a previous paper in *Nature* in which it is claimed that it is possible to destroy a large proportion of bacteria in a liquid suspension at sub-lethal temperatures by the application of high radiofrequency electric fields [27]. Nevertheless, in 1961, H. Doevenspeck [28] describes commercial installations using electrical pulses to break apart cellular components for industrial food related processing of animal meat through non-thermal means, which resemble irreversible electroporation. These involve the electrical discharge of electrical pulses from carbon electrodes through the treated material. It should be emphasized that the paper does not specifically refer to the breakdown of the cell membrane. Neither does it provide specific values for the electrical pulses used. However, the outcome reported is clearly non-thermal ablation of the cell membrane. Furthermore, Doevenspeck also reports results showing that these electrical pulses can inactivate microorganisms with what he considers a non-thermal effect producing a small increase in temperature of at most 30 °C.

The interest in the so-called "bactericidal action of electrical fields" motivated three outstanding and seminal papers by Sale and Hamilton which set the basis for the field of irreversible electroporation and contain the ingredients of many of the future studies in electroporation in general [29-31] . The goal of the first of the three papers was to demonstrate that high field DC electrical pulses can kill cells without a thermal effect. They evaluated the non-thermal bactericidal effect by using ten DC electrical pulses that were very short, between 2 to 20 μs, and separated by long intervals of seconds, to minimize the temperature rise. A systematic study with several types of bacteria and two species of yeast demonstrated that the effect is not related to the stage of growth of the cells, the pH, electrolysis or heating. The measured temperature raise was at most 10 °C. They concluded that the parameters which affect cell killing are, in order of importance, electrical field magnitude first and then the time extent the field is applied. The electrical fields required to completely ablate the cells were found to be quite substantial, (e.g. 6 kV/cm for Saccharomyces cerevisiae and 16 kV/cm for Escherichia coli).

In the second study [30] Hamilton and Sale try to elucidate what is the mechanism by which the electric field pulses kill the cells and they conclude that the irreversible loss of the membrane's function as a semipermeable barrier is what causes cell death. The paper reports leakage of Escherichia Coli cell content in the medium, detected with spectroscopy, as a measure of the loss of cell membrane integrity. It further demonstrates membrane damage leading to the lysis of erythrocytes and protoplasts and suggests that "the electrical field causes an irreversible loss of the membrane's function as the semipermable membrane between the bacterial cell and its environment and that this is the cause of cell death." Electron microscopy of Escherichia Coli and erythrocytes shows that the complete breakdown of the membrane did not occur and suggests that the damage resulting from the DC pulse is confined to particular areas that were not identified.

In the third paper [31] Sale and Hamilton show that the electric field magnitude for inducing the lysis of various organism ranges from 3.1 kV/cm to 17 kV/cm (fields for 50 % population lysis with a protocol of 10 pulses of 20 μs) whereas the equivalent induced transmembrane voltages only range from 0.7 V to 1.15V. And after this result, they suggest that the transmembrane potential induced by the external field may cause "conformational changes in the membrane structure resulting in the observed loss of its semipermeable properties".

In order to compute the induced transmembrane voltages, Sale and Hamilton employed a model in which the cell was considered to be a conductive sphere isolated from the external conductive medium by a thin dielectric layer. Then they obtained equations that were derived from the equations already described by J.C. Maxwell for calculating the conduction through a suspension of spheres [32, 33]. The transmembrane voltage (Vm) has a maximum at the poles facing the electrodes (i.e. direction of the electric field, **E**) and its value at those two points is $Vm = (3/2) \times a \times |\mathbf{E}|$ where a is the radius of the cell. At any point of the cell the transmembrane voltage is $Vm=(3/2) \times a \times |\mathbf{E}| \times \cos(\theta)$ where θ is the polar angle measured from the center of the cell with respect to the direction of the field, this expression is usually referred to as the *Schwan's equation* [34].

In summary, at the end of the 1960s it was known that electrical pulses have a permeabilizing effect on the cell membrane, which can lead to cell lysis through an effect that is non-thermal. It was also known that certain electrical pulses can cause reversible breakdown of the cell membrane. Most of the studies in the field were carried out with neurons or in relation to food processing.

1970 to 1990

The 1970s and 1980s produced some of the key advances in the field of reversible electroporation which brought this field into the mainstream of biotechnology and medicine. The research during these two decades was primarily in the field of reversible electroporation and focused on developing new uses and fundamental understanding of the mechanisms. Since reversible electroporation is not the focus of this review we will discuss only some of the highlights.

In their 1972 paper, E. Neumann and K. Rosenheck, apparently unaware of previous studies on electroporation, show that electrical impulses of about 18 to 24 kV/cm and about 150 μsec long produce reversible permeabilization of the cell membrane of chromaffin granules of bovine-medullary cells used as vesicles for epinedrine, norepinephrine, ATP and proteins [35]. Experiments done at 0 °C show that the largest increase in temperature is 6 °C and that the observed effect of reversible permeabilization is therefore not thermal. However, although all their experiments seem to involve reversible electroporation, Neumann and Rosenbeck fail to recognize that the increase of permeabilization is due to an extreme transmembrane potential induced by the external field and they try to relate such phenomenon to the physiological release of hormones and neurotransmitters in neurons.

From observation of discrepancies in the readings of an electrical Coulter counter for cells, U. Zimmerman and his group, also apparently unaware of previous studies by Sale and Hamilton, determined that the electrical field in the counter induces cell membrane breakdown. The methodology they developed, which combines experimentation in Coulter type counters and between parallel plates with the solution of the Laplace equation has produced some of the first systematic data on the electrical parameters required for electroporation in cells. Working with human and bovine red blood cells they explored the dependence of cell membrane breakdown, as expressed by the presence of intracellular contents in the extracellular solution, on increasing pulse length and amplitude. They found find that the maximal content is reached asymptotically when the pulse length reaches about 50 to 100 μsec and the electrical field strength of about 2.6- 2.8 kV/cm. The critical membrane potential difference leading to membrane breakdown was found to be about 1 V. Results from this group of studies were published in a series of papers, since 1974, e.g. [36, 37]. It should be emphasized that the data in this series of papers is extremely relevant to irreversible electroporation parameters. These parameters are the asymptotic values listed in these papers. It is interesting to note that they found different asymptotic values for human and bovine red blood cells, which may suggest the possibility for differential irreversible electroporation in applications in tissue. They also showed that these effects are not thermal. One practical outcome of their work is the suggestion of employing erythrocytes and lymphocytes as drug and enzyme carrier systems.

In 1977, K. Kinosita and T. Tsong, [38] proposed that the permeabilization of the cell membrane due to the application of electrical pulses is related to the formation of several pores with radii in the range of a few angstroms. In classical osmotic mass transfer experiments with red blood cells they showed that the size of these pores can be varied and that these pores eventually reseal. They also propose the use of permeabilized cells as reservoirs for transport of chemical species through the body circulatory system. In evaluating the time it takes for the cells to reseal as a function of temperature, they find that "At 37 °C the treated membrane rapidly regains its impermeability to cations, whereas, at 3 °C the cells remain highly permeable even after 20 h." To us this suggests that lowering the temperature may produce an effect akin to irreversible electroporation even when using reversible electroporation parameters.

In 1978, S.V. Belov wrote a paper that it is never cited but that may be relevant in the context of irreversible electroporation as it is probably the first case in which IRE of living tissues is intentionally pursued [39]. In an investigation of coagulation type electrosurgical devices he suggests that surgical coagulation is actually related to cell membrane breakdown due to pulses that have a "high ratio of peak to mean voltages". As a reference to the observation that high ratio of peak to mean voltage produce cell membrane breakdown he mentions a 1938 paper by Tatrinov[1]. Belov's research was performed on frog leg muscle. Histological analysis and measurements of changes in electrical resistance were used to show that 500 pulses of about 8.5 kV/cm and pulse length of 1 μsec and 1000 pulses of about 7.5 kV and pulse length of 2 μsec cause the destruction of the cell membrane in tissue, without producing a thermal effect.

The year 1982 brought about two major discoveries which have led reversible electroporation to the forefront of modern biotechnology and medicine: cell fusion with reversible electroporation and introduction of genes into cells using reversible electroporation. The use of reversible electroporation to produce fusion between cells is described in the paper by Zimmermann [40]. A more recent review on cell fusion is by C. Ramos and J. Teissie [41]. In a now classical paper Neumann and his collaborators coin the term *electroporation* to describe the membrane breakdown discussed in this review and introduce the use of reversible electroporation for the insertion of genes into cells [42]. They also present a classical thermodynamic analysis of the formation of pores during electroporation. In relation to the field of irreversible electroporation, it is interesting to note that this paper which focuses on reversible electroporation, naturally, caution against the use of irreversible parameters during the procedure. The reason we bring this up here is because during the two decades following this work, irreversible electroporation was studied primarily as an upper limit to reversible electroporation in the context of applications of reversible electroporation. The researches in the application of reversible electroporation have led to numerous other publications in that field, whose review is not within the scope of this paper.

While until the mid 1980s the research on electroporation was primarily related to cells (with few exceptions, [28, 39]) between 1987 and 1989 the research on electroporation began to deal with tissue. During the late 1980s it pursued three important directions, one of which is directly related to irreversible electroporation.

M. Okino and H. Mohri [43] and S. Orlowski et al., [44] proposed , independently, the use of reversible electroporation to reversible permeabilize cells and thereby introduce more effectively cytotoxic agents into malignant cells, for treatment of cancer. This field has subsequently developed to become an important application of reversible electroporation to treatment of cancer.

In 1989, by performing electrical conductance measurements K.T. Powell et al. [45] demonstrated that frog skin can be reversible electroporated. They did not report how that could influence the passage of drugs through the skin but quite probably this paper was influential on the posterior discovery by M.R. Prausnitz et al. [46] in which transdermal drug delivery was enhanced by electroporation.

An important series of studies in the field of irreversible electroporation of tissue began with the 1987 [47] and 1988 [11] papers of R.C. Lee and his co-investigators on electrical discharge induced trauma in tissue. Lee et al write "While it seems likely that Joule heating causes part of the tissue destruction (due to electrical discharge), particularly near the skin contact points, it does not appear to explain the patter of tissue injury frequently observed at sites distant from contact." . The experiments in [11] were performed on rat muscle cells exposed to

---

[1] V. V. Tatarinov, Arkh. Biol. Nauk, 52, No. 2, 173-177 (1938). Unfortunately we were not able to locate this paper.

electrical fields of 20 to 300 V/cm for periods of 1 millisecond in 30 seconds intervals and 25 V/cm increments. The temperature of the system was monitored during the study. Dissected whole rat muscles were given thirty 1 ms 150 V/cm pulses separated by 5 s. The electrical resistance of the muscle was measured before and after the sequence of pulses. The results show that the cells were irreversible damaged with pulses varying from 50 to 300 V/cm and from 1 to 30 pulses (there are no more precise details), with a temperature rise of less than 0.1 °C. The results with the tissue showed a decrease in electrical resistance following the application of the pulses, which is suggested to be a measure of tissue damage. The non-thermal nature of this damage is proven by measuring an increase in temperature of only 0.7 °C. The study also states that in experiments in which a 100 V/cm electrical field was applied for 1 s a 10 to 20 °C temperature increase is expected. They write, "In this case, cells would be vulnerable to both thermal and nonthermal injury mechanisms. Under these conditions the primary cause of cell rupture would have to be determined since both electroporation and heating may lead to disruption and both may be cytotoxic."

In summary, these two decades end with reversible electroporation and electrofusion becoming important biotechnological and medical tools in cell research and use. A review book that summarizes the state of the art at the end of this period is available [48].

1990 to 2000

In the nineties, many of the ideas related to reversible electroporation that were conceived in the 70s and 80s reach enough maturity for commercial and clinical applications. We will briefly summarize some of those achievements by referencing books and articles, however, herein after we will focus mostly on aspects related exclusively to irreversible electroporation.

After the design by H. Potter et al. [49] of an electroporation cuvette suitable for cells suspensions, microbiology researchers started to employ electrophoresis power supplies in order to perform gene transfection by electroporation. Soon after, multiple commercial generators specifically intended for electroporation were developed and now this transfection technique is very common in microbiology laboratories. Summaries on the technique and its applications can be found in several edited books, e.g. [48, 50].

The first report on the use of reversible electroporation to introduce plasmid DNA into a living tissue was published in 1991 by A.V. Titomirov et al. [51]. Gene delivery to cells in tissue has also become an area of major importance to biotechnology and medicine in which reversible electroporation plays a central role. It has also found applications in treatment of cancer, e.g. [52, 53]. Some of the reviews and edited books written on this topic include [54-56].

In 1991 the group of L.M. Mir published two breakthrough papers on the use of reversible electroporation to treat cancer by facilitating the penetration of anticancer drugs, such as bleomycin, in the malignant cells. They coined the term *electrochemotherapy* to describe this procedure [57] and reported the first clinical trial in the field of electroporation [58]. Electrochemotherapy is now one of the most solid applications of reversible electroporation and it is being used clinically to treat cancer patients. Probably the most updated review information on the topic can be found in [59-61].

During this decade, skin electroporation and its use for drug delivery emerged as an important aspect of reversible electroporation of tissue. The study that established skin electroporation for transdermal drug delivery was published in 1993 [46], [4]. Multiple reviews have been written on this application of tissue electroporation, e.g. [62].

The study of the contribution of irreversible electroporation to tissue damage during electrical shock trauma continued since it was first proposed in the 80s and was led primarily by R.C. Lee [63-65]. An interesting new aspect of this research is the suggestion that irreversible electroporated cell membranes could be therapeutically sealed with surfactants [66].

Related to the above paragraph it is convenient to note that most researchers cite necrosis due to excessive permeabilization and consequent disruption of the osmotic balance as the killing mechanism of electroporation. However, in the late 90s, two independent papers were published in which it was shown *in vitro* that electroporation not only caused necrosis but it also induced cell death with features compatible with apoptosis [67, 68]. In both papers it is reported that electroporation leads to chromosomal DNA fragmentation, which is considered to be an unambiguous indication of late apoptosis. Another interesting outcome of the paper by J. Piñero et al. [67] is the following statement written in the abstract: "The possibility of killing tumour cells by electroporation, as a variant of electrotherapy, constitutes, in our opinion, a promising procedure in cancer therapy, avoiding the undesirable side effects normally derived from treatment with cytotoxic drugs." Quite surprisingly this particular topic is not further developed in the manuscript. In the introduction the authors refer to the use of electrochemotherapy for treating solid tumors and, therefore, we must suppose that the comment in the abstract was actually proposing the use of irreversible electroporation for treating tumors. As far as we know this would be the first time irreversible electroporation is proposed as an ablative method.

In 1998 [69] it was reported a remarkable phenomenon observed during tissue reversible electroporation: blood flow is blocked in the area where the electric field is applied. This phenomenon, referred to as *the vascular lock* [70], has been noticed in muscle [70], liver [69] and tumors [71]. Blood perfusion disruption affects kinetics of drug delivery [70] and is followed by ischemia, which could be beneficial in the treatment of tumors [72]. In fact, it has been proposed the use of electroporation to intentionally interrupt blood flow [73, 74]. J. Gehl et al. [70] suggested two mechanisms that would explain why such vascular lock is produced: 1) the electrical stimulus, or the permeabilization, induces an immediate reflex vasoconstriction of afferent arterioles mediated by the sympathetic system and 2) the permeabilization of endothelial cells causes an increase of the interstitial pressure and a decrease of the intravascular pressure that leads to vascular collapse.

To sum up, the last decade of the 20th century witnessed the consolidation of reversible electroporation as a standard microbiology laboratory technique and the first animal and clinical tries for using reversible *in vivo* were successfully performed. Irreversible electroporation was well characterized but, in biomedical applications, it was perceived as a drawback, or complication, of reversible electroporation.

2000 to 2008

In 1997 K.H. Schoenbach et al. [75, 76] reported the first *in vitro* study on the use of very high voltage pulses of "submicrosecond" duration. Numerous papers have been published since then on the use of pulses with a duration of some nanoseconds or tens of nanosecond [75-80]. The main motivation for this research line came from the believe that those ultra-short pulses, known as *nanosecond Pulsed Electric Field* (nsPEF), could be able to induce electroporation of intracellular membranous structures (e.g. mitochondria) without disturbing the cell membrane. In *conventional* electroporation, that is, with pulses larger than 10 μs, the cell membrane is charged up to a stable transmembrane potential in very few microseconds and the intracellular structures become isolated from the external field and hence cannot experience electroporation. On the other hand, since the charging time constant is proportional to the structure size, it seems reasonable that if very short pulses of high magnitude are applied then the small intracellular membranous structures will be charged to a sufficiently high voltage for electroporation before the cell membrane is barely charged. That is the reason why it was expected that it would be possible to electroporate internal structures without causing electroporation of the plasma membrane. However, recent computer models [81] and experimental results [82] indicate that cell membrane electroporation also occurs when nanosecond pulses are applied. That is, with nsPEF all the membranous structures of the cell are electroporated (some authors refer to this phenomenon as *supraelectroporation*). Still, it is important to point out that researchers in this field found out that nsPEF can induce apoptosis [83] and that they even partially succeeded in inhibiting tumor growth after nanosecond pulses were applied *in vivo* [84, 85].

In 2004, perhaps motivated by the results that researchers in the nsPEF field had been obtaining, C. Yao et al. [86] investigated the use of special pulses, to which they referred to as *Steep Pulsed Electric Field* (SPEF), for killing cells and for *in vivo* tumor growth inhibition. Those SPEF pulses consisted of a fast rising edge (rise time ~ 200 ns) followed by a slow exponential decay (τ ~ 200 μs) which was originated from a capacitance discharge. The authors wanted to combine the intracellular effects from the "high-frequency components" of the rising slope together with the plasma membrane effects from the "low-frequency components" of the falling slope so that it was possible to "destroy both nucleus and membrane". And indeed they succeeded in killing cells and delaying tumor growth. However, taking into account the low amplitude of the applied pulses and the fact that a rise time of 200 ns is quite standard in electroporation, we believe that what they induced and observed does not differ from the effects of *conventional* IRE. Therefore, despite their unawareness, it is very likely that these researchers were the first ones to obtain empirical evidences of the capabilities of IRE as a tissue ablation method for tumors.

Also in 2004, R. Davalos and B. Rubinsky filled a US patent application [87] which followed a provisional application filled in 2003 and that proposed the use of conventional IRE (pulses longer than 5 μs) as a tissue ablation method. In the patent publication, the inventors detail methods for treatment planning by adjusting the electrode configurations and the applied voltages. They also point out the fact that IRE can be easily applied in areas where perfusion is high (e.g. in the vicinity of blood vessels), as opposed to the case of thermal methods for ablation. Nevertheless, probably the most essential aspect of this application, and of the research carried out immediately afterwards by the same authors [81], is the recognition that IRE is an ablation method that can be designed in such a way as to avoid tissue damaging thermal effects and that such feature has important implications in the outcome of the procedure as well as post-treatment healing. The use of IRE in a way that avoids thermal damage is referred to as Non Thermal Irreversible Electroporation (NTIRE) and is the type of IRE that is currently pursued in clinical applications.

The subsequent research efforts carried out by Rubinsky's group were aimed at treating hepatocarcinomas by means of irreversible electroporation. First it was shown *in vitro* that IRE is capable of killing human hepatocarcinoma cells [89]. Afterwards it was demonstrated that IRE can ablate selectively areas of non-pathological rodent livers [90]. In this later study the histological assessment three hours after pulses were applied showed some interesting features: the treated areas exhibited microvascular occlusion, endothelial cell necrosis and diapedeses, resulting in ischemic damage to parenchyma and massive pooling of erythrocytes in the hepatic sinusoids. Hepatocytes displayed blurred cell borders, pale eosinophilic cytoplasm, variable pyknosis and vacuolar degeneration. On the other hand, large blood vessel architecture was preserved.

The last published report of this series of studies on liver ablation by IRE was performed on pigs under experimental conditions relatively close to those of a clinical scenario [91]. After exposing the liver, IRE was applied by using 18 gauge stainless needles that were positioned with the assistance of ultrasound sonography. Previously, the size and shape of the lesions had been designed by means of a treatment planning procedure that will be briefly explained later in the context of the paper by J. Edd and R. Davalos [92]. Following the surgical procedure, animals were sacrificed at 24 hours, 3 days, 7 days and 14 days and liver samples were excised for histopathological analysis. Multiple interesting observations and conclusions emerged from this study: 1) all the 14 animals survived the procedure; 2) upon application of the IRE pulses a variable degree of generalized muscle contraction occurred in each animal and such degree appeared to be related to the administered amount of muscle relaxant (Pancuronium); 3) immediately following pulse application, sonography showed a markedly hypoechoic area in the expected location of the IRE lesion and, at 24 hours, the ultrasound image showed the area had permuted its quality and was now uniformly hyperechoic; 4) histological analysis showed that the IRE ablated area was continuously necrotic and that the transition between this area and the adjacent untreated normal parenchyma was abrupt; 5) macroscopic histological analysis also showed that large vascular structures were mainly unaffected; and 6) all animals manifested peripheral lymphadenopathy in the drainage area of the ablated tissue.

This last paper by Rubinsky et al. [91] on the ablation of pig liver is included in a special issue of *Technology in Cancer Research and Treatment* (TCRT, August 2007) devoted to irreversible electroporation. It is worth to briefly summarize here the results and conclusions communicated in some of the other papers included in that issue:

1) E. Maor et al., from Rubinsky's group, presented the results from a pilot study in which IRE (ten 100 μs pulses of 3800 V/cm) was applied to the carotid artery of rats that were kept alive for 28 days after the procedure [93]. Histology showed that the connective matrix of the blood vessels remained intact whereas the number of vascular smooth muscle cells (VSMC) was decreased very significantly without pathological observable consequences such as aneurysm, thrombus formation or necrosis. These findings seem to indicate that IRE can be applied safely to the vicinity of large blood vessels. Moreover, the fact that VSMC population was significantly reduced suggests that IRE could be the basis for treating pathologies such as restenosis and atherosclerotic processes. As a matter of fact, Maor and co-investigators are now performing research on the use of IRE for treating cardiac restenosis [94].

2) G. Onik et al, also from Rubinsky's group, applied *in vivo* IRE to canine prostates by means of percutaneous needle electrodes placed under ultrasound guidance [95]. Macroscopic observation of the induced lesions revealed a very distinct narrow zone of transition from normal to complete necrosis. Nearby structures such as urethra, vessels, nerves and rectum were apparently not affected by the IRE procedure despite the fact that the areas covered by the high electric field purposely included those structures. This study was preceded by an *in vitro* study in which prostate adenocarcinoma cells were destroyed by IRE [96].

3) E.W. Lee et al. reported an experimental study on pig livers which is quite similar to the one presented by Rubinky et al. [91] but in this case the electrodes were inserted percutaneously, without exposing the liver, and the histological samples were analyzed with apoptotic markers. Equivalent observations were obtained (lesion manifestation by ultrasonography and sharp transition zone between ablated and normal tissue). Regarding the histological analysis the authors conclude: "… confirmed complete apoptotic cell death by PIE (i.e. IRE) on Von Kossa, BAX, and H&E staining. In summary, PIE can provide a novel and unique ablative method with real-time monitoring capability, ultra-short procedure time, non-thermal ablation, and well-controlled and focused apoptotic cell death."

4) B. Al-Sakere et al. [97] performed *in vivo* IRE of tumors subcutaneously inoculated in mice and studied the immune reactions. The objective of the study was to elucidate what was the role of the immune system in the ablation of tumors by means of IRE. Their main conclusion is that the immune system response is not required in order to successfully ablate tumors by IRE and, therefore, IRE is a feasible option to consider for the treatment of immunodepressed cancer patients. This study was preceded by an investigation of multiple electroporation protocols that lead to complete tumor regression of tumors inoculated in mice [98]. Best

results were obtained by using a protocol that consisted of 80 pulses of 100 μs at 0.3 Hz (an interval of 3.3 seconds between pulses) with a field magnitude of 2500 V/cm. With this protocol complete regression was achieved in 12 out of 13 treated tumors and no thermal effects were induced.

5) J. Edd and R. Davalos described how mathematical modeling aided by computer methods can be employed to predict the shape and extent of the lesion created by IRE [92]. The basic principle for such modeling is that any specific tissue region is electroporated if the electric field magnitude is higher than a certain value. Such threshold is specific to the sort of tissue and the features of the IRE pulses (e.g. number of pulses and duration of the pulses). Once this threshold is experimentally obtained, by using numerical methods on computers it is possible to predict the distribution of the field magnitude in a tissue according to the electrode configuration and the applied voltages to those electrodes. This methodology for treatment planning was pioneered before for the case of *in vivo* reversible electroporation by D. Miklavcic's group [99].

Also in 2007, J. Lavee et al. published a study with five pigs in which IRE of the atrium was performed in order to analyze its applicability for the treatment of atrial fibrillation as an alternative to methods based on thermal ablation [100]. Again the demarcation between ablated and normal tissue was clear and sharp. Histological analysis demonstrated complete destruction of atrial cells down to a mean depth of 0.9 cm. The lesions manifested electrical isolation. The authors conclude "we propose and demonstrate here a new and exciting modality to perform atrial ablation, which holds the potential of providing very swift, precise, and complete transmurality with no local heating effects."

The first commercial system approved for clinical irreversible electroporation of soft tissues started to be produced in 2008. It consists of a high voltage pulse generator [101] and single-use disposable electrodes.

Finally we want to point out a recent paper by E. Tekle et al. [102] that demonstrates that electroporation induces the exposure of phosphatidylserine (PS) to the outer surface of the cell membrane and that such externalization results in phagocytic clearance of the cells by macrophages. The authors refer to the cells exposed to the electric fields as *apoptosis-mimetic* cells because they consider that those cells are non-apoptotic but exhibit some features typical of cells undergoing apoptosis. This phenomenon is of interest because it could be employed for removal of pathogenic cells through non-inflammatory phagocytosis. Furthermore, this observation could concur with the non-thermal nature of IRE, and the associated preservation of tissue scaffolding, in order to explain the rapid recovery of tissues after an IRE treatment.

References

[1] Chen, C., Smye, S.W., Robinson, M.P., Evans, J.A.: Membrane electroporation theories: a review. Medical & Biological Engineering & Computing 44, 5–14 (2006)

[2] Nollet, J.A.: Recherches sur les causes particulieres des phénoménes électriques. Chez H.L. Guerin & L.F. Delatour, Paris (1754)

[3] Reilly, J.P.: Applied Bioelectricity: From Electrical Stimulation to Electropathology. Springer, New York (1998)
[4] Prausnitz, M.R.: A practical assessment of transdermal drug delivery by skin electroporation. Advanced Drug Delivery Reviews 35, 61–76 (1999)
[5] Vanbever, R., Préat, V.: In vivo efficacy and safety of skin electroporation. Advanced Drug Delivery Reviews 35, 77–88 (1999)
[6] Noad, H.M.: Lectures on electricity; comprosing galvnism, magnetism, electro-magnetism, magneto- and thermo- electricity, and electo-physiology, 3rd edn. George Knight and Sons, London (1849)
[7] Fuller, G.W.: Report on the investigations into the purification of the Ohio river water at Louisville Kentucky. D. Van Nostrand Company, New York (1898)
[8] Rockwell, A.D.: The Medical and surgical uses of electricity: including the X-ray, Finsen light, vibratory therapeutics, and high-frequency currents. E.B. Treat & Company, New York (1903)
[9] Abidor, I.G., Li, L.H., Hui, S.W.: Studies of cell pellets: II. Osmotic properties, electroporation, and related phenomena: membrane interactions. Biophys. J. 67, 427–435 (1994)
[10] Jex-Blake, A.J.: Death by electric currents and by lightning. The Goulstonian lectures for 1913. British Medical Journal 11, 425–552, 492–498, 548–552, 601–603 (1913)
[11] Lee, R.C., Gaylor, D.C., Bhatt, D., Israel, D.A.: Role of cell membrane rupture in the pathogenesis of electrical trauma. The Journal of Surgical Research 44, 709–719 (1988)
[12] O'Keefe Gatewood, M., Zane, R.D.: Lightning injuries. Emergency Medicine Clinics of North America 22, 369–403 (2004)
[13] Al-Khadra, A., Nikolski, V., Efimov, I.R.: The role of electroporation in defibrillation. Circulation Research 87, 797–804 (2000)
[14] Christie, R.V., Binger, C.A.L.: An experimental study of diathermy: iv. Evidence for the penetration of high frequency currents through the living body. Journal of Experimental Medicine 46, 715–734 (1927)
[15] Weinberg, E.D., Ward, G.E.: Diathermy and regeneration of bone. Archives of Surgery 28, 1121–1129 (1934)
[16] McKinley, G.M.: Short electric wave radiation in biology. In: Duggar, B.M. (ed.) Biological effects of radiation, vol. 1, pp. 541–558. McGraw-Hill Book Co., New York (1936)
[17] Hodgkin, A.L.: The ionic basis of electrical activity in nerve and muscle. Biological reviews of the Cambridge Philosophical Society 26, 339–409 (1951)
[18] Fricke, H.: A mathematical treatment of the electric conductivity and capacity of disperse systems. II. The capacity of a suspension of conducting spheroids surrounded by a non-conducting membrane for a current of low frequency. Physical Review 26 (1925)
[19] Crowley, J.M.: Electrical breakdown of bimolecular lipid membranes as an electromechanical instability. Biophys. J. 13, 711–724 (1973)
[20] Frankenhaeuser, B., Widén, L.: Anode break excitation in desheathed frog nerve. The Journal of Physiology 131, 243–247 (1956)
[21] Biedermann, W.: Electro-physiology, vol. 2. Macmillan, London (1898)
[22] Stämpfli, R., Willi, M.: Membrane potential of a ranvier node measured after electrical destruction of its membrane. Experimentia 13, 297–298 (1957)

[23] Stämpfli, R.: Reversible electrical breakdown of the excitable membrane of a Ranvier node. Anais da Academia Brasileira de Ciencias 30, 57–63 (1957)
[24] Toepfl, S., Mathys, A., Heinz, V., Knorr, D.: Review: Potential of High Hydrostatic Pressure and Pulsed Electric Fields for Energy Efficient and Environmentally Friendly Food Processing. Food Reviews International 22, 405–423 (2006)
[25] Burton, H.: A survey of literature on bacterial effects of short electromagnetic waves. In: National Institute for Research in Dairying, Shinfield, England (1949)
[26] Burton, H.: Effects of Radio-Frequency Voltages on Bacteria. Nature 166, 434 (1950)
[27] Nyrop, J.E.: A Specific Effect of High-Frequency Electric Currents on Biological Objects. Nature 157, 51–52 (1946)
[28] Doevenspeck, H.: Influencing cells and cell walls by electrostatic impulses. Fleishwirtshaft 13, 986–987 (1961)
[29] Sale, A.J.H., Hamilton, W.A.: Effects of high electric fields on microorganisms. 1. Killing of bacteria and yeasts. Biochimica et Biophysica Acta 148, 781–788 (1967)
[30] Hamilton, W.A., Sale, A.J.H.: Effects of high electric fields on microorganisms. 2. Mechanism of action of the lethal effect. Biochimica et Biophysica Acta 148, 789–800 (1967)
[31] Sale, A.J.H., Hamilton, W.A.: Effects of high electric fields on microorganisms. 3. Lysis of erythrocytes and protopasts. Biochimica et Biophysica Acta 163, 37–43 (1968)
[32] Maxwell, J.C.: A Treatise on Electricity and Magnetism, 3rd edn. Clarendon Press, Oxford (1904)
[33] Cole, K.S.: Electric impedance of suspensions of spheres. J. Gen. Physiol. 12, 29–36 (1928)
[34] Miklavcic, D., Kotnik, T.: Electroporation for Electrochemotherapy and Gene Therapy. In: Rosch, P.J., Markov, M.S. (eds.) Bioelectromagnetic Medicine, pp. 637–656. Informa Health Care, New York (2004)
[35] Neumann, E., Rosenheck, K.: Permeability changes induced by electric impulses in vesicular membranes. The Journal of Membrane Biology 29, 279–290 (1972)
[36] Zimmermann, U., Pilwat, G., Riemann, F.: Dielectric Breakdown of Cell Membranes. Biophys. J. 14, 881–899 (1974)
[37] Riemann, F., Zimmermann, U., Pilwat, G.: Release and uptake of haemoglobin and ions in red blood cells induced by dielectric breakdown. Biochimica et Biophysica Acta 394, 449–462 (1975)
[38] Kinosita, K.J., Tsong, T.Y.: Formation and resealing of pores of controlled sizes in human erythrocyte membrane. Nature 268, 438–441 (1977)
[39] Belov, S.V.: Effects of high-frequency current parameters on tissue coagulation. Biomedical Engineering 12, 209–211 (1978)
[40] Zimmermann, U.: Electric field-mediated fusion and related electrical phenomena. Biochimica et Biophysica Acta 694, 227–277 (1982)
[41] Ramos, C., Teissie, J.: Electrofusion: a biophysical modification of cell membrane and a mechanism in exocytosis. Biochimie 82, 511–518 (2000)
[42] Neumann, E., Schaeffer-Ridder, M., Wang, Y., Hofschneider, P.H.: Gene transfer into mouse lymphoma cells by electroporation in high electric fields. EMBO J. 1, 841–845 (1982)
[43] Okino, M., Mohri, H.: Effects of a high-voltage electrical impulse and an anticancer drug on in vivo growing tumors. Japanese Journal of Cancer Research 78, 1319–1321 (1987)

[44] Orlowskim, S., Belehradek, J.J., Paoletti, C., Mir, L.M.: Transient electropermeabilization of cells in culture. Increase of the cytotoxicity of anticancer drugs. Biochemical pharmacology 34, 4727–4733 (1988)

[45] Powell, K.T., Morgenthaler, A.W., Weaver, J.C.: Tissue electroporation. Observation of reversible electrical breakdown in viable frog skin. Biophys. J. 56, 1163–1171 (1989)

[46] Prausnitz, M.R., Bose, V.G., Langer, R., Weaver, J.C.: Electroporation of mammalian skin: a mechanism to enhance transdermal drug delivery. Proc. Natl. Acad. Sci. USA 90, 10504–10508 (1993)

[47] Lee, R.C., Kolodney, M.S.: Electrical Injury Mechanisms: Electrical Breakdown of Cell Membranes. Plastic & Reconstructive Surgery 80, 672–679 (1987)

[48] Neumann, E., Sowers, A.E., Jordan, C.A.: Electroporation and Electrofusion in Cell Biology. Plenum Press, New York (1989)

[49] Potter, H., Weir, L., Leder, P.: Enhancer-dependent expression of human kappa immunoglobulin genes introduced into mouse pre-B lymphocytes by electroporation. Proc. Natl. Acad. Sci. USA 81, 7161–7165 (1984)

[50] Nickoloff, J.A.: Electroporation Protocols for Microorganisms. Humana Press, Totowa (1995)

[51] Titomirov, A.V., Sukharev, S., Kistanova, E.: In vivo electroporation and stable transformation of skin cells of newborn mice by plasmid DNA. Biochimica et Biophysica Acta 1088, 131–134 (1991)

[52] Daud, A.I., DeConti, R.C., Andrews, S., Urbas, P., Riker, A.I., Sondak, V.K., Munster, P.N., Sullivan, D.M., Ugen, K.E., Messina, J.L., Heller, R.: Phase I Trial of Interleukin-12 Plasmid Electroporation in Patients With Metastatic Melanoma. J. Clin. Oncol. 26, 5896–5903 (2008)

[53] Tamura, T., Sakata, T.: Application of In Vivo Electroporation to Cancer Gene Therapy. Current Gene Therapy 3, 59 (2003)

[54] Jaroszeski, M.J., Heller, R., Gilbert, R.: Electrochemotherapy, electrogenetherapy, and transdermal drug delivery: electrically mediated delivery of mollecules to cells, vol. 37. Humana Press, Totowa (2000)

[55] Mir, L.M., Moller, P.H., André, F., Gehl, J.: Electric Pulse-Mediated Gene Delivery to Various Animal Tissues. In: Huang, L., Hung, M.-C., Wagner, E. (eds.) Advances in Genetics, vol. 54, pp. 83–114. Academic Press, London (2005)

[56] Andre, F., Mir, L.M.: DNA electrotransfer: its principles and an updated review of its therapeutic applications. Gene Therapy 11, S33–S42 (2004)

[57] Mir, L.M., Orlowski, S., Belehradek, J.J., Paoletti, C.: Electrochemotherapy potentiation of antitumour effect of bleomycin by local electric pulses. European Journal of Cancer 27, 68–72 (1991)

[58] Mir, L.M., Belehradek, M., Domenge, C., Orlowski, S., Poddevin, B., Belehradek, J.J., Schwaab, G., Luboinski, B., Paoletti, C.: Electrochemotherapy, a new antitumor treatment: first clinical trial. Comptes Rendus de l'Academie des Sciences Serie III Sciences de la Vie 313, 613–618 (1991)

[59] Mir, L.M., Gehl, J., Sersa, G., Collins, C.G., Garbay, J.-R., Billard, V., Geertsen, P.F., Rudolf, Z., O'Sullivan, G.C., Marty, M.: Standard operating procedures of the electrochemotherapy: Instructions for the use of bleomycin or cisplatin administered either systemically or locally and electric pulses delivered by the CliniporatorTM by means of invasive or non-invasive electrodes. European Journal of Cancer Supplements 4, 14–25 (2006)

[60] Marty, M., Sersa, G., Garbay, J.R., Gehl, J., Collins, C.G., Snoj, M., Billard, V., Geertsen, P.F., Larkin, J.O., Miklavcic, D., Pavlovic, I., Paulin-Kosir, S.M., Cemazar, M., Morsli, N., Soden, D.M., Rudolf, Z., Robert, C., O'Sullivan, G.C., Mir, L.M.: Electrochemotherapy - An easy, highly effective and safe treatment of cutaneous and subcutaneous metastases: Results of ESOPE (European Standard Operating Procedures of Electrochemotherapy) study. European Journal of Cancer Supplements 4, 3–13 (2006)

[61] Sersa, G.: The state-of-the-art of electrochemotherapy before the ESOPE study; advantages and clinical uses. European Journal of Cancer Supplements 4, 52–59 (2006)

[62] Prausnitz, M.R., Mitragotri, S., Langer, R.: Current status and future potential of transdermal drug delivery. Nature Reviews. Drug discovery 3, 115–124 (2004)

[63] Bhatt, D.L., Gaylor, D.C., Lee, R.C.: Rhabdomyolysis Due to Pulsed Electric Fields. Plastic and Reconstructive Surgery 86(1), 1–11 (1990)

[64] Abramov, G.S., Bier, M., Capelli-Schellpfeffer, M., Lee, R.C.: Alteration in sensory nerve function following electrical shock. Burns 22, 602–606 (1996)

[65] Bier, M., Hammer, S.M., Canaday, D.J., Lee, R.C.: Kinetics of sealing for transient electropores in isolated mammalian skeletal muscle cells. Bioelectromagnetics 20, 194–201 (1999)

[66] Lee, R.C., River, L.P., Pan, F.S., Ji, L., Wollmann, R.L.: Surfactant-induced sealing of electropermeabilized skeletal muscle membranes in vivo. Proc. Natl. Acad. Sci. USA 89, 4524–4528 (1992)

[67] Piñero, J., Lopez-Baena, M., Ortiz, T., Cortes, F.: Apoptotic and necrotic cell death are both induced by electroporation in HL60 human promyeloid leukaemia cells. Apoptosis 2, 330–336 (1997)

[68] Hofmann, F., Ohnimus, H., Scheller, C., Strupp, W., Zimmermann, U., Jassoy, C.: Electric field pulses can induce apoptosis. The Journal of Membrane Biology 169, 103–109 (1999)

[69] Ramirez, L.H., Orlowski, S., An, D., Bindoula, G., Dzodic, R., Ardouin, P., Bognel, C., Belehradek, J.J., Munck, J.N., Mir, L.M.: Electrochemotherapy on liver tumours in rabbits. British Journal of Cancer 77, 2104–2111 (1998)

[70] Gehl, J., Skovsgaard, T., Mir, L.M.: Vascular reactions to in vivo electroporation: characterization and consequences for drug and gene delivery. Biochimica et Biophysica Acta 1569, 51–58 (2002)

[71] Sersa, G., Cemazar, M., Parkins, C.S., Chaplin, D.J.: Tumour blood flow changes induced by application of electric pulses. European Journal of Cancer 35, 672–677 (1999)

[72] Sersa, G., Jarm, T., Kotnik, T., Coer, A., Podkrajsek, M., Sentjurc, M., Miklavcic, D., Kadivec, M., Kranjc, S., Secerov, A., Cemazar, M.: Vascular disrupting action of electroporation and electrochemotherapy with bleomycin in murine sarcoma. British Journal of Cancer 98, 388–398 (2008)

[73] Rubinsky, B., Edd, J., Horowitz, L.: Electroporation to interrupt blood flow. US 2005/0171574 A1, USPTO, ed. The Regents of the University of California, USA (2004)

[74] Palanker, D., Vankov, A., Freyvert, Y., Huie, P.: Pulsed electrical stimulation for control of vasculature: temporary vasoconstriction and permanent thrombosis. Bioelectromagnetics 29, 100–107 (2008)

[75] Schoenbach, K.H., Peterkin, F.E., Alden, R.W.I., Beebe, S.J.: The effect of pulsed electric fields on biological cells:experiments and applications. IEEE Trans. Plasma Science 25, 284–292 (1997)
[76] Schoenbach, K.H., Beebe, S.J., Buescher, E.S.: Intracellular effect of ultrashort electrical pulses. Bioelectromagnetics 22, 440–448 (2001)
[77] Sun, Y., Vernier, P.T., Behrend, M., Marcu, L., Gundersen, M.A.: Electrode microchamber for noninvasive perturbation of mammalian cells with nanosecond pulsed electric fields. IEEE Trans. NanoBioscience 4, 277–283 (2005)
[78] Jordan, D.W., Uhler, M.D., Gilgenbach, R.M., Lau, Y.Y.: Enhancement of cancer chemotherapy in vitro by intense ultrawideband electric field pulses. Journal of Applied Physics 99, 094701 (2006)
[79] Frey, W., White, J.A., Price, R.O., Blackmore, P.F., Joshi, R.P., Nuccitelli, R., Beebe, S.J., Schoenbach, K.H., Kolb, J.F.: Plasma Membrane Voltage Changes during Nanosecond Pulsed Electric Field Exposure. Biophys. J. 90, 3608–3615 (2006)
[80] Vernier, P.T., Sun, Y., Chen, M.-T., Gundersen, M.A., Craviso, G.L.: Nanosecond electric pulse-induced calcium entry into chromaffin cells. Bioelectrochemistry 73, 1–4 (2008)
[81] Smith, K.C., Gowrishankar, T.R., Esser, A.T., Stewart, D.A., Weaver, J.C.: The Spatially Distributed Dynamic Transmembrane Voltage of Cells and Organelles due to 10-ns Pulses: Meshed Transport Networks. IEEE Trans. Plasma Science 34, 1394–1404 (2006)
[82] Pakhomov, A.G., Kolb, J.F., White, J.A., Joshi, R.P., Xiao, S., Schoenbach, K.H.: Long-lasting plasma membrane permeabilization in mammalian cells by nanosecond pulsed electric field (nsPEF). Bioelectromagnetics 28, 655–663 (2007)
[83] Beebe, S.J., Fox, P.M., Rec, L.J., Somers, K., Stark, R.H., Schoenbach, K.H.: Nanosecond pulsed electric field (nsPEF) effects on cells and tissues: apoptosis induction and tumor growth inhibition. IEEE Trans. Plasma Science 30, 286–292 (2002)
[84] Nuccitelli, R., Pliquett, U., Chen, X., Ford, W., James Swanson, R., Beebe, S.J., Kolb, J.F., Schoenbach, K.H.: Nanosecond pulsed electric fields cause melanomas to self-destruct. Biochemical and Biophysical Research Communications 343, 351–360 (2006)
[85] Garon, E.B., Sawcer, D., Vernier, P.T., Tang, T., Sun, Y., Marcu, L., Gundersen, M.A., Koeffler, H.P.: In vitro and in vivo evaluation and a case report of intense nanosecond pulsed electric field as a local therapy for human malignancies. International Journal of Cancer 121, 675–682 (2007)
[86] Yao, C., Sun, C., Mi, Y., Xiong, L., Wang, S.: Experimental studies on Killing and inhibiting effects of steep pulsed electric field (SPEF) to target cancer cell and solid tumor. IEEE Trans. Plasma Science 32, 1626–1633 (2004)
[87] Davalos, R.V., Rubinsky, B.: Tissue ablation with irreversible electroporation. US 2007/0043345 A1, USPTO, ed. The Regents of the University of California, USA (2004)
[88] Davalos, R.V., Mir, L.M., Rubinsky, B.: Tissue Ablation with Irreversible Electroporation. Ann. Biomed. Eng. 33, 223 (2005)
[89] Miller, L., Leor, J., Rubinsky, B.: Cancer cells ablation with irreversible electroporation. Technology in Cancer Research and Treatment 4, 699–706 (2005)
[90] Edd, J., Horowitz, L., Davalos, R.V., Mir, L.M., Rubinsky, B.: In-Vivo Results of a New Focal Tissue Ablation Technique: Irreversible Electroporation. IEEE Trans. Biomed. Eng. 53, 1409–1415 (2006)

[91] Rubinsky, B., Onik, G., Mikus, P.: Irreversible electroporation: a new ablation modality – clinical implications. Technology in Cancer Research and Treatment 6, 37–48 (2007)

[92] Edd, J.F., Davalos, R.V.: Mathematical modeling of irreversible electroporation for treatment planning. Technology in Cancer Research and Treatment 6, 275–286 (2007)

[93] Maor, E., Ivorra, A., Leor, J., Rubinsky, B.: The effect of irreversible electroporation on blood vessels. Technology in Cancer Research and Treatment 6, 307–312 (2007)

[94] Maor, E., Ivorra, A., Leor, J., Rubinsky, B.: Irreversible electroporation attenuates neointimal formation after angioplasty. IEEE Trans. Biomed. Eng. 55, 2268–2274 (2008)

[95] Onik, G., Rubinsky, B., Mikus, P.: Irreversible Electroporation: Implications for Prostate Ablation. Technology in Cancer Research and Treatment 6, 295–300 (2007)

[96] Rubinsky, J., Onik, G., Mikus, P., Rubinsky, B.: Optimal Parameters for the Destruction of Prostate Cancer Using Irreversible Electroporation. The Journal of Urology 180, 2668–2674 (2008)

[97] Al-Sakere, B., Bernat, C., Andre, F., Connault, E., Opolon, P., Davalos, R.V., Mir, L.M.: A study of the immunological response to tumor ablation with irreversible electroporation. Technology in Cancer Research and Treatment 6, 301–306 (2007)

[98] Al-Sakere, B., André, F., Bernat, C., Connault, E., Opolon, P., Davalos, R.V., Rubinsky, B., Mir, L.M.: Tumor ablation with irreversible electroporation. PLoS ONE 2, e1135 (2007)

[99] Miklavcic, D., Beravs, K., Semrov, D., Cemazar, M., Demsar, F., Sersa, G.: The Importance of Electric Field Distribution for Effective in Vivo Electroporation of Tissues. Biophys. J. 74, 2152–2158 (1998)

[100] Lavee, J., Onik, G., Mikus, P., Rubinsky, B.: A Novel Nonthermal Energy Source for Surgical Epicardial Atrial Ablation: Irreversible Electroporation. The Heart Surgery Forum 10, E162–E167 (2007)

[101] Bertacchini, C., Margotti, P.M., Bergamini, E., Lodi, A., Ronchetti, M., Cadossi, R.: Design of an irreversible electroporation system for clinical use. Technology in Cancer Research and Treatment 6, 313–320 (2007)

[102] Tekle, E., Wolfe, M.D., Oubrahim, H., Chock, P.B.: Phagocytic clearance of electric field induced 'apoptosis-mimetic' cells. Biochemical and Biophysical Research Communications 376, 256–260 (2008)

Tissue Electroporation as a Bioelectric Phenomenon: Basic Concepts

Antoni Ivorra

Department of Mechanical Engineering, University of California, Berkeley, CA 94720, USA
Corresponding author E-mail: antoni.ivorra@gmail.com

Abstract. Electroporation is the phenomenon in which cell membrane permeability to ions and macromolecules is increased by exposing the cell to short (microsecond to millisecond) high electric field pulses. In living tissues, such permeabilization boost can be used in order to enhance the penetration of drugs (electrochemotherapy) or DNA plasmids (electrogenetherapy) or to destroy undesirable cells (irreversible electroporation). The main purpose of the present chapter is to provide an overview of the electrical concepts related to electroporation for those not familiar with electromagnetism. It is explained that electroporation is a dynamic phenomenon that depends on the local transmembrane voltage and it is shown how a voltage difference applied though a pair of electrodes generates an electric field which in turn induces the required transmembrane voltage for electroporation to occur. Quite exhaustive details are given on how electroporation changes the passive electrical properties of living tissues. Furthermore, some remarks are given about the effects of electroporation on other bioelectric phenomena such as cardiac arrhythmias.

1 Introduction

Bioelectricity typically refers to the electromagnetic energy produced by living organisms. Such energy is usually manifested as ionic currents at nerves or muscles due to the propagation of the so called *action potentials*. However, in a broader sense, bioelectricity also refers to the study of the interaction of externally applied electromagnetic energy with living organisms, as it is the case of the electroporation phenomenon.

Like in other bioelectricity topics, electroporation can be approached from a life sciences perspective or from a physical sciences perspective. This duality enriches the topic but also creates some confusion and hinders a comprehensive view of the phenomenon. The present chapter is intended to help bridge both approaches: its main purpose is to provide an overview of the electrical concepts related to electroporation for those not familiar with electromagnetism. Some advanced aspects are presented that may be of interest for engineers or physicists working in the field (particularly in sections 5 and 6) but electrical terminology and concepts are simplified as much as possible.

Going back to the duality of bioelectricity it is interesting to note that discoveries in electricity and biology have been interlinked throughout the history. For instance, in a series of experiments started in 1780, Luigi Galvani, a physician,

B. Rubinsky (Ed.): Irreversible Electroporation, BIOMED, pp. 23–61.
springerlink.com

discovered that when a dead frog was placed on an iron grating and a bronze hook touched the spinal cord then the frog's muscle twitched. His explanation to the phenomenon was based on what he called *animal electricity*. Later, it was Alessandro Volta, a physicist, who found the correct explanation: the presence of two different metals in the same electrolyte (frog's body fluids) had created a DC current that had stimulated the frog's muscles. That lead to the invention of the voltaic pile (Volta replaced the frog's fluids by brine-soaked paper), which was the first device able to produce steady electric current and that became a basic element for later discoveries in electromagnetism.

In 1925, Fricke [1] produced a result in the opposite direction: two decades before the lipid bi-layer was understood, he was able to hypothesize a reasonable value for the membrane thickness (30 nm instead of the actual 7 nm) by analyzing the passive electrical properties of red blood cells. His calculations were based on an electrical model for the cell, and its environment, in which the cell membrane was modeled as a dielectric layer. Interestingly, from such a model an additional hypothesis could be drawn: some sort of dielectric rupture phenomenon could exist in the case of living tissues as it is manifested in most dielectrics: when a dielectric is subjected to a sufficiently high electric field, some bound electrons are freed and accelerated, then those electrons can liberate additional electrons during collisions in a process called avalanche breakdown. This process leads to a dramatic conductivity increase and, in some cases, to permanent physical damage of the dielectric material. Now it is accepted that electroporation is not due to dielectric rupture by electron avalanche [2] but the idea that membrane breakdown could be caused by excessive transmembrane voltage surely helped to understand some experimental observations that are related to electroporation. At that time, neurophysiology researchers knew that when a certain electrical stimulus threshold was surpassed the electrophysiological manifestations were disrupted. And, for instance, A. L. Hodgkin, who won the 1963 Nobel Prize in Physiology or Medicine for his work with A. F. Huxley on the basis of nerve action potentials, noted in 1951 that the *Ritter's opening tetanus* might be due to the fact that "the insulating properties of the membrane break down under the influence of the abnormally high potential difference" [3]. In 1957, Stämplfi [4] correctly demonstrated the phenomenon as being a disturbance of the cell membrane dielectric properties and noticed that this effect could be reversible under certain circumstances. Since then, multiple scientific discoveries and practical applications of electroporation have emerged. In 1967, Sale and Hamilton [5-7] observed that leakage of intracellular ions and molecules occurs immediately the application of the electric field pulses and, ten years later, Kinosita and Tsong [8] demonstrated that reversible electroporation could also be employed to allow cellular uptake. Then Neumann et al. in 1982 [9] were the first ones to facilitate gene transfer into cells by means of electroporation. This created a powerful method for gene transfection that has been used in microbiology labs for more than 25 years. In the early 90s, Mir pioneered the use of *in vivo* electroporation for the introduction of chemotherapy drugs in solid tumors [10, 11]. And now we are witnessing how *in vivo* irreversible electroporation (IRE) is becoming a new surgical tool able to perform tissue ablation without the drawbacks of thermal methods.

This chapter is organized into seven sections. The first three sections, including the present introduction section, are intended to provide a general overview of some basic concepts that are required to understand electroporation from an electrical point of view. In section 4 it is explained that electroporation occurs when the transmembrane voltage reaches a certain threshold and it is shown how an external electric field induces such transmembrane voltage in cell suspensions and tissues. Furthermore, some basics concepts are given about the use of the *finite element method* for simulating the electric field distribution in tissues. The fifth section deals with the dynamic changes that occur in membrane conductance during the application of the electroporation pulse. It is shown how these changes influence the electric current waveform and how relevant these changes can be for computing the electric field distribution. Then, the sixth section describes the changes in conductance that occur after pulse termination. And, finally, the last section very briefly introduces a topic that would require another full chapter and that will probably deserve further future research: the effect of electroporation on other bioelectric phenomena.

2 Overview of Basic Electricity Concepts

This subsection is intended to provide a brief overview of the main concepts on electromagnetism that are required to understand electroporation from an electrical perspective. Those familiar with basic electromagnetism or circuit theory can skip this subsection without hesitation. Readers who feel they require another source to learn the concepts presented here are encouraged to look for the electromagnetism section in any general physics textbook at undergrad level.

Electric charge (Q) is the fundamental property of some subatomic particles that determines their electromagnetic interaction. Positive and negative charges have been defined. The unit used to express the amount of charge is the coulomb (C) and the value of the elementary charge is +1.602 × 10–19 C which is the charge of a proton (charge of an electron = -1.602 × 10–19 C). The electric charge of larger particles, such as ions or molecules, is an integer multiple of the elementary charge. Particles with charges with the same sign repel one another, whereas opposite sign charged particles attract. The magnitude of this force is proportional to the product of their charges and the inverse square of the distance between them. It is useful to employ the concept of the electric field (**E**): at a given point the electric field is a vector (i.e. magnitude and direction)[1] that expresses the force that would be exerted

[1] In this chapter, bold symbols are used to denote vectors; other authors prefer to use a small arrow over the symbol for the same purpose. A vector is a geometric object that has both a magnitude (length) and a direction. It is usually graphically represented by an arrow connecting two points in a bidimensional or a tridimensional space. In physics, vectors are employed to represent physical entities such as forces and flows.

on a positive charge of 1 coulomb placed at that point. Each charged particle creates an electric field which is proportional to its charge, inverse to the square of the distance and that points towards, or against, the particle depending on the sign of the charge. The convenience of the electric field concept comes from the fact that at any point the electric field is calculated as the summation of the electric fields created by each one of the charged particles independently.

Electric charges also interact with magnetic fields. The velocity and trajectory of moving electric charges are modified by magnetic fields which in turn are modified by the moving electric charges. Moreover, moving electric charges generate magnetic fields and alternating magnetic fields cause electric charges to move. As a matter of fact, electric forces and magnetic forces are both manifestations of the *electromagnetic force*, which is one of the four fundamental forces identified in nature. To some extent, the interactions between magnetic fields and electric fields are present in the electroporation phenomenology. However, in order to understand most of the electroporation concepts presented here it is not necessary to comprehend those interactions. Therefore, for the sake of simplicity, this chapter intentionally skips magnetic fields and their interactions with electric fields.

Voltage (or potential) value in a point A indicates the energy that a unitary electric charge would have in this point compared to the energy that the unitary charge would have in point B. In other words, voltage is defined as a difference in energy values due to the electric field. When the electric field is constant through the trajectory A-B (e.g. in an infinitesimal trajectory) then the voltage equals the electric field times the distance. For that reason the electric field (**E**) units are V/m. The direction of **E** indicates in which direction the maximum drop in voltage is produced and the magnitude of **E** represents the value of this voltage drop. These concepts are probably better understood with graphical illustrations in which voltages are represented as heights (Figure 1).

If an electric path exists between two points with different voltages then free electric charges will move from the high energy position to the low energy position. The electric current value indicates the flow of electric charge through the cross-section of the electric path in a second. Traditionally, an analogy with fluidic elements has been displayed in order to explain the voltage and the electric current concepts in circuits: two water tanks at different heights are connected through a pipe and the hydrostatic pressure difference (analogy for the voltage difference) causes the water (analogy for the electrical charge) to flow through the pipe. This analogy is also useful to explain the electrical resistance concept: water flow will not only depend on pressure difference between the tanks but also on pipe diameter and length; the shorter and wider the pipe is, the larger the flow is. A lot of materials exhibit a linear relationship between the electric current and the voltage difference. This relationship is known as the *Ohm's law* and the constant that relates both parameters is the resistance:

$$V_1 = V_1 - V_2$$

$$R = \frac{V}{I}$$

where R is the resistance (units: ohms, Ω), V is the voltage difference (units: volts, V) and I is the current that flows through the resistance (units: amperes, A). The inverse of the resistance (G = 1/R = I/V) is called conductance and is expressed in siemens, S.

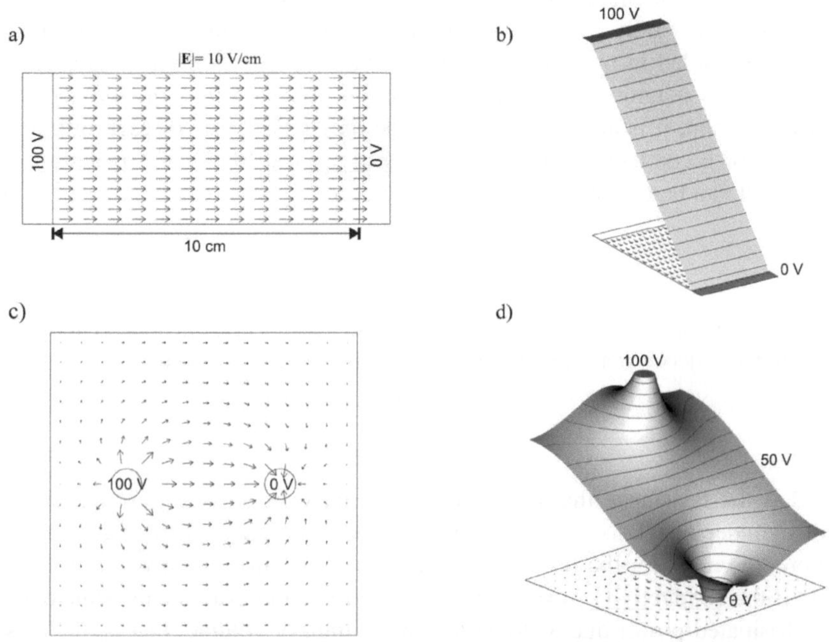

Fig. 1. Electric fields and potentials in two bidimensional structures. a) a rectangular slab of homogeneous material with two rectangular electrodes at opposite sides; an electrode is held at 100 V whereas the other is maintained at 0 V; in this case the electric field (represented by arrows) is uniform and its magnitude is equal to the ratio between the voltage difference and the distance (10 V/cm = 1000 V/m), b) voltage values in the previous structure are depicted as height values, c) a square slab of homogeneous material contains two round electrodes, an electrode is held at 100 V whereas the other is maintained at 0 V; in this case the electric field is not uniform, d) voltage values in the previous structure are depicted as height values.

Fig. 2. Resistance symbol

It is important to note that power is dissipated as heat at any conductor. This phenomenon is known as *Joule heating* and it is also referred to as *ohmic heating* or *resistive heating* because of its relationship with the Ohm's law:

$$P_{dissipated} = VI = I^2R = \frac{V^2}{R} = V^2G$$

The above definitions for the resistance and the electric current are intended for two terminal components embedded in a circuit. In the case of tridimensional domains it is possible to define equivalent parameters at infinitesimal portions. First it is possible to define the concept of current density (**J**, units: A/m^2) which is the electric current per unit area of cross section. Note that the symbol for current density, **J**, is bolded in order to show that current density is defined as a vector (length indicates magnitude and vector direction indicate direction and sense of the current). Then, it is also possible to define the "resistance" of each infinitesimal portion of a conductive medium: the resistivity (ρ; "rho") of a material at a specific point can be formally defined as:

$$\rho = \frac{\mathbf{E}}{\mathbf{J}}$$

and this expression is completely equivalent to the Ohm's law introduced above.

For a homogeneous material the resistivity can also be defined as:

$$\rho = R\frac{S}{L}$$

where R is the resistance that a cylinder of the material (S is the cross-sectional area of cylinder and L is its length) exhibits between its opposite flat surfaces. Therefore, the units for the resistivity are Ω.m (= Ω.m^2/m). The inverse of the resistivity is the conductivity (σ = 1/ρ) and it is expressed in S/m (siemens/meter).

The dissipated power due to Joule heating at unitary volume, that is, the dissipated power density (units are watts/cubic meter), can now be written[2]:

$$p_{dissipated} = \frac{|\mathbf{E}|^2}{\rho} = \sigma|\mathbf{E}|^2$$

Here an example can be useful to illustrate the significance of the Joule heating: if a pulse of 1000 V/cm (100,000 V/m) is applied to a living tissue, the maximum possible dissipated power density that will be produced will be about 15×10^9 W/m^3 since the maximum conductivity for tissues is about 1.5 S/m. If the pulse is 100 microseconds (μs) long, then the dissipated energy density (energy = U = power × time) will be 1.5 ×10^6 J/m^3 (joules/cubic meter). If we assume that the thermal properties of the tissue are equal to those of liquid water and that no heat

---

2 |**E**| indicates **E** magnitude; direction is not relevant.

is lost by radiation, diffusion or convection, which is the worst plausible scenario, then the temperature rise (ΔT) due to the pulse would be:

$$\Delta T = \frac{U}{c\,d} = 0.36\ K$$

where c is the specific heat capacity (c_{water} = 4.184 J/(g.K), joules/(grams × kelvin)) and d is the mass density ($d_{water} = 0.997 \times 10^6$ g/m^3). Therefore, although the pulse can produce a measurable temperature increase of 0.36 degrees Celsius (°C), it does not seem likely that such ΔT could produce any effect on the viability of the tissues. On the other hand, it is perfectly plausible that some electroporation protocols with multiple pulses (or longer or larger than the 100 μs 1000 V/cm pulse) could result in thermal damage to the tissues. This issue has been addressed in detail by some researchers in the field [12-16] and the main conclusion from those studies is that electroporation, even when it is irreversible, is not necessarily accompanied by thermal damage in the treated region. Nevertheless, thermal damage can appear in specific regions where the electric field is too large, for instance, at the edges of the electrodes. This is an issue to keep in mind when assaying new applications, protocols or electrode setups[3].

All the electrical concepts presented above were referred to conductive materials and continuous signals. It is convenient now to explain different notions regarding dielectrics and time dependant signals (e.g. alternating voltages).

Dielectrics (i.e. electrical insulators) are materials that are not able to conduct charge. On the other hand, dielectrics support charge accumulation. The physical principles that explain such feature are beyond the scope of this section. Yet it is necessary to introduce here the concept of capacitance. A capacitance is a circuit element able to store, and to release, electric charge. It is created by the combination of conductors and dielectrics. One of its most basic implementations consists of two conductive plates separated by a dielectric material. The amount of charge (Q) that this structure is able to store is determined by its dimensions and by a fundamental parameter of the dielectric: the permittivity (ε, units: C/V.m). Charge stored in the capacitance, at any time, is:

$$Q = C\,V$$

where C (not to be confused with the unit for charge, coulombs) is the capacitance value (units: farads, F). For the conductor-dielectric-conductor structure (Figure 3) the value of the capacitance is:

$$C = \varepsilon \frac{A}{d} = \varepsilon_r \varepsilon_0 \frac{A}{d}$$

where ε_0 is the permittivity of vacuum (= 8.9×10^{-12} C/V.m), ε_r is the relative permittivity of the material ($\varepsilon_r = \varepsilon / \varepsilon_0$), A is the area of the plates and d is the distance between the plates.

---

[3] This topic is covered in detail in another chapter of this book by J. Edd, R. Davalos and P. Garcia.

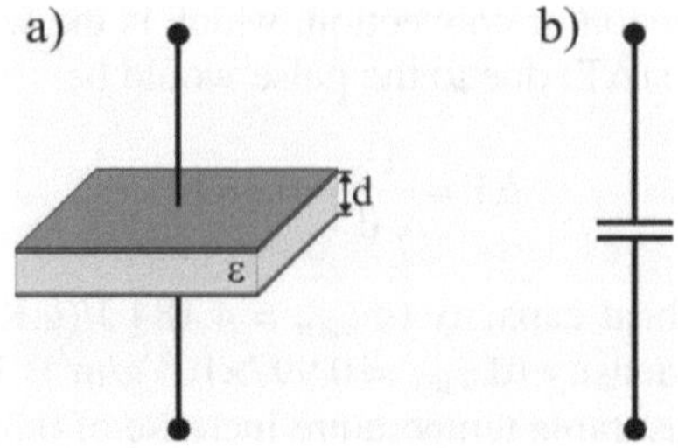

Fig. 3. Capacitance. a) simple implementation based on two conductive plates and a dielectric. b) Electrical symbol for a capacitance.

An interesting feature of the capacitance is that it can conduct *displacement currents*:

$$\frac{dQ}{dt} = C\frac{dV}{dt}$$

$$\Downarrow$$

$$I = C\frac{dV}{dt}$$

that is, since the time derivative of the stored or released charge (I=dQ/dt) depends on the time derivative of the voltage, a capacitance acts as a conducting element for fast changing currents. Actual charges do not flow through the capacitance but current appears to do so (i.e. *displacement currents*). Examples of fast changing currents are alternating currents (AC), which are currents whose direction reverses cyclically with a specific frequency. For an alternating signal (voltage or current), it can be said that a capacitance behaves similarly to a resistance with a value that depends on the frequency of the signal. The "resistance" of a capacitance (actually its impedance magnitude or modulus[4]) is:

$$"R_{CAPACITANCE}" = |\mathbf{Z}_C| = \frac{1}{2\pi fC}$$

where C is the capacitance value, f is the frequency of the alternating signal and π is the mathematical constant pi. Note that for high frequencies a capacitance will act as a short circuit whereas for low frequencies it will act as an open circuit.

In all practical cases, capacitances are accompanied by resistances in series. The following example (Figure 4) shows what happens when a voltage pulse is

[4] Impedance is a common term in electronics. Usually, it is simply described as the opposition to the flow of an alternating electric current through an electric element or system. Therefore in this sense it is equivalent to a resistance. However, impedance is a broader concept that includes the *phase shift* (delay measured in parts of cycle) between the voltage and the current signals. Impedance is graphically represented with a bidimensional vector whose length represents the magnitude and whose angle represents the phase shift. For mathematical convenience, impedance values are represented with complex numbers.

applied to a circuit that consists of a resistance in series with a capacitance (*RC circuit*).

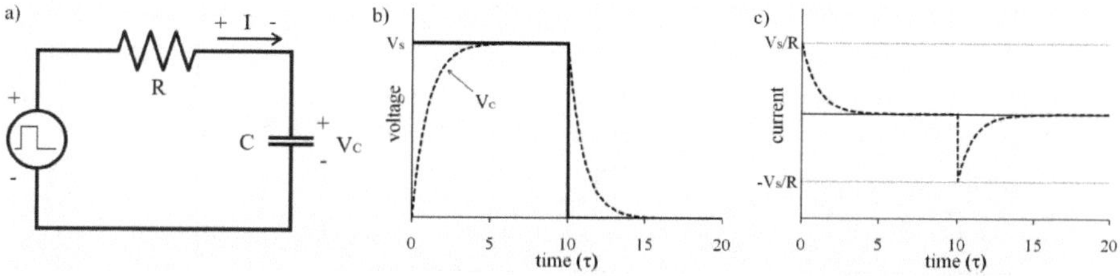

Fig. 4. Voltages (b) and currents (c) in a RC circuit (a) when a single rectangular voltage pulse is applied

At the beginning the capacitance is discharged (Q= 0 coulombs) and, consequently, the voltage difference between its terminals is 0 V. The applied voltage pulse (continuous line) causes current to flow through the resistance and starts to charge the capacitance so that the capacitance voltage (dashed line) increases and the current decreases (voltage difference at resistance terminals drops). After a while ($5\tau = 5RC$ [5]) the capacitance is completely charged and almost no current flows through the circuit because the voltage difference at the resistance terminals is null. Then, when the voltage source comes back to 0 V, the capacitance starts to return the accumulated charge through the resistance and the voltage at the capacitance terminals drops slowly down to 0 V. Both, charging and discharging capacitance voltages follow exponential functions:

$$V_C(t) = V_S\left(1 - e^{-(t-t_0)/\tau}\right) \qquad \text{(charging)}$$

$$V_C(t) = V_S\left(e^{-(t-t_0)/\tau}\right) \qquad \text{(discharging)}$$

where V_S is the amplitude of the applied pulse and t_0 is the time at which the charging or discharging starts (in the above example $t_{0\_charging} = 0$, $t_{0\_discharging} = 10\tau$).

3 Cells and Tissues in Electrical Terms

The electrical model that Fricke used in the 20s [1] is still considered to be a good approximation of the passive electrical properties of a single cell for frequencies up to several megahertz. In this model every infinitesimal portion of the extracellular and intracellular media is modeled as a resistance and every infinitesimal

[5] The product RC (resistance × capacitance) has units of time and is known as the time constant, τ. In a "RC circuit" it is considered that stable voltages and currents are reached after 5τ.

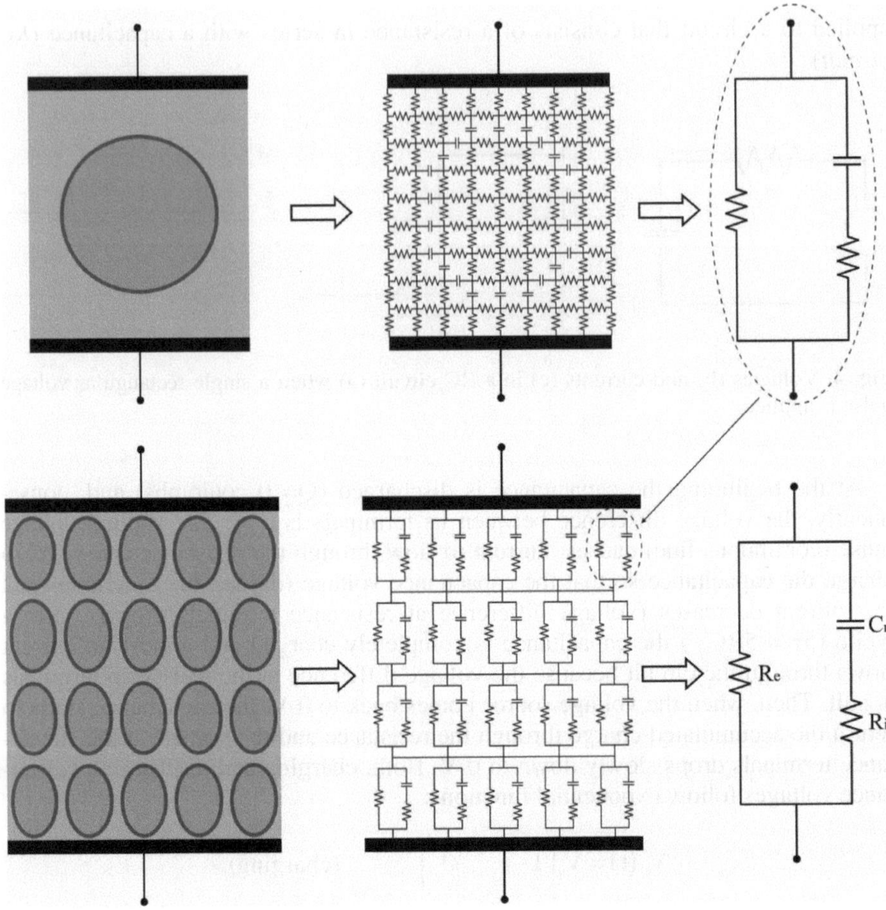

Fig. 5. Electrical models for cell (top) and tissue (bottom) as seen from the electrodes. Infinitesimal portions of the extracellular and intracellular media are modeled as resistances and infinitesimal portions of the membrane are modeled as capacitances. All those elements can be combined in an *extracellular resistance* (R_e) in parallel with the series combination of a *membrane capacitance* (C_m) and an *intracellular resistance* (R_i). The same three-element model can be employed to represent the behavior of tissues.

portion of the membrane is modeled as a capacitance. Circuit theory allows all those elements to be combined to form a simple equivalent circuit as seen from the electrodes (Figure 5): a resistance representing the extracellular medium (R_e) in parallel with the series combination of a capacitance (C_m), which represents the membrane, and another resistance which represents the intracellular medium (R_i). In a tissue or cell suspension the impedance contribution from all cells is combined so that the same electrical model (obviously with different values) can be employed to characterize the impedance behavior as seen from the measurement electrodes.

The resistive behavior of the extracellular and intracellular media is basically due to their contents of ions; both media are in fact ionic solutions. Most abundant ions in the extracellular medium are Na^+ and Cl^- whereas in the intracellular media K^+ is the most abundant ion. Blood plasma conductivity at 37 °C is 1.5 S/m (resistivity = 0.66 Ω.m). This is the value that most researchers chose for representing the extracellular conductivity. In some cases this same value is also employed for the intracellular conductivity, although most researchers prefer significantly lower values around 0.6 S/m [17].

The cell membrane consists primarily of a thin lipid bilayer. This film (~ 7 nm thick) is partially permeable to lipids and water molecules to pass through but it is almost impermeable to ions. Its intrinsic electrical conductance is very low and can be considered as a good dielectric. Therefore, the structure formed by the extracellular medium, the lipid bilayer and the intracellular medium is a conductor-dielectric-conductor and it behaves as a capacitance. Experimentally it has been found that such capacitance has a value of about 0.01 F/m^2.

Now, taking into account what has been learnt in the previous subsection about capacitance behavior in the frequency domain (i.e. it behaves as an open circuit at low frequencies and as a short-circuit at high frequencies) it can be understood that low frequency currents will not penetrate into the cell whereas high frequency currents will flow freely through it (figure 6.a). Hence the impedance magnitude (i.e. opposition to current flow) will be higher at lower frequencies (i.e. $|\mathbf{Z}|=R_e$) than at higher frequencies (i.e. $|\mathbf{Z}|= R_e//R_i$) because the electrical paths at low frequencies are narrower[6]. For intermediate frequencies a transitional behavior is

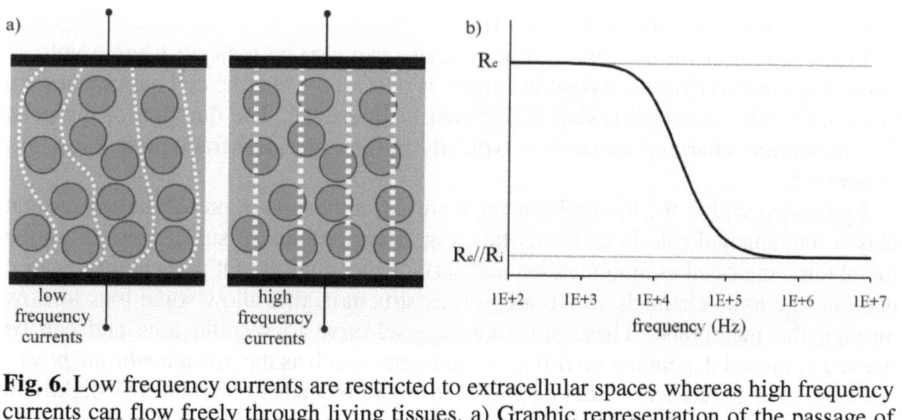

Fig. 6. Low frequency currents are restricted to extracellular spaces whereas high frequency currents can flow freely through living tissues. a) Graphic representation of the passage of low frequency and high frequency currents through a cell suspension or tissue. b) Idealized graph of impedance magnitude versus frequency in a living tissue. Note that the scale for the frequency is logarithmic. $R_e//R_i$ represents the value that results from the parallel combination of R_e and R_i which is equal to $R_eR_i/(R_e+R_i)$.

---

[6] Instead of $R_e//R_i$ (i.e. the parallel combination of R_e and R_i), the impedance magnitude at high frequencies is also denoted as R_∞ (where ∞ indicates infinite frequency) and instead of using R_e it is noted R_0 to indicate the impedance magnitude at low frequencies (ideally 0 Hz).

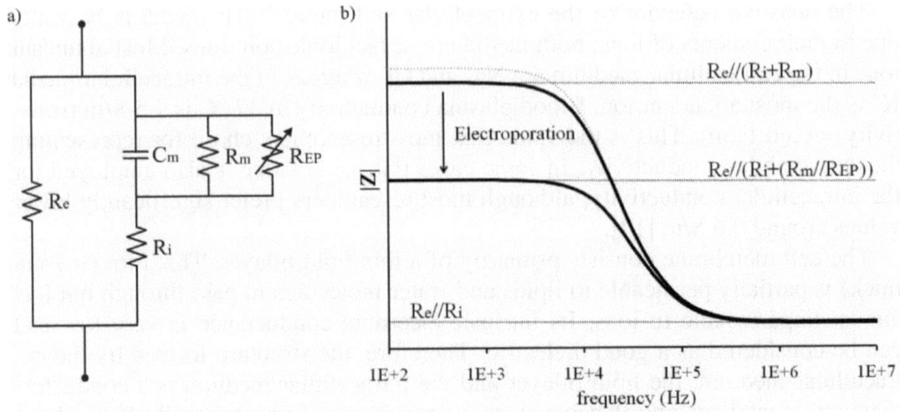

Fig. 7. The cell membrane is not purely dielectric, it has some residual conductivity that is increased by the electroporation phenomenon. a) Enhanced electrical model for a living tissue: R_m represents the short-circuiting resistance across the membrane due to ionic channels and R_{EP} represents the increased conductance due to the electroporation phenomenon (the arrow denotes that it is variable). b) Graph of impedance magnitude versus frequency for the enhanced electrical model. The impedance magnitude at low frequencies is slightly lower than R_e (figure 6.c). Electroporation causes a significant drop in impedance magnitude at low frequencies.

manifested (figure 6.b). Typically, for most animal tissues, the transition between the low frequency behavior and the high frequency behavior occurs at frequency band from about 10 kHz to about 1 MHz.

In the time domain the capacitance behavior can also be noticed: when a voltage pulse is applied to a tissue, a peak in current is observed while the cell membranes are being charged; similarly to what is depicted in figure 4.c. The duration of the peak (i.e. membrane charging process) is typically in the order of a fraction or a few microseconds.

Embedded within the lipid bilayer there are different sorts of protein structures that play a fundamental role in cell activities (e.g. transportation of substances across the membrane, inter-cell communication and surface recognition). Of particular relevance here are the ionic channels which are porous structures that allow some ions to flow through the membrane. These structures are selective to specific ions and can be opened or closed depending on different parameters such as the *transmembrane potential*[7]. Hence the electrical model displayed in figure 6 needs to be modified slightly in order to include the short-circuiting effect caused by these structures. In other words, the cell membrane is not a perfect dielectric and some leakage current through it can exist (Figure 7).

When electroporation occurs, cell membrane becomes more permeable and, therefore, the cell conductance (i.e. the inverse of the cell resistance) increases. At high frequencies this effect is not manifested in the impedance magnitude but at

[7] The *transmembrane potential*, or *transmembrane potential* or, simply, *membrane voltage*, is the voltage difference between the interior and the exterior of the cell.

low frequencies a significant impedance magnitude drop can be observed. This issue will be covered in more detail in subsection 6.

All what has been said above refers to the linear[8] passive electrical properties of biological samples. These properties are usually grouped with the term *bioimpedance*. The bioimpedance model described here is the simplest one for cells and tissues. Such model is appropriate to understand the electroporation phenomenon but it must be said that current bioimpedance knowledge is significantly more complicated. Readers interested in this topic will find further details in [18] or [19].

Cells also exhibit non-linear and active electrical properties. Nerve and muscle cell membranes contain specific ion channels that make them electrically excitable: an adequate disturbance of the transmembrane resting potential will trigger a sequence of fast changes in the membrane permeability to specific ions known as an *action potential* (AP). The AP in a spot of the membrane will affect adjacent portions of the membrane and will propagate as a voltage impulse through the nerve or the muscle fiber. This is the basic mechanism that allows *nerve impulses* (actually another name for AP) to propagate through the axons of afferent (i.e. sensory) nerves or efferent (i.e. motor) nerves. This also explains how the contraction wave is propagated through the myocardium so that the heart performs its pumping action.

AP propagation can be considered as a non-linear passive phenomenon. However, it depends to a large extent on the existence of a transmembrane resting potential. This resting potential between the interior and the exterior of the cell (approximately -70 mV for neurons) is kept constant thanks to an active mechanism in which specific ions are pumped inside and outside of the cell. Action potentials can be initiated when the resting transmembrane is cancelled or surpassed (i.e. *membrane depolarization*) by the application of an external electrical stimulus.

Additional knowledge on AP generation and its propagation may be convenient to fully comprehend subsection 7 in this chapter, however, no further details are included here because there is a large variety of sources that explain those topics from a natural sciences perspective, without assuming a strong background in electromagnetism; textbooks on cell physiology and human physiology are probably proper choices for most readers.

4 Electroporation Threshold

There are numerous evidences from experiments on cell suspensions [7, 20, 21], on isolated cells [4, 22, 23] and on artificial membranes [24-26] that electroporation occurs when the transmembrane voltage reaches a specific threshold. The value of such threshold depends on the characteristics of the applied pulses (number, duration and shape) and also on how electroporation is assessed (e.g. by noticing an increase of membrane conductance, by detecting intracellular contents

[8] The word *linear* in this context implies that these properties have the same value regardless of the magnitude of the test signal. When certain stimulus thresholds are surpassed, those properties are not linear anymore.

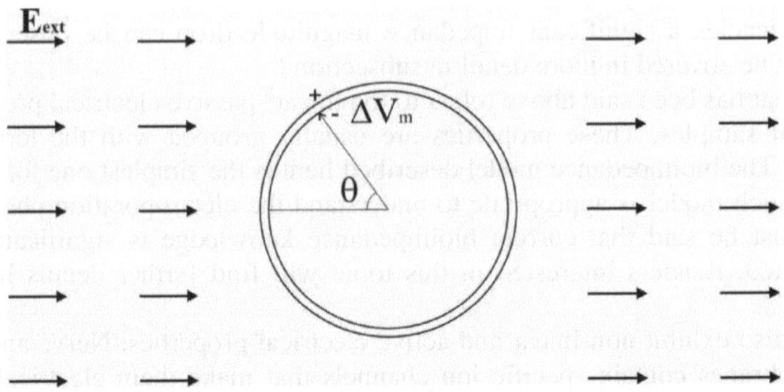

Fig. 8. An external electric ($\mathbf{E}_{ext}$) field induces a modification of the transmembrane potential. The electric field is intentionally not drawn in the vicinity of the cell because there it is not uniform (see Figure 9).

release or by observing cell lysis). Most authors report threshold values in the range from 200 mV to 1 V.

When an electric field is applied to a biological sample (e.g. a cell suspension or a tissue), after a short delay[9], transmembrane voltages are induced on the cells membranes. If the magnitude of the field is large enough, the induced transmembrane potentials will cause electroporation.

Before electroporation occurs, or for low electric field magnitudes, the induced transmembrane voltages are proportional to the magnitude of the applied field[10]. In the case that a uniform field is applied to a single cell in suspension, or to a diluted suspension[11], it is possible to employ a simple model in order to predict when electroporation will occur: for a spherical cell of radius r with negligible membrane conductivity (Figure 8) the induced transmembrane potential (ΔV_m) at each membrane point is [22]:

$$\Delta V_m = \frac{3}{2}|\mathbf{E}_{ext}|\, r\cos(\theta)$$

where θ is the angle between the radius (from cell center to evaluation point) and the applied external field ($\mathbf{E}_{ext}$). This expression is sometimes referred to as the *Schwan's equation* [28].

---

[9] This delay is due to the cell membrane charging process. See the next section 5.

[10] When electroporation occurs the conductance of the membrane changes abruptly and the relationship between the electric field magnitude and the transmembrane voltage is not linear anymore (this is analyzed in the next section 5). Actually, the cell membrane behavior can be nonlinear before reaching the electroporation threshold; some of the ion channels are voltage gated and they will switch at low transmembrane voltages. Nevertheless, for the sake of simplicity, this voltage gating phenomenon is ignored by most researchers working on electroporation models.

[11] According to [27] this model is valid when the volume percentage occupied by the cells is lower than 0.6%. For larger fractions the same paper gives an expression to correct the predicted transmembrane voltage values.

Therefore, since ΔV_m is proportional to the cell radius, lower fields will be required to achieve electroporation in larger cells. This fact has significant consequences when cell suspensions are electroporated in a cuvette[12]. For instance, the magnitude of the applied electric field needs to be optimized for each cell type and better results will be obtained for cell lines with small variations in cell sizes. On the other hand, in some cases it may be interesting to electroporate selectively the large cells in a heterogeneous sample [30].

Another interesting consequence of the above equation is that electroporation will not occur uniformly across the cell membrane; some areas will be easily electroporated (i.e. large $|\Delta V_m|$) whereas other will remain intact ($|\Delta V_m| \sim 0$ V). In particular, cell areas facing the electrodes, that is, perpendicular to the field direction ($\theta \sim 0$) will experience larger transmembrane voltages and therefore will become more easily electroporated. This phenomenon is nicely illustrated in [23]. In that paper, the researchers employed voltage-sensitive fluorescence dyes combined with fast microscopy and they were able to observe that 1) large $|\Delta V_m|$ occurred at the cell poles facing the electrodes and 2) at those areas it was observed a huge increase in membrane conductivity (i.e. electroporation). In a more recent paper by the same research group [31], the effect of the resting transmembrane potential is discussed: since this physiological potential (approximately +70 mV as defined[13] in figure 8) is added to the transmembrane voltage change induced by the external field, the cell pole facing the positive electrode (i.e. the anode, that is, left side in figure 8) reaches the transmembrane voltage threshold required for electroporation before it does the negative pole. This interesting observation, however, has little practical impact on electroporation since the induced transmembrane voltages will be typically much larger than the resting potential (e.g. 500 mV against 70 mV), particularly in irreversible electroporation.

For more complex geometries than that shown in figure 8 the equations that describe the induced transmembrane potential are much more intricate or do not exist. For those cases it is convenient to make use of numerical methods implemented on computers. For instance, in [32] the *finite element method* was employed to conclude that the maximum transmembrane potential induced in erythrocytes (biconcave shape) must be 22 % lower than the maximum transmembrane potential that is reached in an spherical cell of the same size.

There are different numerical methods for calculating physical variables in scenarios with arbitrary geometries. One of those methods suitable for computer simulations is the *finite element method* (FEM). The key idea of the FEM is the decomposition of the geometry into small simple elements (i.e. the *mesh* of elements) in which it is possible to solve differential equations related to the phenomenon under study. Boundary conditions impose the constraints that allow the method to generate a single approximate solution for the complete geometry. Up

[12] This electroporation method has been employed for more than 20 years in microbiology labs after the design by Potter et al. [29].

[13] In neurophysiology this potential is represented with negative sign because the potential is defined between the interior and the exterior of the cell, and not between the exterior and the interior of the cell as it is the case here.

to a point, FEM is equivalent to the decomposition in a matrix of simple elements shown in figure 5.

Figure 9 displays results from tridimensional FEM simulations of the induced transmembrane potential for different modeled cells under an external uniform field of 1000 V/cm (100 kV/m). The first three cases (a, b and c) correspond to the spherical cell model depicted in figure 8 and show the two main features that were expected from the above equation: 1) induced transmembrane potential is proportional to cell size and 2) transmembrane potential is not uniform across the cell membrane; the poles facing the electrodes experience a larger voltage. As matter of fact, the cell poles in the subfigure b (radius = 5 μm) reach the transmembrane voltage predicted by the equation, that is, 750 mV. The last three cases (d, e and f) correspond to an ellipsoidal cell (20 μm × 10 μm × 10 μm) that is oriented in different directions relative to the electric field. Although it is not easily observable in the figure, in the parallel orientation higher voltages are reached. Therefore, a general rule that can be intuited from these results is that induced transmembrane potentials depend on cell size, cell shape and cell orientation. It is s worth noting that a recent paper by Towhidi et al. [33] reports magnificently a experimental study in which multiple cells exhibit different electroporation thresholds depending on their sizes, shapes and orientations.

Notice in the simulations displayed in figure 9 that the electric field is significantly distorted in the vicinity of cells. This fact implies that in dense cell suspensions and tissues, where the spaces between cells are narrow, the electric field will be far from uniform at microscopic level. Hence the above equation for the induced transmembrane voltage in the case of an isolated cell will be useless for tissues. However, given that the resistivity of some tissues can be assumed to be homogeneous at macroscopic level, it is possible to consider a *macroscopic* electric field distribution. Then it is reasonable to hypothesize that cells in a specific tissue region will be electroporated if the macroscopic electric field magnitude at that region reaches or surpasses a specific field magnitude threshold. Such threshold will not only be dependent on electroporation parameters but also on tissue type, similarly to what happens with suspensions of cells of different lines. Besides cell sizes, shapes and orientations, here the intercellular distances will also play a significant role. That is, the separation between cells will modulate the transmembrane potential induced by the external macroscopic field [34].

The above hypothesis is now commonly employed to predict the extension and shape of tissue volume that will be electroporated with a specific electrode setup. Figure 10 illustrates how this is done. It is first created a geometrical model of the tissues and the electrodes in which tissues conductivities are modeled as being homogeneous (in Figure 10 a single tissue type is modeled). Then the electric field distribution is computed by analytical or computer assisted methods[14]. And,

[14] In some simple cases it is possible to find analytical solutions (i.e. equations) that describe the field distribution [35], however, in most cases computer tools based on numerical methods such as FEM will be necessary.

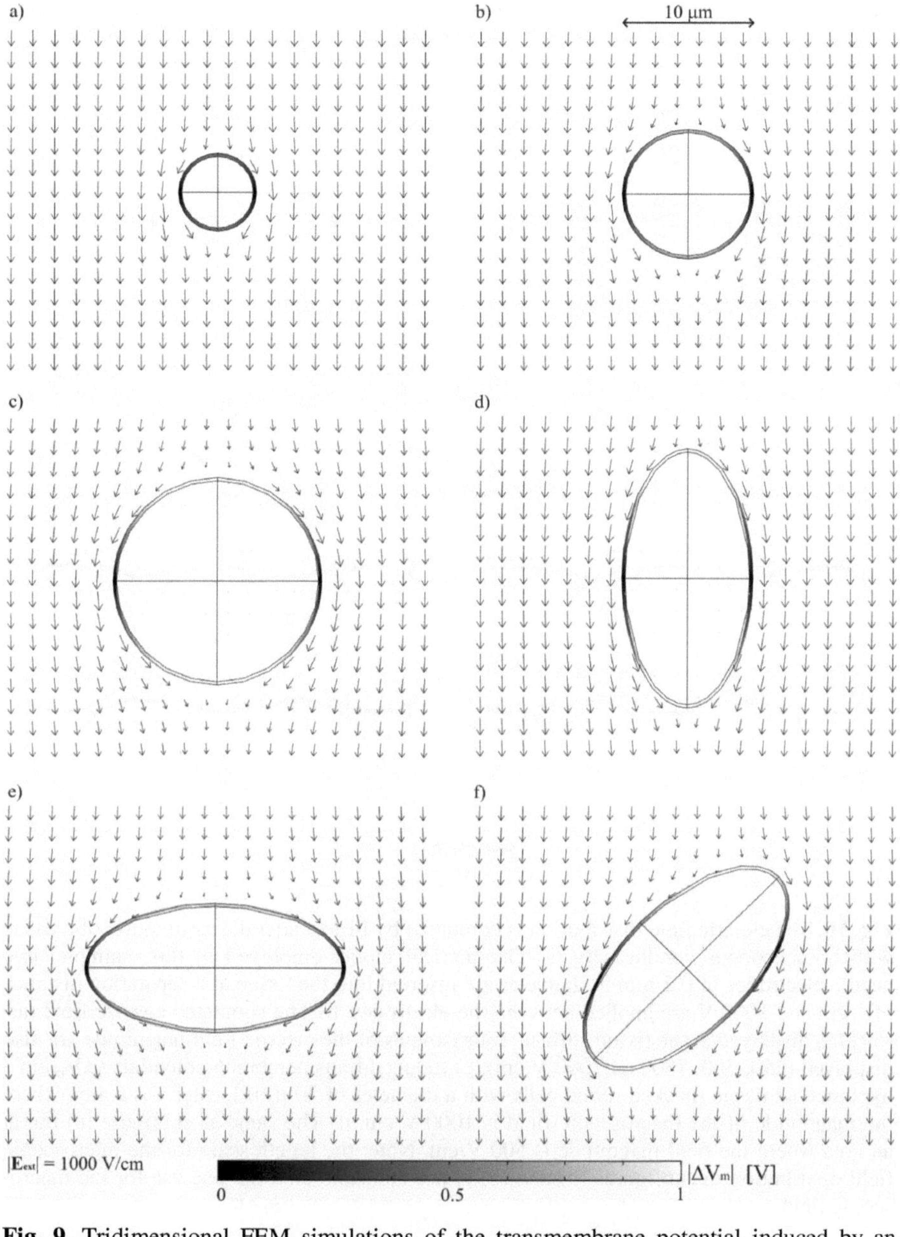

Fig. 9. Tridimensional FEM simulations of the transmembrane potential induced by an external electric field of 100 kV/m. Only a single plane (across the center) is shown for each cell case. Arrows indicate field magnitude and direction. a) Spherical cell of 6 microns in diameter: no membrane region is above 0.5 V .b) Spherical cell of 10 μm in diameter: top and bottom poles reach 0.75 V (in absolute value). c) Spherical cell of 16 μm in diameter: large regions are above 1 V. d) Ellipsoidal cell (20 μm, 10 μm, 10 μm) parallel to the field direction. e) Ellipsoidal cell (20 μm, 10 μm, 10 μm) perpendicular to the field direction. f) Ellipsoidal cell (20 μm, 10 μm, 10 μm) oblique to the field direction.

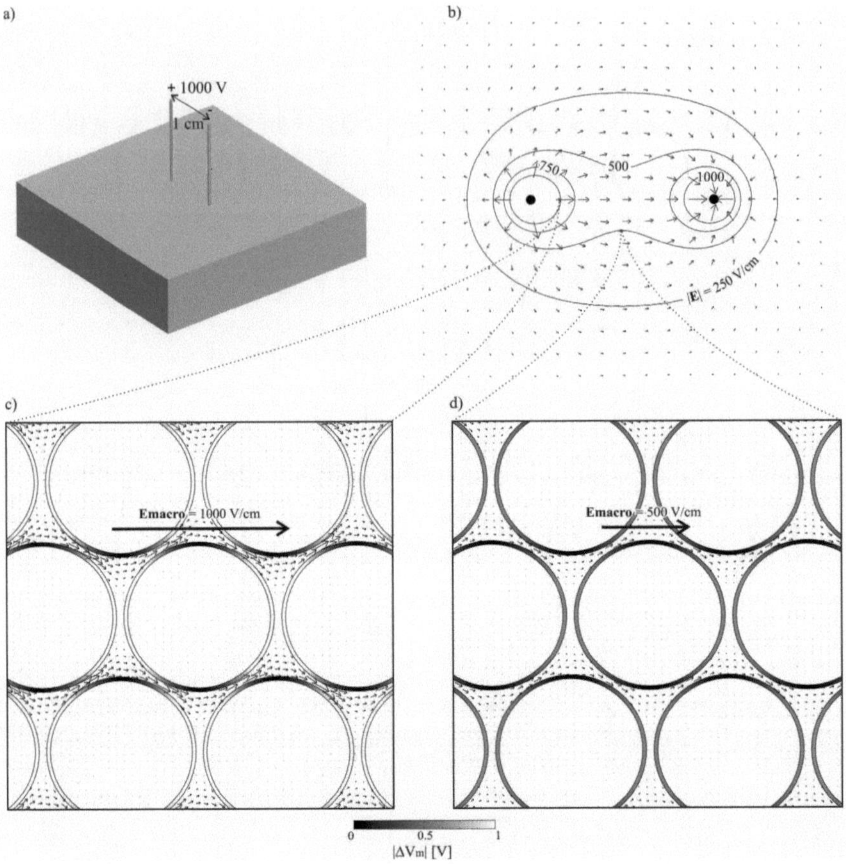

Fig. 10. The electric field distribution is computed by FEM under the assumption of a tissue with homogeneous conductivity. a) Geometrical model employed in this example: two needle electrodes of 0.5 mm in diameter are inserted into the tissue at a separation distance of 1 cm and 1000 V are applied between the electrodes. b) The simulated electric field (arrows) is displayed at the tissue surface; four isolines of the electric field magnitude are also displayed (250, 500, 750 and 1000 V/cm). c) Induced transmembrane potentials (ΔV_m) in a hypothetical tissue (packed round cells with a diameter of 20 μm) located at an area where the magnitude of the macroscopic field is 1000 V/cm. d) The same as subfigure (c) but at an area where the field magnitude is 500 V/cm. Note: the length scale for the microscopic field distribution in subfigures (c) and (d) is not the same than the one use for the macroscopic field.

finally, it is considered that only the cells at areas where the field magnitude surpasses a specific threshold reach a sufficiently high transmembrane voltage for successful treatment. This sort of *treatment planning* has been validated in multiple experimental studies [36-39].

At least three different electric field thresholds can be defined that are of interest in electroporation. The lowest of them would be the threshold for the manifestation of reversible electroporation (E_{rev}); only the cells within areas where $E \geq E_{rev}$ are electroporated. If a second threshold (E_{irrev}) is reached or surpassed, electroporation will compromise the viability of the cells. A larger threshold can also be defined ($E_{thermal}$) for the manifestation of thermal damage caused by the Joule effect. This is particularly relevant in the case of IRE ablation techniques: if irreversibility threshold is surpassed but thermal threshold is not reached then cells are destroyed but tissue scaffold is spared and that facilitates post-treatment healing [40].

Therefore, the goal of simulations in electroporation treatment planning must be to guarantee that a rather homogeneous electric field magnitude ($E_{min} \leq E < E_{max}$) is created in the region of interest and that an electric field magnitude as low as possible is induced in the regions not to be treated. Note that here it has been indicated that the electric field magnitude in the region of interest must be between E_{min} and E_{max} instead of E_{rev} and E_{irrev} or E_{irrev} and $E_{thermal}$. The E_{rev} and E_{irrev} labels are usually employed to denote the minimum electric field magnitudes at which reversible electroporation and irreversible electroporation can be manifested, but that does not imply that all the cells will experience reversible or irreversible electroporation. In other words, if the electric field magnitude at a certain area is slightly larger than E_{rev} that does not guarantee that all the cells in that area will experience enough reversible electroporation. Again, multiple factors, such as cell size, cell shape and intercellular spacing, modulate the induced transmembrane voltage. By performing animal and preclinical experiments it is possible to obtain an E_{target} that represents the optimal electric field magnitude for a specific treatment. Then it is also possible to define a tolerance around this value. That is, in the region to be treated, the electric field magnitude must be within this tolerance interval in order to achieve successful treatment.

Up to now, electroporation treatment planning has been focused on designing suitable electric field distributions by choosing: 1) the location, and shape of the electrodes and 2) in some cases, the order and sequence in which high voltage pulses are applied between the involved electrodes. For instance, Gilbert et al.[41] proposed the use of needle structures in which the pulses were applied between opposite electrodes in a rotating sequence. Recently, in [42], we introduced a third mechanism of control for the distribution of the electric field: the use of electrolytic gels and liquids which have specific conductivities so that the field distribution can be modulated. For instance, in the particular case of tissue electroporation by using plate electrodes, a gel with a conductivity similar to that of the tissue can be employed in order to homogenize the field distribution as it is illustrate in figure 11. In [43] we demonstrate *in vivo* the effectiveness of this specific technique and we show that it is tolerant to conductivity mismatching errors between the gel and the tissue.

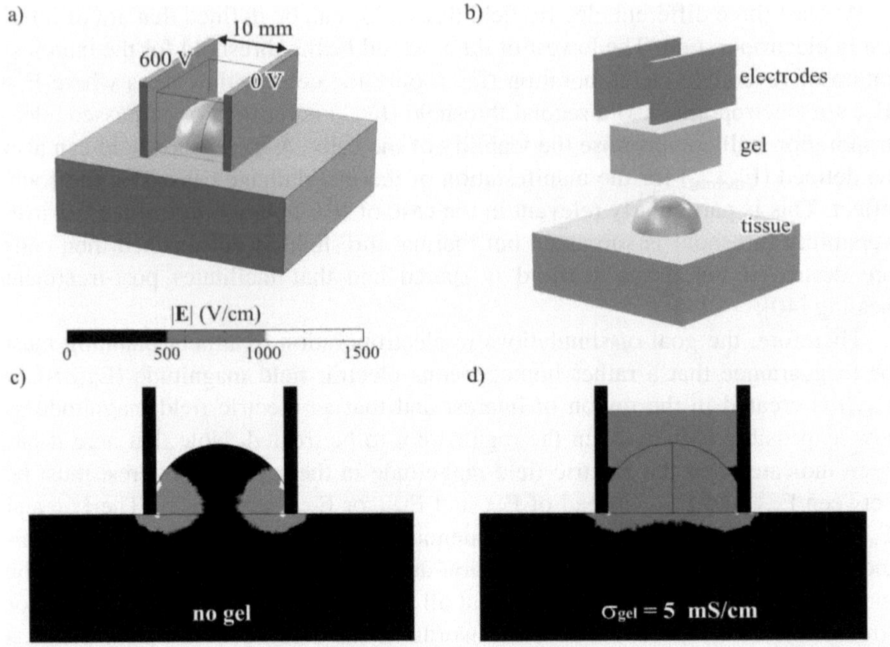

Fig. 11. Use of conductive gels, matched to the conductivity of the tissue, to fill dead spaces between plate electrodes gripping the tissue so that the electric field distribution becomes less heterogeneous. a) and b) Geometrical model employed for the simulation. c) Simulation result when no gel is used; the field distribution is too heterogeneous. d) Simulation result when a gel with conductivity of 0.5 S/m is applied. This technique is tolerant to conductivity mismatches between the gel and the tissue [43].

5 Time Course of Currents and Electric Fields during Electroporation Pulses

As it has been mentioned above, due to the membrane charging process there is a delay between the application of the electric field and the materialization of the induced transmembrane voltage. Actually, it is more accurate to say that transmembrane voltage is not delayed but increases slowly up to a stable value. For the spherical cell model of figure 8 it is possible to write the equations that describe such behavior [23]:

$$\Delta V_m = \frac{3}{2}|\mathbf{E}_{\mathbf{ext}}|\, r\cos(\theta)\left(1-e^{-t/\tau}\right)$$

$$\tau = rC_m\left(\rho_i + \frac{\rho_e}{2}\right)$$

where C_m is the capacitance of the cell membrane per area unit and ρ_i and ρ_e are the resistivities of the intracellular and the extracellular media respectively.

Therefore, if a very short pulse is applied (duration in the order of τ)[15] the transmembrane voltage will not reach its maximum possible value ($\Delta V_{m\_max}=1.5\times|\mathbf{E_{ext}}|\times r$ at time >> τ) and probably the cell will not experience electroporation unless $|\mathbf{E_{ext}}|$ is extremely high. Another interesting consequence is that larger cells will experience larger transmembrane voltages but they will require more time to reach the stable voltage. At this point it is opportune to mention the research that multiple groups are carrying out regarding the use of very high voltage pulses of very short duration; in the order of some nanoseconds or tens of nanoseconds [44-49]. The main motivation for this research comes from the believe that those ultra-short pulses, known as nanosecond Pulsed Electric Field (nsPEF), could be able to induce electroporation of intracellular membranous structures (e.g. mitochondria) without disturbing the cell membrane. In *conventional* electroporation, that is, with pulses larger than 10 μs, the cell membrane reaches its maximum transmembrane potential and when this happens the intracellular structures become isolated from the external field and hence cannot experience electroporation[16]. On the other hand, it seems reasonable that if very short pulses of high magnitude are applied then the intracellular membranous structures will be charged to a sufficiently high voltage for electroporation before the cell membrane is barely charged. That is the reason why it was expected that it would be possible to electroporate internal structures without causing electroporation of the plasma membrane. However, recent computer models [51] and experimental results [52] indicate that cell membrane electroporation also occurs when nanosecond pulses are applied. That is, with nsPEF all the membranous structures of the cell are electroporated (some authors refer to this phenomenon as *supraelectroporation*). This is illustrated in figure 12: it is simulated the evolution of the transmembrane potential in the cell membrane (ΔV_{m_out}) and in the membrane of an internal structure (ΔV_{m_in}). After two microseconds (subfigure a) the cell membrane potential reaches the predicted stable value ($\Delta V_{m\_max}=1.5\times|\mathbf{E_{ext}}|\times r$) and the potential across the membrane of the internal structure drops to zero. However, during the first nanoseconds after the pulse is applied (subfigure b) both transmembrane potentials are very similar. Hence if a very short electric field pulse is applied (~ 20 ns) and it has a much larger magnitude than the pulse applied in the example, it is reasonable to expect that both membranes will reach a sufficient transmembrane voltage for electroporation.

The fact that nanosecond pulses are able to achieve electroporation implies that the electroporation process occurs very rapidly once the transmembrane voltage threshold is reached; in a matter of a few nanoseconds at most. In fact, by performing membrane conductance measurements it was demonstrated more than twenty-five years ago that "membrane breakdown" occurs in less than 10 ns [53].

---

[15] A round cell of 20 μm in diameter will have a charging time constant, τ, of about 0.18 μs (Cm = 0.01F/m, ρ_e = (1/1.5) Ω.cm, ρ_i =(1/0.66) Ω.cm).

[16] This statement, given by researchers in the nsPEF field [50], is not necessarily true: when the cell membrane transmembrane voltage is large enough to induce electroporation, the membrane resistivity drops dramatically and internal structures are not isolated anymore from the external field. Therefore, if the magnitude of the field is extremely large (as it may be the case in IRE) the smaller internal structures might also be electroporated.

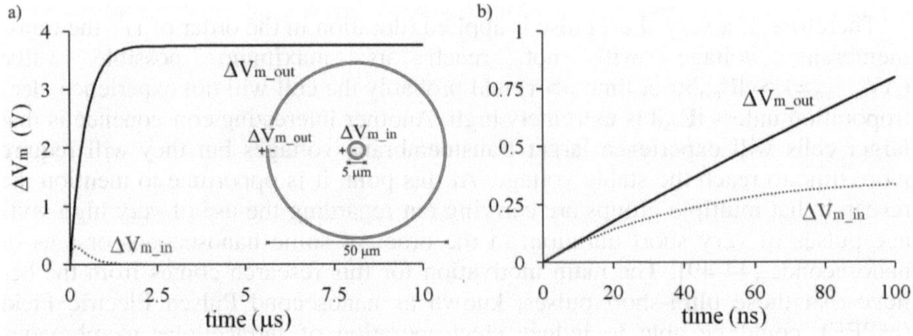

Fig. 12. Simulation of the evolution of the transmembrane potential in a cell membrane (ΔV_{m_out}) and in the membrane of an internal structure (ΔV_{m_in}). The magnitude of the applied electric field is 1000 V/cm.

In all the above equations and examples for the transmembrane potential it has been assumed that the conductivity of the membrane is always extremely low. However, once the transmembrane potential reaches the critical value and electroporation occurs, this assumption is no longer valid and it is easy to understand that the short-circuiting effect caused by the permeabilization will somehow neutralize the transmembrane voltage rise. This is illustrated experimentally in [23]. There it is shown that the cos(θ) is not held when the electric field magnitude goes above a certain threshold: the profile of the transmembrane voltage across the membrane flattens at the poles facing the electrodes. That is, the transmembrane potential does not seem to go above a critical value which is only slightly larger than the threshold to initiate electroporation.

In fact, one could expect that the transmembrane voltage would go back to values much lower than the electroporation threshold (imagine the hypothetical case in which the electroporation effect is so intense that the membrane is fully short-circuited; that would imply $\Delta V_m = 0$). And indeed it has been noticed experimentally [31] and it has been predicted by electroporation models [54] that the transmembrane potential can slowly decrease in a matter of a microseconds or tens of microseconds. Nevertheless, a dramatic and sudden drop in transmembrane potential is not observed. The transmembrane potential remains quite stable after the critical value has been reached. This would be an indication that the sort of permeabilization that occurs immediately (in a matter of nanoseconds) is somehow reversible also in an immediate fashion: the permeabilization tends to reduce the transmembrane voltage but if that happens then the conductivity decreases immediately and hence voltage goes up again, and, therefore, the voltage remains stable. On the other hand, the slow decrease of the voltage that comes later during the pulse would indicate that membrane conductivity is increasing in a way that cannot be reversed immediately after removing the electric field.

To sum up all the above concerning single cell studies: 1) electroporation does not occur until the transmembrane voltage reaches, in about a few microseconds, a critical value (between 200mV and 1V), 2) once it reaches this value the conductivity of the membrane increases immediately (< 10 ns) to a value that keeps the

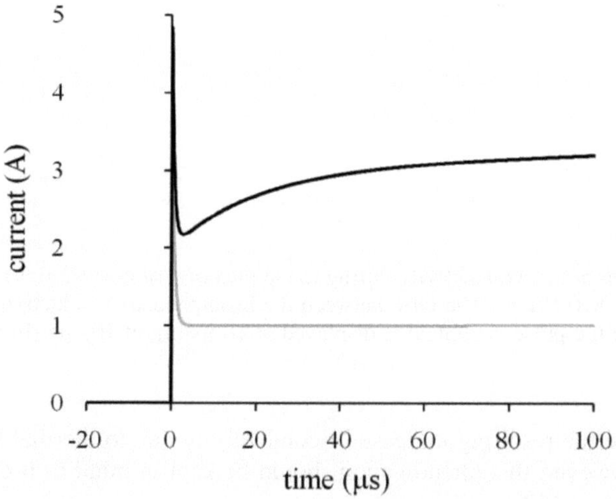

Fig. 13. Typical recording (black line) of current during the application of a 100 μs high voltage pulse. The gray line indicates what would be the response if electroporation did not occur: after a peak in current due to cell membrane charging, current would increase very slightly (almost unappreciable in the graph) because of resistive heating.

transmembrane voltage close to the critical value and 3) then the membrane conductivity keeps rising slowly during the application of the pulse.

In figure 13 it can be observed a typical behavior of the current signal when a rectangular electroporation pulse is applied to a living tissue or a very dense cell suspension. This example does not correspond to an actual experiment but it shows the main features that can be noticed in actual measurements in tissues [55, 56] or dense cell suspensions [57]: after an initial peak due to cell membrane charging, current increases exponentially and afterwards it seems to increase much slower in a linear fashion or it stops increasing. The initial abrupt change in conductance[17], masked by the membrane charging process, is probably the manifestation of the immediate and reversible membrane permeabilization mentioned before. The later exponential rise shows that membrane conductance increases slowly and moderately during the pulse.

The gray line in figure 13 shows what would be the behavior of the tissue in case the electroporation phenomenon did not exist. After the peak due to the membrane charging process, tissue conductivity would increase moderately, almost inappreciably, due to resistive heating. This tissue conductivity increase due to Joule heating is related to the fact that the conductivity of ionic solutions shows a positive dependence on the temperature. In the particular case of physiological fluids is estimated that the conductivity increases 2% per each °C [58]. In the

[17] G=I/V and in this case V is constant, therefore, conductance is directly proportional to the current.

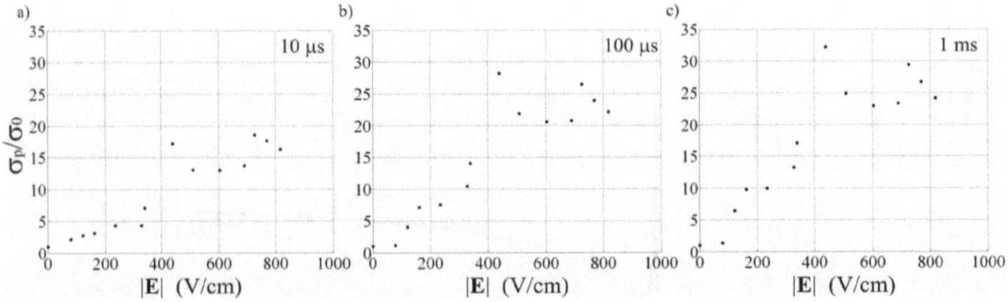

Fig. 14. Relative potato conductivity during the application of electric field pulses ranging from 0 V/cm to 820 V/cm . The ratio between the instantaneous conductivity and the conductivity before the pulse is applied is displayed at 10 μs (a), at 100 μs (b) and at 1000 μs (c).

above example the resulting increase of conductivity due to thermal heating is not significant, however, this phenomenon should be kept in mind as it can be significant in other cases [57].

Following the replacement concept of the 3 Rs approach for animal testing[18], it is convenient to note here that some vegetables can be a proper alternative for studying bioelectrical aspects of tissue electroporation. In particular, raw potato tuber is a good choice because any irreversibly electroporated area will be distinctively darker 5 hours after electroporation[19]. Figure 14 displays the ratio between the conductivity at specific times of a 5 ms pulse (10 μs, 100 μs and 1000 μs) and the original potato conductivity (σ_0); multiple field magnitudes are tried. Observe that the immediate (10 μs) change in conductivity is quite significant (15 times the original conductivity), that it follows an almost linear relationship with the field magnitude up to 400 V/cm and that for larger field values it seems to saturate. During the pulse the conductivity keeps rising, but at a slower pace (values at 5 ms are virtually equal to those at 1 ms), and the maximum conductivity that it is reached is about $30\times\sigma_0$. It is interesting to note that this maximum ratio ($\sigma_p/\sigma_0 = 30$) is really close to the impedance ratio R_0/R_∞ that is measured before electroporation[20]. This is in agreement with the electrical model of electroporation depicted in figure 7: in principle, the minimum resistance that can be reached by electroporation must be equal to the minimum impedance magnitude that can be obtained by applying high frequencies (R_∞); in both cases the cell membrane is "short-circuited". Therefore, at 5 ms, for field

[18] The 3Rs approach was introduced by Russell and Burch [59]. The 3 Rs represent: reduction of the number of animals used, refinement of techniques and procedures to reduce pain and distress, and replacement of animal with non-animal techniques.

[19] When the electroporation protocol consists of 8 pulses of 100 μs at 10 pulses/s then the electric field threshold for irreversible electroporation is about 300 V/cm (not published data obtained by the author; methods equivalent to those in [60]). Potato darkening after 5 hours is probably due to an accelerated oxidation of chemical constituents caused by a decompartmentalization of certain enzymes and substrates [61] that occurs at cell lysis caused by electroporation.

[20] R_0/R_∞ is the ratio between the impedance magnitude at low frequencies and the impedance magnitude at high frequency; see subsection 3.

magnitudes larger than 400 V/cm, cell membrane has become so permeable to ions that its resistance is insignificant when compared to the resistance of the intracellular and extracellular media. Such statement does not imply that membrane is completely disrupted; only a tiny fraction of the membrane area needs to be opened (<1%) in order to achieve such irrelevance in terms of conductivity [31].

In comparison with animal and human soft tissues the ratio R_0/R_∞ of potato tuber is extremely high. Probably the only exception to this statement is the skin, which has a very low conductivity, particularly when it is dry[21]. Hence it is not too surprising that electroporated skin shows a much higher ratio σ_p/σ_0 at 100 μs than other animal tissues (Figure 15).

In figure 15 it is also interesting to note that muscle seems to reach σ_p/σ_0 saturation for field magnitudes above 600 V/cm whereas at this same field the ratio σ_p/σ_0 in the case of tumors has barely gone above 1. In fact, tumor σ_p/σ_0 only rises above 3 when field magnitude is larger than 2500 V/cm (not published data obtained in collaboration with Prof. L. Mir). These observations indicate how large the difference in electroporation thresholds can be between different tissues. The same can be noticed when the effects of electroporation are analyzed: for the same protocol, 8 rectangular pulses of 100 μs at 10 Hz, irreversible electroporation in muscle is observed at field magnitudes lower than 500 V/cm [56] whereas for sarcoma tumors the threshold seems to be larger than 2000 V/cm [65].

The fact that tissue conductivity depends on the magnitude of the electric field must be taken into account when calculating field distributions: once a high voltage pulse is applied some tissue areas will be electroporated very fast, their conductivities will increase and a redistribution of the electric field magnitude will occur. Since electroporation takes place almost immediately (<10 ns) when the critic transmembrane potential is reached (in about 1μs after the pulse), it can be presumed that the whole process of field redistribution will happen in a few microseconds, that is, in an almost insignificant portion of the pulse (typically 100 μs). This phenomenon has been incorporated in recent FEM simulations of the electric field distribution [66-69]. In these simulations the electric field magnitude distribution is computed in an iterative process in which in each step the conductivities are adjusted according to the electric field computed in the previous step. Figure 16 shows an example that shows the importance of taking into account the dependence of the conductivity on the field magnitude: when the iterative method is not applied the computed field distribution indicates that the field magnitude within the tumor is very low because of the skin resistance; it would not be capable of electroporation. A lot of experimental and clinical studies on electrochemotherapy [70] indicate that that cannot be the case; electroporation indeed occurs within tumors. Moreover, in a recent paper [71] it was measured the electric field magnitude within tumors and it was found that, although the skin indeed has a significant shielding effect, the actual inner electric field value was in the same order of magnitude than the inner electric field predicted without the influence of the skin.

---

[21] A comprehensive collection on tissue impedance data can be found in http://niremf.ifac.cnr.it/tissprop/ which is based on the data collected in [62-64].

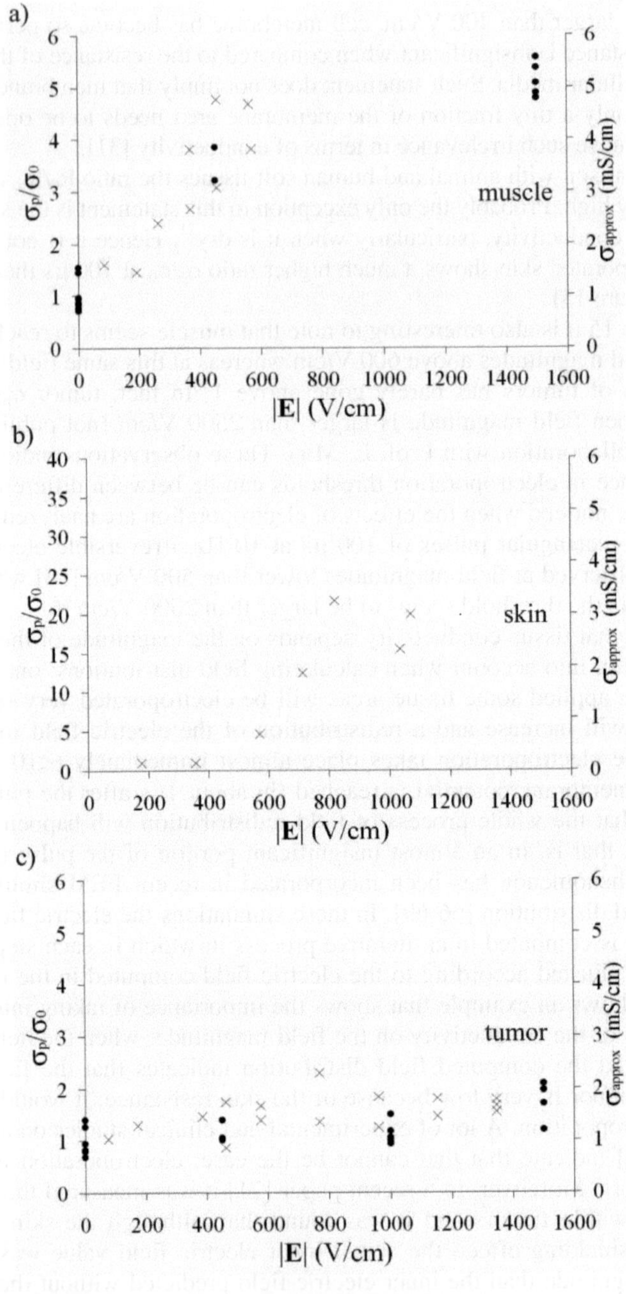

Fig. 15. For three different animal tissues (rat transversal skeletal muscle, rat skin and sarcoma tumor implanted in mice), relative conductivity (σ_p/σ_0) at 100 μs after the beginning of the pulse. The approximate absolute conductivity is also represented on the secondary vertical axis (right). The symbol × indicates data from [67] whereas the symbol • indicates data collected by the author (muscle data partially published in [60]).

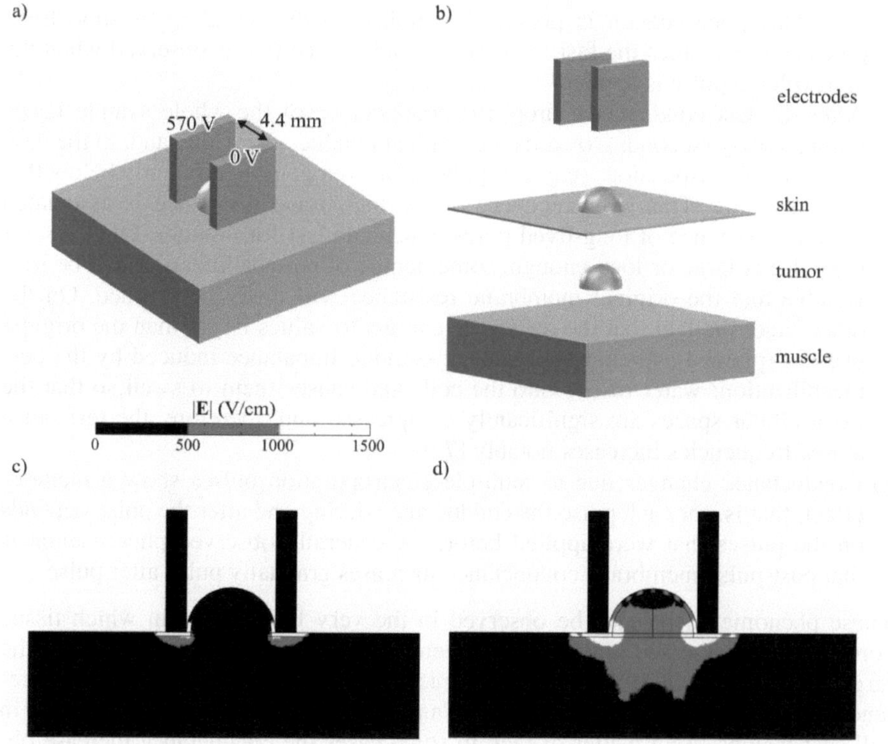

Fig. 16. Comparison between the simulation results of the electric field magnitude (|**E**|) distribution when it is assumed that tissue conductivity does not change (c) and when it is assumed that tissue conductivity depends on the electric field and the distribution is computed with an iterative process (d). The first case (c) is completely unrealistic as it would imply that only the skin is electroporated.

6 Conductivity Changes Induced by Electroporation

There are quite a few studies on isolated membranes [26] and on cells [23, 57, 72-77] in which the electrical conductance is measured after the application of the electroporation pulses or in between the application of multiple pulses. From all these studies some interesting general observations can be extracted:

1) After cessation of the high voltage pulse, membrane conductance[22] drops very fast (<<1 ms) to a value which is much smaller than the conductance during the pulse (σ_p) but significantly larger than the conductance value before the pulse

[22] Conductance before and after the pulse is typically measured by means of low frequency (<10 kHz) AC signals of very low amplitude. If the frequency is sufficiently low then the impedance magnitude is very similar to the resistance value at "0 Hz" (see figure 6.b). DC signals are generally avoided because they have an effect on the sample (electrochemical reactions and possible effects on the electroporation process).

(σ_0). This phenomenon is presumably linked to the resealing of short-lived pores and resembles the fast increase in conductance that is observed when the high voltage pulse is applied.

2) After the fast conductance drop, the conductance of the whole sample keeps falling slowly (seconds) towards the original conductance value and, in the case of dense cell suspensions (e.g. cell pellets), it even goes significantly below that original value. This slow recovery of the membrane resistance is associated with the existence of long-lived pores which can last for minutes. Of course, if the pulse is large or long enough, some degree of permeabilization will be irreversible and the original membrane resistance will never be reached. On the other hand, the fact that the resistance can get to values larger than the original one is explained as being the result of osmotic imbalance induced by the permeabilization: water rushes into the cells and causes them to swell so that the extracellular spaces are significantly compressed and, therefore, the resistance at low frequencies increases notably [72].
3) Conductance changes due to multiple electroporation pulses show a memory effect, that is, for each pulse the conductance during and after the pulse depends on the pulses that were applied before. A generally observed phenomenon is that post-pulse membrane conductance increases gradually pulse after pulse.

These phenomena can also be observed in the very few studies in which tissue conductance is assessed after or in between the pulses [55, 60, 78-80]. Due to the large ratio between cell volume and extracellular volume in tissues, the conductance increase caused by electroporation in tissues is much more noticeable than in cell suspensions. As a matter of fact, in some cases the conductance increase observed in cell suspensions is not related to an increase of membrane conductance but to secondary effects such as the release of intracellular contents [57].

Typical conductivity behaviors during tissue electroporation (8 pulses of 100 μs at 10 Hz) can be observed in figure 17. These examples (obtained from [80]) correspond to *in vivo* experiments with rat livers. When the electric field magnitude is 450 V/cm (subfigure a) conductivity slightly increases pulse after pulse up to a value $\sigma/\sigma_0 = 1.08$. The conductivity during the pulses (represented with a single circle) is significantly higher than the conductivity just before or just after each pulse. This *in-pulse* conductivity (σ_p) also shows a quite linear progressive increase. On the other hand, when the magnitude is 1500 V/cm (subfigure b) the in-pulse conductivity seems to saturate at a value (3.2 mS/cm) which is quite close to the pre-treatment admittance value at high frequencies [80]. In this case the conductivity increase induced by each pulse is quite significant and it accumulates up to a conductivity value a 42% higher than the original conductivity ($\sigma/\sigma_0 = 1.42$).

As it would be expected, in all tissue studies reported up to the current date tissue conductivity immediately after the electroporation sequence is larger than the original conductivity. The in-pulse conductivity also seems to increase pulse after pulse, or at least reaches a saturation level and remains stable (Figure 17.b). However, there are two experimental cases in which a decrease of the in-pulse conductivity was observed: irreversible electroporation of rat muscle [60] and rat artery [81, 82]. A plausible, but not demonstrated, explanation to this unexpected

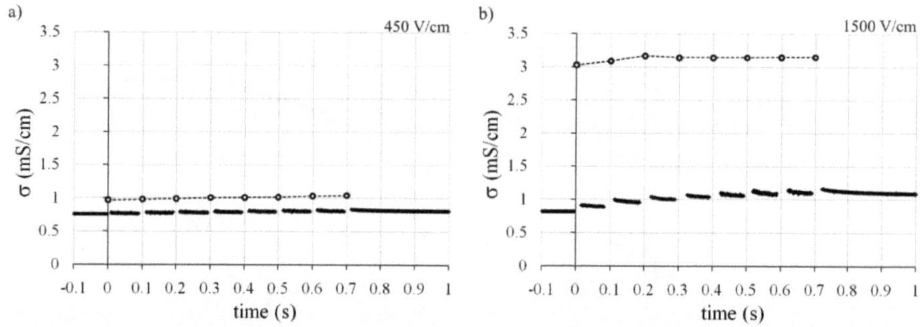

Fig. 17. Two examples of conductivity evolution during *in vivo* electroporation of rat liver. Points marked with circles and joined by a dashed line represent the conductivity values obtained at the end of the electroporation pulses (*in-pulse* conductivity, σ_p). The electroporation protocol consisted of 8 rectangular pulses of 100 μs at 10 Hz. In a) the magnitude of the electric field was 450 V/cm and in b) it was 1500 V/cm.

phenomenon could be that the voltage pulses cause contraction of the arteries by stimulating the muscle cells (vascular smooth muscle cells in the case of the artery) and that such contraction results in an increase of the electrical impedance [83-85].

The eight 100 μs pulses at 10 Hz with a field magnitude of 450 V/cm is considered to be a reversible electroporation protocol for the rat liver case. On the other hand, when the same protocol is applied with a magnitude of 1500 V/cm the effects are unmistakably irreversible. For instance, in less than two hours after electroporation the area exposed to the electric field shows a very distinctive dark red color that is due to the accumulation of erythrocytes after the destruction of the endothelium of the liver vasculature system. Furthermore, microscopic observations show a very significant increase of the interhepatocyte space that is probably caused by massive cell lysis [80].

Conductance monitoring after the electroporation sequence yields interesting observations (figure 18). The "reversible" protocol (450 V/cm, gray line) induces first a fast, but small (<10%), increase in conductance and then conductance decreases slowly down to a minimum, at about 2 minutes, that is lower than the original conductance value. As indicated earlier, such decrease is thought to be related to the osmotic imbalance induced by the permeabilization [72]. Besides permeabilization, it is possible that another source of osmotic imbalance concurs in this case: tissue electroporation is usually accompanied by reversible blood occlusion [86] which causes ischemia and in turn induces cell swelling [87]. After the minimum, conductance goes up very slowly and reaches values higher than the original value, probably due to some degree of irreversible electroporation (cell lysis) around the electrode edges. Understandably, the "irreversible" protocol (1500 V/cm. black line) shows a much larger immediate increase in conductance (>35%). However, the pattern up to 10 minutes is quite similar to the 450 V/cm case: a minimum after one minute and a tendency to reach quite stable values

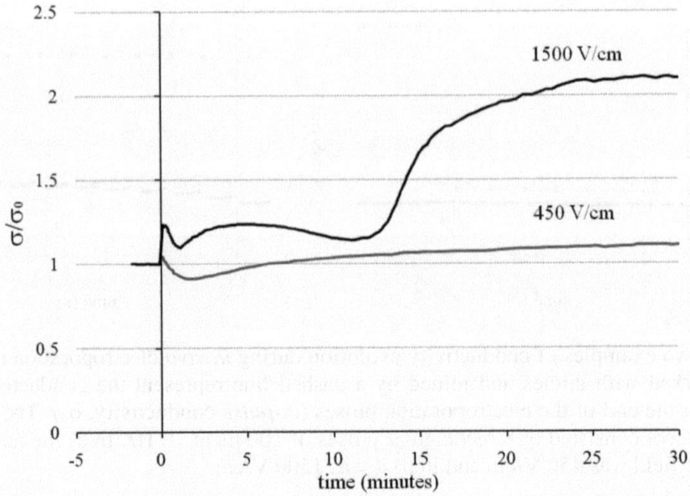

Fig. 18. Examples of *in vivo* relative conductivity evolution after electroporation of rat liver. Electroporation (8 rectangular pulses of 100 μs at 10 Hz) is produced at time = 0 minutes. Grey trace is for the case in which the magnitude of the electric field is 450 V/cm whereas the black trace is for a magnitude of 1500 V/cm and it clearly manifests the effect of irreversible electroporation, that is, a remarkable increase in conductivity.

Table 1. Experimentally measured increases (mean ± standard deviation) in tissue conductivity after an electroporation sequence that consists of 8 pulses of 100 μs at 10 Hz. Rat liver results are from [80] and rat skeletal muscle results are from [60]. Results from sarcoma tumor implanted in mice and results from potato have been obtained by the author in collaboration with Prof. L. Mir's lab and have not been published yet.

Δσ (50 ms after EP)	EP protocol: 8 pulses of 100 μs separated by 100 ms		
	|E| = 450 V/cm	|E| = 1500 V/cm	|E| = 2500 V/cm
Rat liver	9 ± 3 %	43 ± 5 %	-----
Rat muscle	70 ± 15 %	95 ± 23 %	-----
Mice sarcoma	8 ± 2 %	12 ± 3 %	39 ± 5 %
Potato	78 ± 15 %	-----	-----

afterwards. Then, between 10 and 15 minutes, a quite dramatic increase in conductivity is observed that is probably related massive lysis of liver cells.

Table 1 summarizes some experimental results in which tissue conductivities are increased by applying an electroporation protocol that consists of 8 pulses of 100 μs at 10 Hz. It is interesting to note the disparity of electric field magnitudes needed to get a certain conductivity increase. For instance, to reach a 40 % increase, 2500 V/cm are needed for tumors whereas for liver only 1500 V/cm are needed and at 450 V/cm in the case of the muscle the conductivity increase is 70 %. Up to a point, irreversible field thresholds and conductivity increases seems to be correlated: in all the examples of the table, 30% or larger conductivity increases coincide with irreversible conditions. However, it must be said that no such a thing as a universal conductivity increase threshold for irreversibility has been found.

7 Effect of Electroporation on Other Bioelectric Phenomena

In all bioelectric phenomena the cell membrane permeability (or better said, its selective permeability) plays a crucial role. Therefore, it seems quite obvious that electroporation can have a significant impact on such phenomena, reversibly or irreversibly. Moreover, the high voltage pulses employed to achieve electroporation can initiate action potentials in muscles and nerves within the region under treatment or in distant areas; it must remembered that the thresholds for excitation are significantly lower than the thresholds for electroporation. As a matter of fact, one of the few drawbacks of electroporation based therapies is that they can cause acute pain and muscle contraction in the vicinity of area where the electrodes are applied. In some cases local anesthesia is enough to minimize discomfort to the patient but in other cases general sedation may be preferred [88] combined or not with muscle relaxants [89].

However, the main immediate concern of clinicians dealing with electroporation is whether the electroporation voltage pulses can elicit ventricular fibrillation (VF) or other cardiac arrhythmias. Such concern is perfectly justified if it is taken into account that cardiac muscle is extremely sensitive to a large variety of electrical stimuli [90]. For instance, it is quite astonishing to learn that currents above 10 A may be used in irreversible electroporation treatments whereas the residual breaker circuit breakers[23] are designed to open its contacts when the leakage current goes above 30 mA. Hence it sounds reasonable that since the beginning of the use of *in vivo* electroporation researchers were worried with the possibility that an electroporation treatment in any part of the body could result in fatal fibrillation due to residual currents flowing through the myocardium. As a matter of fact, a technical solution to this problem was envisioned quite early [91]: voltage pulses can be synchronized with the electrocardiogram signal so that they are delivered when all myocardium cells are in the *absolute refractory period*; in this period it is not possible to initiate action potentials regardless of the amplitude of the excitation signal. This sort of solution has been studied more recently by other researchers [92, 93]. Nevertheless, it appears that the topic has been quite ignored during the 90s and the beginning of the 2000s, when *in vivo* electroporation started to be used clinically for superficial treatments and in animal research for multiple

[23] Residual current circuit breakers, also known as differential breakers or ground fault interrupters, are electrical wiring devices intended to prevent ventricular fibrillation in the case in which a grounded person touches the energized part of a circuit so that a leakage current flows through the body. These devices disconnect the circuit when the electric current is not balanced between the phase conductor and the neutral conductor.

internal treatments. The reason for such oblivion is also understandable: it is not possible to find a single report in which ventricular fibrillation occurred after electroporation, neither in human trials nor in animal experimentation. In the cases in which electroporation was applied to human limbs residual current densities through the heart were probably too weak to induce any effect. On the other hand, when tumors on the chest were treated, and particularly during experimentation with small animals, it is very plausible that current densities above 0.5 mA/cm^2 flowed through the heart[24] and, therefore, VF would have been very likely to occur. However, it must be taken into account that this specific 0.5 mA/cm^2 current density threshold for ventricular fibrillation is applicable for prolonged (> 1s) 50 Hz or 60 Hz exposures but it is meaningless for short pulses as those employed in electroporation. It is a known fact that long exposures at around 50 Hz are the most dangerous ones in terms of VF; lower or higher frequencies imply much larger thresholds and that could explain why VF was never observed in electroporation *in vivo* treatments. Unfortunately, little experimental data exists on stimuli with features similar to those of electroporation pulses and it is not possible to provide here a reliable threshold value. Mali et al. [93] employed a VF threshold from [94] for single DC pulses of 500 μs: total current of 100 mA flowing through the entire heart. This seems a reasonable value as in most of the cases electroporation pulses will be shorter and, in principle, less prone to induce VF. Even so, it is clear that further research is required in this area. For instance, it is necessary to analyze how the fact that multiple pulses are applied can increase the chances of VF and what is the importance of the pulse duration and of the interval between pulses.

In relation to the above paragraph it is interesting to note that Lavee et al. [95] carried out a study with five pigs in which IRE of the atrium was performed for the treatment of atrial fibrillation, as an alternative to methods based on thermal ablation. Pulse sequences with an amplitude of 1500 V or 2000 V, a frequency of 5 Hz and up to 32 pulses were applied and no permanent arrhythmia or any significant rhythm disturbance was observed apart from the rapid pacing imposed momentarily by the sequence of pulses. That study does not contain information regarding electric current measurements but, taking into account the electrode setup geometry (two parallel 4 cm long cylindrical rods of 2 mm in diameter at opposite sides of the tissue with a thickness ranging from 5 mm to 14 mm) and if we assume that the myocardium resistivity during IRE can go down to 300 Ω.cm, we estimate that the current applied in some cases was larger than 20 A. This astonishing result confirms that further research is required in this area.

In relation also with the heart it must be mentioned that quite a few recent research articles are pointing out at the possible role of electroporation on defibrillation therapy [96-104]. In this case the research is not focused on the effect of the electrical stimulus associated with electroporation but on electroporation itself, that is, on the effect of reversible, or irreversible, permeabilization. This topic has been recently reviewed by Nikolski and Efimov in [105] and they conclude that electroporation indeed can affect the outcome of defibrillation and that it can have

---

24 Reilly, in (Reilly 1998), compiled results from numerous experimental studies on myocardium stimulation and fibrillation.

both pro- and anti-arrhythmic consequences. It is also interesting to note here that it is suspected that electroporation could play a significant role in radiofrequency myocardium ablation [106].

It is generally accepted that electroporation influences action potential generation and transmission by causing a nonspecific increase of cell membrane permeability to ions. The main observable effects of electroporation on excitable cells seem to be: 1) reduction of action potential amplitude and 2) reduction of conduction velocity. For the case of rat sciatic nerve Abramov et al. [107] found that after 75 V/cm shocks (12 pulses of 4 ms at 0.1 Hz) the normal action potential amplitude and conduction velocity were not recovered until 3 hours after treatment, whereas 37 V/cm pulses did not produce significant changes and the changes produced by 150 V/cm were permanent. Comparable transitory muscle incapacitation has been achieved by applying eight 100 μs 300 V/cm pulses [108]. Therefore, it seems very plausible to modulate or to block transitorily, or permanently, conduction in nerves by applying electroporation pulses [109].

References

[1] Fricke, H.: A mathematical treatment of the electric conductivity and capacity of disperse systems. II. The capacity of a suspension of conducting spheroids surrounded by a non-conducting membrane for a current of low frequency. Physical Review 26 (1925)

[2] Crowley, J.M.: Electrical breakdown of bimolecular lipid membranes as an electromechanical instability. Biophys. J. 13, 711–724 (1973)

[3] Hodgkin, A.L.: The ionic basis of electrical activity in nerve and muscle. Biological reviews of the Cambridge Philosophical Society 26, 339–409 (1951)

[4] Stämpfli, R.: Reversible electrical breakdown of the excitable membrane of a Ranvier node. Anais da Academia Brasileira de Ciencias 30, 57–63 (1957)

[5] Sale, A.J.H., Hamilton, W.A.: Effects of high electric fields on microorganisms. 1. Killing of bacteria and yeasts. Biochimica et Biophysica Acta 148, 781–788 (1967)

[6] Hamilton, W.A., Sale, A.J.H.: Effects of high electric fields on microorganisms. 2. Mechanism of action of the lethal effect. Biochimica et Biophysica Acta 148, 789–800 (1967)

[7] Sale, A.J.H., Hamilton, W.A.: Effects of high electric fields on microorganisms. 3. Lysis of erythrocytes and protopasts. Biochimica et Biophysica Acta 163, 37–43 (1968)

[8] Kinosita, K.J., Tsong, T.Y.: Formation and resealing of pores of controlled sizes in human erythrocyte membrane. Nature 268, 438–441 (1977)

[9] Neumann, E., Schaeffer-Ridder, M., Wang, Y., Hofschneider, P.H.: Gene transfer into mouse lymphoma cells by electroporation in high electric fields. EMBO J. 1, 841–845 (1982)

[10] Mir, L.M., Orlowski, S., Belehradek, J.J., Paoletti, C.: Electrochemotherapy potentiation of antitumour effect of bleomycin by local electric pulses. European Journal of Cancer 27, 68–72 (1991)

[11] Mir, L.M., Belehradek, M., Domenge, C., Orlowski, S., Poddevin, B., Belehradek, J.J., Schwaab, G., Luboinski, B., Paoletti, C.: Electrochemotherapy, a new antitumor treatment: first clinical trial. Comptes Rendus de l'Academie des Sciences Serie III Sciences de la Vie 313, 613–618 (1991)

[12] Becker, S.M., Kuznetsov, A.V.: Numerical Modeling of In Vivo Plate Electroporation Thermal Dose Assessment. Journal of Biomechanical Engineering 128, 76–84 (2006)

[13] Davalos, R.V., Rubinsky, B., Mir, L.M.: Theoretical analysis of the thermal effects during in vivo tissue electroporation. Bioelectrochemistry 61, 99–107 (2003)

[14] Pliquett, U.F., Martin, G.T., Weaver, J.C.: Kinetics of the temperature rise within human stratum corneum during electroporation and pulsed high-voltage iontophoresis. Bioelectrochemistry 57, 65–72 (2002)

[15] Edd, J.F., Davalos, R.V.: Mathematical modeling of irreversible electroporation for treatment planning. Technology in Cancer Research and Treatment 6, 275–286 (2007)

[16] Maor, E., Ivorra, A., Rubinsky, B.: Intravascular irreversible electroporation: Theoretical and experimental feasibility study. Presented at 30th Annual International Conference of the IEEE Engineering in Medicine and Biology Society, EMBS 2008, Vancouver, BC, Canada (2008)

[17] Pilwat, G., Zimmermann, U.: Determination of intracellular conductivity from electrical breakdown measurements. Biochimica et Biophysica Acta 820, 305–314 (1985)

[18] Ivorra, A.: Annex A - Bioimpedance monitoring for physicians: an overview, http://www.tdx.cbuc.es/TESIS_UPC/AVAILABLE/TDX-0302105-135356//08Aic08de11.pdf; Contributions to the measurement of electrical impedance for living tissue ischemia injury monitoring, PhD Thesis. Universitat Politècnica de Catalunya, Barcelona, 131–176 (2005)

[19] Grimnes, S., Martinsen, O.G.: Bioimpedance and Bioelectricity Basics. Academic Press, London (2000)

[20] Riemann, F., Zimmermann, U., Pilwat, G.: Release and uptake of haemoglobin and ions in red blood cells induced by dielectric breakdown. Biochimica et Biophysica Acta 394, 449–462 (1975)

[21] Teissie, J., Rols, M.P.: An experimental evaluation of the critical potential difference inducing cell membrane electropermeabilization. Biophys. J. 65, 409–413 (1993)

[22] Zimmermann, U., Pilwat, G., Riemann, F.: Dielectric Breakdown of Cell Membranes. Biophys. J. 14, 881–899 (1974)

[23] Kinosita, K.J., Ashikawa, I., Saita, N., Yoshimura, H., Itoh, H., Nagayama, K., Ikegami, A.: Electroporation of cell membrane visualized under a pulsed-laser fluorescence microscope. Biophys. J. 53, 1015–1019 (1988)

[24] Abidor, I.G., Arakelyan, V.B., Chernomordik, L.V., Chizmadzhev, Y.A., Pastushenko, V.F., Tarasevich, M.R.: 246 - Electric breakdown of bilayer lipid membranes I. The main experimental facts and their qualitative discussion. Bioelectrochemistry and Bioenergetics 6, 37–52 (1979)

[25] Melikov, K.C., Frolov, V.A., Shcherbakov, A., Samsonov, A.V., Chizmadzhev, Y.A., Chernomordik, L.V.: Voltage-Induced Nonconductive Pre-Pores and Metastable Single Pores in Unmodified Planar Lipid Bilayer. Biophys. J. 80, 1829–1836 (2001)

[26] Glaser, R.W., Leikin, S.L., Chernomordik, L.V., Pastushenko, V.F., Sokirko, A.I.: Reversible electrical breakdown of lipid bilayers: formation and evolution of pores. Biochimica et Biophysica Acta 940, 275–287 (1988)

[27] Canatella, P.J., Karr, J.F., Petros, J.A., Prausnitz, M.R.: Quantitative Study of Electroporation-Mediated Molecular Uptake and Cell Viability. Biophys. J. 80, 755–764 (2001)

[28] Miklavcic, D., Kotnik, T.: Electroporation for Electrochemotherapy and Gene Therapy. In: Rosch, P.J., Markov, M.S. (eds.) Bioelectromagnetic Medicine, pp. 637–656. Informa Health Care, New York (2004)
[29] Potter, H., Weir, L., Leder, P.: Enhancer-dependent expression of human kappa immunoglobulin genes introduced into mouse pre-B lymphocytes by electroporation. Proc. Natl. Acad. Sci. USA 81, 7161–7165 (1984)
[30] Sixou, S., Teissié, J.: Specific electropermeabilization of leucocytes in a blood sample and application to large volumes of cells. Biochimica et Biophysica Acta (BBA) - Biomembranes 1028, 154–160 (1990)
[31] Hibino, M., Itoh, H., Kinosita, K.J.: Time courses of cell electroporation as revealed by submicrosecond imaging of transmembrane potential. Biophys. J. 64, 1789–1800 (1993)
[32] Miller, C.E., Henriquez, C.S.: Three-dimensional finite element solution for biopotentials: erythrocyte in an applied field. IEEE Trans. Biomed. Eng. 35, 712–718 (1988)
[33] Towhidi, L., Kotnik, T., Pucihar, G., Firoozabadi, S.M., Mozdarani, H., Miklavcic, D.: Variability of the minimal transmembrane voltage resulting in detectable membrane electroporation. Electromagnetic biology and medicine 27, 372–385 (2008)
[34] Pavlin, M., Pavselj, N., Miklavcic, D.: Dependence of induced transmembrane potential on cell density, arrangement, and cell position inside a cell system. IEEE Trans. Biomed. Eng. 49, 605–612 (2002)
[35] Dev, S.B., Dhar, D., Krassowska, W.: Electric field of a six-needle array electrode used in drug and DNA delivery in vivo: analytical versus numerical solution. IEEE Trans. Biomed. Eng. 50, 1296 (2003)
[36] Miklavcic, D., Beravs, K., Semrov, D., Cemazar, M., Demsar, F., Sersa, G.: The Importance of Electric Field Distribution for Effective in Vivo Electroporation of Tissues. Biophys. J. 74, 2152–2158 (1998)
[37] Miklavcic, D., Semrov, D., Mekid, H., Mir, L.M.: A validated model of in vivo electric field distribution in tissues for electrochemotherapy and for DNA electrotransfer for gene therapy. Biochimica et Biophysica Acta 1523, 73–83 (2000)
[38] Sugibayashi, K., Yoshida, M., Mori, K., Watanabe, T., Hasegawa, T.: Electric field analysis on the improved skin concentration of benzoate by electroporation. International Journal of Pharmaceutics 219, 107–112 (2001)
[39] Edd, J., Horowitz, L., Davalos, R.V., Mir, L.M., Rubinsky, B.: In-Vivo Results of a New Focal Tissue Ablation Technique: Irreversible Electroporation. IEEE Trans. Biomed. Eng. 53, 1409–1415 (2006)
[40] Rubinsky, B.: Irreversible electroporation in medicine. Technology in Cancer Research and Treatment 6, 255–260 (2007)
[41] Gilbert, R.A., Jaroszeski, M.J., Heller, R.: Novel electrode designs for electrochemotherapy. Biochimica et Biophysica Acta (BBA) - General Subjects 1334, 9 (1997)
[42] Ivorra, A., Rubinsky, B.: Electric field modulation in tissue electroporation with electrolytic and non-electrolytic additives. Bioelectrochemistry 70, 551–560 (2007)
[43] Ivorra, A., Al-sakere, B., Rubinsky, B., Mir, L.M.: Use of conductive gels for electric field homogenization increases the antitumor efficacy of electroporation therapies. Physics in Medicine and Biology 53, 6605–6618 (2008)
[44] Schoenbach, K.H., Peterkin, F.E., Alden, R.W.I., Beebe, S.J.: The effect of pulsed electric fields on biological cells:experiments and applications. IEEE Trans. Plasma Science 25, 284–292 (1997)

[45] Schoenbach, K.H., Beebe, S.J., Buescher, E.S.: Intracellular effect of ultrashort electrical pulses. Bioelectromagnetics 22, 440–448 (2001)
[46] Sun, Y., Vernier, P.T., Behrend, M., Marcu, L., Gundersen, M.A.: Electrode microchamber for noninvasive perturbation of mammalian cells with nanosecond pulsed electric fields. IEEE Trans. NanoBioscience 4, 277–283 (2005)
[47] Jordan, D.W., Uhler, M.D., Gilgenbach, R.M., Lau, Y.Y.: Enhancement of cancer chemotherapy in vitro by intense ultrawideband electric field pulses. Journal of Applied Physics 99, 094701 (2006)
[48] Frey, W., White, J.A., Price, R.O., Blackmore, P.F., Joshi, R.P., Nuccitelli, R., Beebe, S.J., Schoenbach, K.H., Kolb, J.F.: Plasma Membrane Voltage Changes during Nanosecond Pulsed Electric Field Exposure. Biophys. J. 90, 3608–3615 (2006)
[49] Vernier, P.T., Sun, Y., Chen, M.-T., Gundersen, M.A., Craviso, G.L.: Nanosecond electric pulse-induced calcium entry into chromaffin cells. Bioelectrochemistry 73, 1–4 (2008)
[50] Schoenbach, K.H., Nuccitelli, R., Beebe, S.J.: Zap. IEEE Spectrum 43, 20–26 (2006)
[51] Smith, K.C., Gowrishankar, T.R., Esser, A.T., Stewart, D.A., Weaver, J.C.: The Spatially Distributed Dynamic Transmembrane Voltage of Cells and Organelles due to 10-ns Pulses: Meshed Transport Networks. IEEE Trans. Plasma Science 34, 1394–1404 (2006)
[52] Pakhomov, A.G., Kolb, J.F., White, J.A., Joshi, R.P., Xiao, S., Schoenbach, K.H.: Long-lasting plasma membrane permeabilization in mammalian cells by nanosecond pulsed electric field (nsPEF). Bioelectromagnetics 28, 655–663 (2007)
[53] Benz, R., Zimmermann, U.: Pulse-length dependence of the electrical breakdown in lipid bilayer membranes. Biochimica et Biophysica Acta 597, 637–642 (1980)
[54] Krassowska, W., Filev, P.D.: Modeling electroporation in a single cell. Biophysical Journal 92, 404–417 (2007)
[55] Pliquett, U., Elez, R., Piiper, A., Neumann, E.: Electroporation of subcutaneous mouse tumors by trapezium high voltage pulses. Bioelectrochemistry 62, 83–93 (2004)
[56] Cukjati, D., Batiuskaite, D., Andre, F., Miklavcic, D., Mir, L.M.: Real time electroporation control for accurate and safe in vivo non-viral gene therapy. Bioelectrochemistry 70, 501–507 (2007)
[57] Pavlin, M., Kanduser, M., Rebersek, M., Pucihar, G., Hart, F.X., Magjarevic, R., Miklavcic, D.: Effect of Cell Electroporation on the Conductivity of a Cell Suspension. Biophys. J. 88, 4378–4390 (2005)
[58] Gersing, E.: Monitoring temperature-induced changes in tissue during hyperthermia by impedance methods. Ann. NY. Acad. Sci. 873, 13–20 (1999)
[59] Russell, W.M.S., Burch, R.L.: The Principles of Humane Experimental Technique. Methuen & Co. Ltd., London (1959)
[60] Ivorra, A., Miller, L., Rubinsky, B.: Electrical impedance measurements during electroporation of rat liver and muscle. In: Scharfetter, H., Merva, R. (eds.) 13th International Conference on Electrical Bioimpedance. IFMBE Proceedings, vol. 17, pp. 130–133. Springer, Berlin (2007)
[61] Ashie, I.N.A., Simpson, B.K.: Application of high hydrostatic pressure to control enzyme related fresh seafood texture deterioration. Food Research International 29, 569–575 (1996)
[62] Gabriel, C., Gabriel, S., Corthout, E.: The dielectric properties of biological tissues: I. Literature survey. Physics in Medicine and Biology 41, 2231–2249 (1996)

[63] Gabriel, S., Lau, R.W., Gabriel, C.: The dielectric properties of biological tissues: II. Measurements in the frequency range 10 Hz to 20 GHz. Physics in Medicine and Biology 41, 2251–2269 (1996)
[64] Gabriel, S., Lau, R.W., Gabriel, C.: The dielectric properties of biological tissues: III. Parametric models for the dielectric spectrum of tissues. Physics in Medicine and Biology 41, 2271–2293 (1996)
[65] Al-Sakere, B., André, F., Bernat, C., Connault, E., Opolon, P., Davalos, R.V., Rubinsky, B., Mir, L.M.: Tumor ablation with irreversible electroporation. PLoS ONE 2, e1135 (2007)
[66] Sel, D., Cukjati, D., Batiuskaite, D., Slivnik, T., Mir, L.M., Miklavcic, D.: Sequential finite element model of tissue electropermeabilization. IEEE Trans. Biomed. Eng. 52, 816–827 (2005)
[67] Pavselj, N., Bregar, Z., Cukjati, D., Batiuskaite, D., Mir, L.M., Miklavcic, D.: The course of tissue permeabilization studied on a mathematical model of a subcutaneous tumor in small animals. IEEE Trans. Biomed. Eng. 52, 1373 (2005)
[68] Sel, D., Macek-Lebar, A., Miklavcic, D.: Feasibility of Employing Model-Based Optimization of Pulse Amplitude and Electrode Distance for Effective Tumor Electropermeabilization. IEEE Trans. Biomed. Eng. 54, 773–781 (2007)
[69] Ivorra, A., Rubinsky, B.: Optimum Conductivity of Gels for Electric Field Homogenization in Tissue Electroporation Therapies. In: Müller-Karger, C., Wong, S., La Cruz, A. (eds.) IV Latin American Congress on Biomedical Engineering, Bioengineering Solutions for Latin America Health. IFMBE Proceedings, vol. 18, pp. 619–622. Springer, Berlin (2007)
[70] Marty, M., Sersa, G., Garbay, J.R., Gehl, J., Collins, C.G., Snoj, M., Billard, V., Geertsen, P.F., Larkin, J.O., Miklavcic, D., Pavlovic, I., Paulin-Kosir, S.M., Cemazar, M., Morsli, N., Soden, D.M., Rudolf, Z., Robert, C., O'Sullivan, G.C., Mir, L.M.: Electrochemotherapy - An easy, highly effective and safe treatment of cutaneous and subcutaneous metastases: Results of ESOPE (European Standard Operating Procedures of Electrochemotherapy) study. European Journal of Cancer Supplements 4, 3–13 (2006)
[71] Mossop, B.J., Barr, R.C., Henshaw, J.W., Zaharoff, D.A., Yuan, F.: Electric fields in tumors exposed to external voltage sources: implication for electric field-mediated drug and gene delivery. Annals of Biomedical Engineering 34, 1564–1572 (2006)
[72] Abidor, I.G., Li, L.H., Hui, S.W.: Studies of cell pellets: II. Osmotic properties, electroporation, and related phenomena: membrane interactions. Biophys. J. 67, 427–435 (1994)
[73] Kinosita, K.J., Tsong, T.Y.: Voltage-induced conductance in human erythrocyte membranes. Biochimica et Biophysica Acta 554, 479–497 (1979)
[74] Schmeer, M., Seipp, T., Pliquett, U., Kakorin, S., Neumann, E.: Mechanism for the conductivity changes caused by membrane electroporation of CHO cell-pellets. Phys. Chem. Chem. Phys. 6, 5564–5574 (2004)
[75] Suehiro, J., Hatano, T., Shutou, M., Hara, M.: Improvement of electric pulse shape for electropermeabilization-assisted dielectrophoretic impedance measurement for high sensitive bacteria detection. Sensors and Actuators B: Chemical 109, 209–215 (2005)
[76] Wegener, J., Keese, C.R., Giaever, I.: Recovery of adherent cells after in situ electroporation monitored electrically. BioTechniques 33, 348–357 (2002)
[77] Glahder, J., Norrild, B., Persson, M.B., Persson, B.R.: Transfection of HeLa-cells with pEGFP plasmid by impedance power-assisted electroporation. Biotechnology and Bioengineering 92, 267–276 (2005)

[78] Lee, R.C., River, L.P., Pan, F.S., Ji, L., Wollmann, R.L.: Surfactant-induced sealing of electropermeabilized skeletal muscle membranes in vivo. Proc. Natl. Acad. Sci. USA 89, 4524–4528 (1992)
[79] Pliquett, U., Langer, R.S., Weaver, J.C.: Changes in the passive electrical properties of human stratum corneum due to electroporation. Biochimica et Biophysica Acta 1239, 111–121 (1995)
[80] Ivorra, A., Rubinsky, B.: In vivo electrical impedance measurements during and after electroporation of rat liver. Bioelectrochemistry 70, 287–295 (2007)
[81] Maor, E., Ivorra, A., Leor, J., Rubinsky, B.: The effect of irreversible electroporation on blood vessels. Technology in Cancer Research and Treatment 6, 307–312 (2007)
[82] Maor, E., Ivorra, A., Leor, J., Rubinsky, B.: Irreversible electroporation attenuates neointimal formation after angioplasty. IEEE Trans. Biomed. Eng. 55, 2268–2274 (2008)
[83] Liao, T.J., Nishikawa, H.: The variation of action potential and impedance in human skeletal muscle during voluntary contraction. The Tohoku Journal of Experimental Medicine 173, 303–309 (1994)
[84] Shiffman, C.A., Aaron, R., Rutkove, S.B.: Electrical impedance of muscle during isometric contraction. Physiol. Meas. 24, 213–234 (2003)
[85] Jackson, V.M., Trout, S.J., Cunnane, T.C.: Regional variation in electrically-evoked contractions of rabbit isolated pulmonary artery. British Journal of Pharmacology 137, 488–496 (2002)
[86] Gehl, J., Skovsgaard, T., Mir, L.M.: Vascular reactions to in vivo electroporation: characterization and consequences for drug and gene delivery. Biochimica et Biophysica Acta 1569, 51–58 (2002)
[87] Haemmerich, D., Ozkan, O.R., Tsai, J.Z., Staelin, S.T., Tungjitkusolmun, S., Mahvi, D.M., Webster, J.G.: Changes in electricel resistivity of swine liver after occlusion and postmortem. Med. Biol. Eng. Comput. 40, 29–33 (2002)
[88] Mir, L.M., Gehl, J., Sersa, G., Collins, C.G., Garbay, J.-R., Billard, V., Geertsen, P.F., Rudolf, Z., O'Sullivan, G.C., Marty, M.: Standard operating procedures of the electrochemotherapy: Instructions for the use of bleomycin or cisplatin administered either systemically or locally and electric pulses delivered by the CliniporatorTM by means of invasive or non-invasive electrodes. European Journal of Cancer Supplements 4, 14–25 (2006)
[89] Onik, G., Rubinsky, B., Mikus, P.: Irreversible Electroporation: Implications for Prostate Ablation. Technology in Cancer Research and Treatment 6, 295–300 (2007)
[90] Reilly, J.P.: Cardiac Sensitivity to Electrical Stimulation. In: Reilly, J.P. (ed.) Applied Bioelectricity: From Electrical Stimulation to Electropathology, pp. 194–239. Springer, New York (1998)
[91] Okino, M., Tomie, H., Kanesada, H., Marumoto, M., Esato, K., Suzuki, H.: Optimal electric conditions in electrical impulse chemotherapy. Japanese Journal of Cancer Research 83, 1095–1101 (1992)
[92] Mali, B., Jarm, T., Jager, F., Miklavcic, D.: An algorithm for synchronization of in vivo electroporation with ECG. Journal of Medical Engineering & Technology 29, 288–296 (2005)
[93] Mali, B., Jarm, T., Corovic, S., Paulin-Kosir, M.S., Cemazar, M., Sersa, G., Miklavcic, D.: The effect of electroporation pulses on functioning of the heart. Medical & Biological Engineering & Computing 46, 745–757 (2008)

[94] Antoni, H.: Electrical Properties of the Heart. In: Reilly, J.P. (ed.) Applied Bioelectricity: From Electrical Stimulation to Electropathology, pp. 148–193. Springer, New York (1998)
[95] Lavee, J., Onik, G., Mikus, P., Rubinsky, B.: A Novel Nonthermal Energy Source for Surgical Epicardial Atrial Ablation: Irreversible Electroporation. The Heart Surgery Forum 10, E162–E167 (2007)
[96] Fogelson, L.J., Tung, L., Thakor, N.V.: Electrophysiologic depression in myocardium by defibrillation-levelshocks. Presented at Proceedings of the Annual International Conference of the IEEE Engineering in Medicine and Biology Society (1988)
[97] Tovar, O., Tung, L.: Electroporation of cardiac cell membranes with monophasic or biphasic rectangular pulses. Pacing and Clinical Electrophysiology 14, 1887–1892 (1991)
[98] Tovar, O., Tung, L.: Electroporation and recovery of cardiac cell membrane with rectangular voltage pulses. The American Journal of Physiology 263, H1128–H1136 (1992)
[99] Knisley, S.B., Smith, W.M., Ideker, R.E.: Prolongation and shortening of action potentials by electrical shocks in frog ventricular muscle. The American Journal of Physiology 266, H2348–H2358 (1994)
[100] Krassowska, W.: Effects of electroporation on transmembrane potential induced by defibrillation shocks. Pacing and Clinical Electrophysiology 18, 1644–1660 (1995)
[101] DeBruin, K.A., Krassowska, W.: Electroporation and shock-induced transmembrane potential in a cardiac fiber during defibrillation strength shocks. Annals of Biomedical Engineering 26, 584–596 (1998)
[102] Al-Khadra, A., Nikolski, V., Efimov, I.R.: The role of electroporation in defibrillation. Circulation Research 87, 797–804 (2000)
[103] Ashihara, T., Yao, T., Namba, T., Ito, M., Ikeda, T., Kawase, A., Toda, S., Suzuki, T., Inagaki, M., Sugimachi, M., Kinoshita, M., Nakazawa, K.: Electroporation in a model of cardiac defibrillation. Journal of Cardiovascular Electrophysiology 12, 1393–1403 (2001)
[104] Cheek, E.R., Fast, V.G.: Nonlinear changes of transmembrane potential during electrical shocks: role of membrane electroporation. Circulation Research 94, 208–214 (2004)
[105] Nikolski, V.P., Efimov, I.R.: Electroporation of the heart. Europace 7, 146–154 (2005)
[106] Wu, C.C., Fasciano, R.W.N., Calkins, H., Tung, L.: Sequential change in action potential of rabbit epicardium during and following radiofrequency ablation. Journal of Cardiovascular Electrophysiology 10, 1252–1261 (1999)
[107] Abramov, G.S., Bier, M., Capelli-Schellpfeffer, M., Lee, R.C.: Alteration in sensory nerve function following electrical shock. Burns 22, 602–606 (1996)
[108] Clausen, T., Gissel, H.: Role of Na,K pumps in restoring contractility following loss of cell membrane integrity in rat skeletal muscle. Acta Physiol. Scand. 183, 263–271 (2006)
[109] Joshi, R.P., Mishra, A., Hu, Q., Schoenbach, K.H., Pakhomov, A.: Self-consistent analyses for potential conduction block in nerves by an ultrashort high-intensity electric pulse. Phys. Rev. E. Stat. Nonlin. Soft Matter Phys. 75, 061906 (2007)

Experimental Studies on Irreversible Electroporation of Cells

Avigail Ben-Or and Boris Rubinsky*

Center for Bioengineering in the Service of Humanity and Society,
School of Computer Science and Engineering, Hebrew University of Jerusalem,
Givat Ram Campus, Jerusalem, Israel
* Corresponding author E-mail: rubinsky@cs.huji.ac.il

Introduction

In the following chapter we wish to explore the fundamental physical properties of electroporation that have emerged from experiments conducted with cells. Since, up to date, a uniform comprehensive theory that explains the mechanism that underlies the electroporation phenomena has not yet been proposed, it is interesting to look into the experimental facts that such a theory needs to encompass.

Many studies have been conducted in order to assess the efficiency of a certain electroporation method. These studies are usually meant to help adjust the electroporation parameters in a way that will promise the highest viability of the cells and the highest total uptake of DNA or drugs into the cells. We will try to focus here on the physics that studies of this sort reveal. We will also bring forth early studies (from the late 1950's and onwards) that were the first attempts to describe the electroporation phenomena.

In the first part of this chapter we will present studies that have to do with the process of pore formation. This part will also include the analysis of the transport through the electropores.

The second part deals with the resealing process of reversible electroporation that follows the pore opening process, once the electric field of the electroporating pulse is removed. We will discuss the time frame and stages of resealing that the cell membrane goes through until it reaches its full recovery.

In the third part we will look at factors that affect the uptake, efficiency and viability of the cells. Such factors are the voltage pulse (height, duration and shape), temperature and geometric factors (cell shape, electric field shape and the shape of the surroundings).

The fourth part examines part of the changes in the cell caused by electroporation. We will take notice of the conductivity changes and the mechanical stress of the cells. We will conclude with a short discussion.

In no way does this chapter fully grasp the entire work that has been done by many researchers over the years in this field. Our goal was to point out a few studies that bring forth the basic phenomena that have been encountered in experimental work on cell electroporation.

B. Rubinsky (Ed.): Irreversible Electroporation, BIOMED, pp. 63–83.
springerlink.com

Pore Formation

It has been observed as early as 1957 by R. Stämpfli [1] that applying an electric field above a certain threshold creates pores in the membrane of a cell. The mechanism that underlies the formation of these pores is not yet fully understood. Many experiments have been carried out in order to elucidate this phenomenon. One basic question is what controls the size of the pores and their quantity in the membrane.

Lipid bilayers are a simple system to study electroporation in comparison with biological cells. The similarity of the electroporation phenomena may point to a common mechanism in the lipid bilayers and the biological cell membrane. As Chernomordik et al. [2] concluded in their study, the pores arise in the lipid matrices of the erythrocytes membranes and therefore the properties of this procedure are very close to those of pores forming in lipid bilayer membranes. One has to be careful, of course, when drawing conclusions from one of these systems to another.

Abidor et al. [3] tested the electric breakdown of bilayer lipid membranes of various compositions in a solution of 0.1M NaCl. The dependence of the logarithm of the mean life time of the membranes on the voltage applied to the membrane is close to linear in the range of 300 to 600 *mV*. At higher potentials, the decrease of the mean lifetime of membranes became less pronounced with the increasing potential. The measurements of the current when applying a step voltage pulse reveal that after the membrane is charged it reaches a steady state current at first, which depends on the applied voltage. In the next stage, current fluctuations appear followed by an onset of irreversible growth of current to a saturation value in which the membrane ruptures. The character and duration of the fluctuating stage are different in repeated measurements. The membrane conductance and lifetime both exhibit stochastic behavior. As Weaver and Chizmadzhev [4] pointed out, these results imply that a purely deterministic mechanism cannot fully describe electroporation. Abidor et al. [3] regard the membrane as a metastable phase which, at a given voltage, is characterized by a mean lifetime. They believe that the membrane breakdown takes place as a result of the development of defects, which play the same role as nuclei in the theory of formation of new phases.

The stochastic nature of the voltage fluctuations through the nano electro-pores was shown by Kotulska et al. [5]. They used planar lipid bilayer membranes in a current clamp experiment and concluded that the voltage fluctuations are a stochastic fractal process with long memory.

In adherent cells, Truel and Meyer [6] found that surprisingly the calcium entry sites formed by electroporation were not a stochastic process. When the same pulse protocol was repeated, the entry sites were the same. This suggests that the local composition of the plasma membrane, cytoskeleton components or local cell adhesion define a specific number of sites where hyperpolarizing membrane potentials can lead to preferential membrane disruption and transient formation of entry sites.

It is a known experimental fact that the poration occurs in an asymmetric fashion in the cell, unless a reverse polarity pulse is used which causes a symmetric poration on the cathode and anode sides. The physical explanation to this fact is that the potential across the membrane, in its steady state, plays an important role. The field induced potential difference is added to the resting one, therefore the side of the cell facing the anode is hyperpolarized while the side facing the cathode is depolarized (illustrated in Fig. 1).

The transport of various molecules across the membrane through these pores gives reason to believe that the bilipid polar layer of the membrane changes its structure at the pore site to form a hydrophilic pore. Above a certain pore radius it is believed that the system energetically favors a hydrophilic structure than the hydrophobic first stage [7, 8].

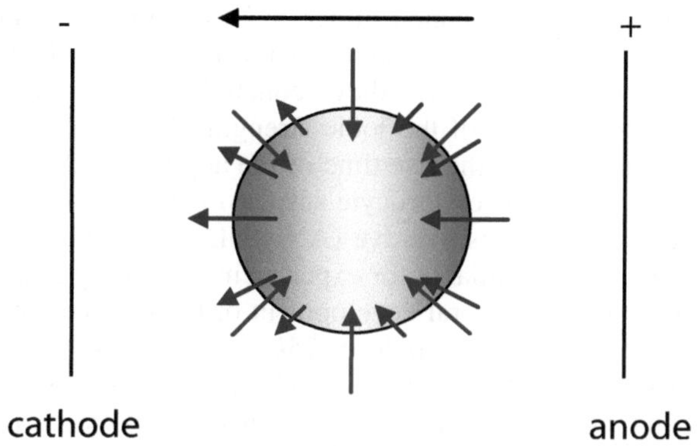

Fig. 1. Effect of electric field on the modulation of the transmembrane electric potential difference of a cell. Blue arrows represent the resting potential, red arrows represent induced potential. Their length is indicative of their magnitude. Adopted from [15].

A comprehensive study on the formation and evolution of the pores in lipid bilayers was conducted by Glaser et al. [9] in 1988. The first stage of electroporation is described as the formation of short-living small hydrophobic pores. In the next step an inversion of the pore edge occurs. As a result, hydrophilic pores determining the membrane conductivity are formed. The first stage happens due to thermal motion of the lipid molecules; hence, the lifetime is in the order of the lipid fluctuations. After the pore reaches a critical radius, the inversion occurs. They assume that the critical radius is independent of the membrane voltage. The conductivity of the small pores rapidly grows when the voltage rises, but the large pores are nearly ohmic. This indicates that it might be possible to estimate the number and size of the pores from conductivity measurements at different voltages. They found a linear dependence of $ln\ (\frac{\Delta I}{\Delta t})$ on U^2 for pulses below 1.5 V. The critical pore radius was estimated to be $R^* \approx 0.3 - 0.5nm$. The conductivity of the membrane is not a linear function of the voltage because ions that cross the membrane need to overcome an energy barrier that is voltage dependent as well. The voltage influences the number, mean size, and the conductivity of the pores.

Apart from the long lifetime of the pores after breakdown, Glaser et al. [9] have found more evidence to the hydrophilic nature of the pores. After adding less than $0.5\mu g/ml$ of lysophosphatidylcholine to the membrane bathing solution a significant acceleration of the current increase was observerd, for identical voltage pulses. This acceleration is explained by a higher rate of pore creation resulting from a decrease of the energetic barrier for hydrophilic pore creation. Involvement of lysophosphatidylcholine in hydrophilic pores is energetically favorable because the hydrophobic tail is smaller in comparison to the hydrophilic head so that the lipid fits to the high curvature in the pore. When the energy of small hydrophilic pores is decreased, the energy barrier for the formation of these pores is also decreased.

In an early study done by Kinosita and Tsong [10] on human erythrocytes, they found that larger pores are obtained either by using higher field intensity, by increasing the pulse duration or by reducing the ionic strength of the pulsation medium. In another paper [11] they concluded that the threshold for electroporation is independent of the ionic strength of the medium, and changes little with temperature or with the rise time of the applied field. They estimate the number of pores per cell in the erythrocyte suspension to be $\sim 10^2$.

Recently, a new voltage-sensitive dye (ANNINE-6) that has a sub-nanosecond temporal response became available for experimentation. Frey et al. [12] used this new technique to monitor temporal development of the transmembrane voltage of Jurkat cells exposed to an electric field of 95*kV/cm* for 60 *nsec*. They found that the voltage across the membrane at the anodic pole rises in less than 5nsec to values of 1.2 *V*, and then increases at a slower rate of 30*mV/nsec* to values of 1.6 *V* – 15*nsec* after pulse application. During the same time period, the voltage across the membrane at the cathodic pole rises with a rate of ~40*mV/nsec* to a value of 0.6*V*. From 15 *nsec* onwards, up to the end of the 60 *nsec* high voltage pulse, the membrane voltages decrease slightly. After the electrical pulse, the voltages at both hemispheres return to the resting voltage. The authors suggest that this difference between the two hemispheres indicates a strong asymmetry in conduction mechanisms in the membranes of the two opposite cell hemispheres.

After the initial rise in voltage at the anodic pole, the cell membrane becomes permeable but only for small monovalent or divalent ions. A possible explanation might be that during the first 15 *nsec* the electric field creates a multitude of small nanopores across the cell membrane.

The pore formation or expansion was observed by Ryttsén et al. [13] of electroporated single NG108-15 cells to happen within a few milliseconds after pulse application. In some experiments, however, a delay of up to 30 *msec* was observed. This delay is probably caused by the time it takes for the pores to expand to a sufficient size for ion to pass. Because of the delay between application of the electric field and the formation of pores, the applied field does not affect directly the transport of ions over the membrane. In their study they cannot determine what process caused larger transmembrane currents, but it is evident that when potentials higher than the threshold value were applied, transmembrane currents increased in magnitude. This increase may be attributed to an increase in the number of pores or the expansion of the existing pores without a change in their number.

- **Transport**

Transport across electroporated cell membranes is in wide use for many applications ([14-17] for example), but it is also a useful tool to study the nature of the created pores – their size, composition and so forth. In this part we will not address the entire issue of transport through the electro-pores, but focus on the transport studies that elucidate the nature of the pores themselves.

Tekle et al. [18] took four types of molecules (divalent cation (Ca^{2+}), ethidium bromide (EB), propidium iodide (PI), ethidium homodimer (EthD-1)) and examined their transport across three electroporated cell membrane types (wild-type NIH 3T3, CHO and HeLa cells). The results show that a preferential transport into the cells after electroporation occurs. The different molecules enter either through the anode facing side or the cathode facing side, depending on their type, and by different amounts. From the data they conclude that asymmetric pores are created on both sides of the membrane that face the electrodes – smaller size and greater number on the anode side, large and low population on cathode side. The mechanism they suggest is that the electric field is first slightly greater than threshold on the anode facing side, so this is where initially the pores form. The radius of the pores does not increase rapidly since the field is only just above the threshold value. As the pores continue to form, the membrane becomes more conductive. This amounts to a potential drop of the entire cell falling across the membrane on the cathode facing side, since the anode side became electrically conductive. This high potential is well above threshold, so the first pores that form expand very rapidly. As they become large, the potential is depleted, and there is no further opportunity for many more pores to form.

In another study trying to understand this asymmetric transport property, Tekle et al. [19], looked into membrane permeabilization of NIH3T3 cells. When a critical membrane breakdown potential was applied to the cells, using unipolar AC fields or single DC square pulse, the cell membrane became permeable only on the hemisphere facing the positive electrode. A symmetrical permeabilization of the cells occurred only with bipolar AC fields. Their results show that the uptake rate of a fluorescent indicator by NIH3T3 cells is concentration dependent. This suggests that diffusion is more likely as a molecular exchange process than electrophoresis or electroosmosis in electroporation. Neumann et al. [20] observed that post-field diffusion is the dominant mechanism of electroporative dye transport across mouse B cell membrane.

Visualizing the DNA transfer at the single-cell level by an electric field, Phez et al. [21], saw that although the membrane permeabilization occurs at the two sides of the cell, only one side is "competent" for gene transfer. DNA transfer is a complex process that involves electrophoretically driven association of DNA only with the destabilized membrane facing the cathode. Pulses of reversed polarity allow DNA-membrane interaction to occur at the sides of the cells facing the two electrodes, but with partial loss of DNA at the membrane side facing the anode. Subsequent increases in the number of pulses led to a global increase in the amount of DNA associated to the membrane up to a plateau value.

The asymmetrical transport has also been reported by Gabriel and Teissié [22]. They electroporated CHO cells and used fluorescent dye to image the flow into them. Their results revealed that past a certain electric field intensity, permeabilization of the cell always happened on the anode facing side. Passing a higher, second, threshold value the two sides of the cells facing the electrodes are permeabilized but the side facing the anode is more permeable. This asymmetric pattern did not depend on the nature (propidium iodide or ethidium bromide) or concentration of the dye, the ionic strength of the pulsing buffer or the duration of the pulse. Their conclusion from the results is that using square-wave electric pulses on a single CHO cell the permeabilization remains local on the cell surface. Furthermore, the part of the cell membrane globally affected by the treatment depends on the electric field intensity and the extent of permeability depends on the duration of the pulse. Another conclusion is that the molecular interaction of the dye with the local permeabilized membrane area depends on the concentration gradient of the molecule, as for free diffusion (Fick's law), and is homogeneous on the permeabilized area.

Sukharev et al. [23] attempted in their study to answer the following questions – which force pushes DNA through the membrane and what kind of rearrangement in the structure of the plasma membrane mediates DNA translocation during electroporation. They electroporated Cos-1 cells by a single or dual pulse. They found that when DNA crosses the membrane it interacts with the membrane causing the membrane permeability to indifferent water-soluble dyes to become markedly higher. If a membrane is subjected to a short electroporative pulse in the presence of DNA, subsequent DNA translocation through this membrane can be driven by a relative low electric field, which is unable to form electroporation by itself. Their conclusion is that DNA does not penetrate the plasma membrane of mammalian cells through 'ready-to-use' electropores. The electric field creates primary pores, and these pores undergo size adjustment as a result of mechanical interaction between the pores and the DNA driven by elecrophoretic force.

The analysis of ion diffusion and cell permeabilization has also been conducted by Pavlin et al. [24] via a dense cell suspension of mouse melanoma cell line (B16F1). The fraction of the transient short-lived pores was determined based on measured transient conductivity changes during the electric pulses, whereas the fraction of long-lived transport pores – related to the increased membrane permeability after the pulse, was determined from the increase in conductivity between the pulses due to the ion efflux. They show that the fraction of long-lived transport pores increases with higher electric field due to larger area of the cell membrane exposed to above critical voltage, as well as due to higher energy which is available for pore formation. Therefore formation of the long-lived pores is governed also by the electrostatic energy of the pores. However, since the total amount of uptake (number of molecules) is an integral over a time period which depends on the time constant of resealing, it is clear that this simple dependence on the electric field strength or energy is lost when the transport of molecules is considered.

Another result they present in this study is an almost linear dependence of the efflux coefficient and the fraction of transport pores on the number of pulses, which suggests that each pulse increases the formation (stabilization) of the transport pores. Each new pulse leads to an increased number of long-lived stable pores that enable diffusion of ions and molecules, in contrast to short-lived pores that due to fast relaxation do not contribute to diffusion. The fraction of transient short-lived pores remains the same during each pulse, while the fraction of long-lived transport pores and efflux coefficient increase with each consecutive pulse.

The transient conductivity changes (short-lived pores) are related to permeabilization (long-lived transport pores) only indirectly, hence they claim we have to distinguish between the two types of pores. The process of formation of stable pores could be either stabilization of already present transient pores or an independent process.

The first clear demonstration of the asymmetric permeabilization during, not after, the pulse treatment was by Hibino et al. [25]. They found that the membrane conductance on the negative-electrode side increases faster than the conductance on the positive side. In the induction stage of the sea urchin eggs they tested, up to $\sim 1\mu sec$, poration on the positive side was more extensive than on the negative side. Part of the reason for this asymmetry appears to be the resting potential, since the critical value for poration is first reached on the positive side. The progress of poration is faster on the negative side – beyond 2 μsec the extent of the poration surpasses that on the positive side.

They think that the major cause of the asymmetry at the later stage is probably in the structure of the cell membrane. Apparently, the sidedness of the asymmetric dye penetration is different among different cell types. For example, with protoplasts or NIH3T3 cells, dye molecules enter the electroporated cells predominantly from the positive-electrode side, in sharp contrast with the results on sea urchin eggs. Such a difference points to the importance of asymmetric membrane structures which vary among different cell types. Electroosmosis alone cannot explain the asymmetry of slow dye permeation because elecroosmosis operates only in the presence of an electric field.

Resealing Process

Electroporation of cells may cause irreversible damage to the cells. Many applications use this to kill unwanted cancer cells, for example [26-29]. However, our main interest in this chapter lies in the reversible process. In order for the cell to survive the poration process, one must learn to adjust the electric field parameters. On the one hand the field strength needs to exceed a certain threshold for the poration to occur, but on the other hand it should not be strong enough to cause irreversible damage.

The sealing of the pores takes a much longer time than the opening of the pores. Setting the membrane back to equilibrium has been reported to take from tens of seconds to whole minutes. A research with human red blood cells done by Saulis et al. [30] concluded that there may be three stages involved in the resealing process: (1) a rapid decrease of the pore size (less than 1 sec); (2) slow pore decrease (a few minutes); (3) complete pore closing (more than 10 minutes). For the appearance of the first stage the pores must have a sufficiently large radius. They believe that this stage is caused by the strong dependence of the pore energy on the membrane voltage. Once the pulse has ended, the membrane voltage drops rapidly to zero. Then, under the influence of the strong restoring force the pore size diminishes to a smaller value.

In the second stage the pore radius decreases slowly until a local minimum of the energy is reached. Large molecules (4-5 Å) can no longer pass through the membrane, but small molecules (such as sodium ions) can. The hypothesis they suggest as to why this stage is much slower is the existence of an additional energy barrier for small pores, or that the restoring force that acts on large pores is greater than for the small pores. During the first and second stages the number of the pores stays constant, only their size changes.

The full closure of the pores occurs in the third and final stage. An energy barrier that is due to hydration and electrostatic repulsion of hydrophilic parts of the radially opposed pore wall molecules and the energy of deformation of the lipid molecules prevent the closing of the pores. That is the reason for their long lifetime.

In an earlier study, by Chernomordik et al. [2], it was reported that the recovery of the initial properties of the membrane following the reversible electroporation of human erythrocyte cells consists of two stages. The first is a very rapid stage ($\tau \ll 1msec$) which reflects the lowering of the conductance of small pores with decreasing voltage across the membrane. The diminishing of the number and mean radii of the pores, resulting in their complete disappearance, occurs only at the second stage of the membrane healing, which lasts several seconds or even minutes.

Glaser et al. [9] conducted experiments with lipid bilayers which indicated that the number of the pores in the membrane remains constant immediately after the pulse. The decrease in conductivity results from a reduction of pore size rather than from a decrease of the number of pores. The initial decrease of pore size happens in a time frame of 1-10*msec*. The decrease of the number of pores in the membrane has a rather long time constant that reaches up to more than 10*sec*. The long time constant is due to an energy barrier which prevents the gradual decrease of pore size leading to its disappearance. The existence of this barrier is connected with the increase of the energy of a hydrophilic pore at small radii.

The recovery process of NG108-15 single cells after electroporation usually occurs within 0.5-1 *sec*, as reported by Ryttsén et al. [13]. The recovery process was longer at higher voltages. They also report two time constants that seem to be involved in the recovery process – a fast recovery process (milliseconds) which returns the conductance to the approximate pre-pulse level but with a noisier baseline, and a slow process (seconds or longer) which returns this low-conductance state into a state with closed pores.

In a study by Bier et al. [31] done on skeletal muscle cells it was found that the resealing time does not significantly depend on the imposed transmembrane potential. Transient pores began to form in the cells when the imposed transmembrane potential exceeded 300 to 350 *mV*. The restoration of the membrane barrier to cationic flux occurred in ~9 minutes, and did not appear to be voltage dependent. When the transmembrane potential was above ~600 *mV* stable pores dominated the equilibration kinetics. One must keep in mind that the differences between the geometries of skeletal muscle cells and erythrocytes result in different transmembrane potentials being imposed for a given electric field. Therefore one has to be careful when comparing the results discussed above.

When electroporating a densely packed pellet of CHO cells, Schmeer et al. [32], found that the characteristic post-field pore resealing time (45 ± 3 *sec*) is independent of the field strength of the pulse. The pore resealing time is also independent of the distance between the two electrodes.

The temperature appears to be a significant factor in the resealing process. Benz and Zimmermann [33] conducted experiments with lipid bilayer membranes and concluded that the time constants of the exponential decay process of the pores may vary from about 10*μsec* at 2°C to about 2 *μsec* at 20°C. Above 25°C the resealing process is controlled by two exponential processes. They applied a pulse of 8 *V* for a duration of 300*nsec* and 500 *nsec*, followed by longer voltage pulse of low amplitude to measure the conductance. For duration of the high pulse of the order of several microseconds, the time constant for the resealing process was considerably larger than that for the short pulse.

Kinosita and Tsong [10] also found a temperature dependence of the resealing process. The human erythrocytes they tested rapidly regained their impermeability to cations at 37°C, whereas at 3°C the cells remain highly permeable even after 20 hours. Neumann et al. [20] have also observed that low temperatures slow down the resealing process of electroporated mouse B cells.

The fluidity of the membrane is different in different types of cells. Kandušer et al. [34] electroporated three different cell lines in order to investigate whether the fluidity of the membrane is connected to the resealing process. They have found that although temperature has a significant effect on the cell membrane fluidity, they could not establish a connection between the resealing process and the fluidity of the membrane. The differences in the time course of resealing could be explained by biological differences among cell lines, which masks any other effect. An interesting result in their study is that less fluid membranes were permeabilised at lower voltages than the more fluid ones.

In the study by Tekle et al. [18], mentioned in the previous section, they found that the rate of resealing is significantly enhanced in high ionic strength medium. They suggest that instability of small pores in high ionic media is the reason for the accelerated rate.

Huiqi He et al. [35] fabricated a MEMS device to study the resealing dynamics of human cancer cells (HeLa cells) after electroporation. They applied several pulse amplitudes (4-6 *V*) and widths (200 *μsec* – 2 *msec*) on the cells. They concluded that when a 2 *msec* pulse width is applied, for all amplitudes, the recovery time of the membrane was longer than 1.5 *sec*. A high amplitude pulse with a short width can be more quickly recovered. A high electric field with a relatively long exposure time will cause, as expected, cell lysis.

Factors That Affect the Uptake, Efficiency and Viability of the Cells

- **Voltage – Height, Pulses (Time, Shape)**

One of the basic parameters that control electroporation is the externally applied voltage (or current). Learning how to control this parameter might lead to a better understanding of the mechanism that underlies electroporation. In a work done by Huang and Rubinsky [36] they managed for the first time to see a typical current-voltage correlation in a single cell micro-chip. In their devised micro-chip a cell is captured in a micro-hole and electric currents are forced to pass through the cell, thereby producing electrically measurable information about the electroporation state of the cell. They found that the current increases with the applied voltage, suggesting that the pore size increases with the applied voltage. As they were able to look at single cells, they could conclude that the threshold voltage can vary by 20% in different cells. Exceeding the optimal voltage by 20% can induce irreversible electroporation. Therefore a careful preparation is needed when adjusting the electric pulse when conducting an electroporation experiment.

Kramar et al. [37] used a linear rising voltage signal for electroporating planar lipid bilayers. They learned that a gentle slope of the signal enables electroporation at lower voltages than steep voltage signals. Thus, a minimal breakdown voltage is determined by decreasing the slope (in their study: 0.49 *V*). Their results show that the lifetime of a lipid bilayer also depends on the slope of the signal – as the slope decreases and becomes steeper, the lifetime decreases.

The critical membrane potential difference which induces membrane permeabilization in intact cell suspensions has been reported by Tiessié et al. [38] to be 200-250 *mV*. This is a much smaller value than the value of 1 V, assumed in previous studies by this group.

The role of the pulse shape in electroporation of a cell suspension of DC3F cells (spontaneously transformed Chinese hamster fibroblasts) has been studied by Kotnik et al. [39]. A unipolar trapezoidal pulse of 1 *msec* with rise and fall times of 2*μsec*, 10*μsec* and 100 *μsec* was applied on the cells. The derivative of the pulse, meaning its rise and fall times, did not change the efficiency of electropermeabilization. Three different pulse shapes were examined – bipolar triangular, sine and rectangular, all of 1 *msec* duration and 100*V* peak amplitude. The results revealed that electropermeabilization, cell death, and peak of uptake all occur at the lowest voltage-to-distance (between electrodes) ratio. All these occurred at the highest voltage-to-distance ratio for the triangular pulses. These results are explained by the amount of time during which the pulse amplitude exceeds a certain critical value. Another aspect of the pulse shape is its modulation. In this study they looked at a unipolar sine-modulated rectangular pulse of 1 *msec* duration and 100*V* peak amplitude with a modulation of 10% and 90% of the peak amplitude, at 50 kHz. A shift towards higher voltage-to-distance ratio in the 90% modulated signal, as with the triangular pulse, was observed. Again, the explanation is that despite of the increase in amplitude, the highly modulated signal spends less time above the critical value.

There are several phenomena that occur when applying high electric field strength (5 to 7 *kV/cm*) on cell membranes that are different than the observed phenomena at the usual, relatively low to moderate, electric field (200 *V/cm* to 5 *kV/cm*) used for electroporation. Pliquett et al. [40] have listed them as follows: (1) strong conductivity increase of the cell solution at the very beginning of a high field pulse; (2) fast decrease in conductance which is not compatible with pore shrinking or sealing after the pulse; (3) loss of the phospholipid membrane during high-voltage pulsing, in addition to the pore formation process; (4) penetration of large molecules which implies the creation of large pores.

The fast initial conductivity increase cannot be explained by electroporation alone, since the increase is beyond that expected for completely electroporated membranes. Hence, the measured increase in conductivity of the cell suspension suggests that either new charge carriers are created in a field-dependent manner; or existing immobile charges are being liberated by the applied voltage pulse.

Pliquett et al. [40] suggest the following scenario: the high external field causes electroporation and forces water inside the bilayer structure. The confinement of water dipole molecules within the nano-pores effectively reduces the liquid permittivity. The high electric field near the membrane surface also facilitates the movement of charged molecules causing further structural rearrangement. The permittivity decreases with the electric field and leads to further increases in the local electric field, since the displacement vector $D = \varepsilon E$ is roughly constant. This local increase will produce further reductions in permittivity. The presence of high electric field could also alter the electronic states and cause energy-level mixing through the 'Stark effect'. It could become possible for the hybrid bonding states of one site to align with anti-bonding energies at adjacent sites. Field induced electronic transfer – hopping, leading to the weakening of the molecular bonds may follow. Hence, field-assisted fragmentation and micelle formation might be facilitated in the presence of high fields.

Kennedy et al. [41] electroporated HL60 human leukemia cells and also witnessed PI (propidium iodide) high level accelerating uptake signature associated with high field intensities – 2 to 3 *kV/cm*. They claim that this accelerated uptake signature is unexplainable by the traditional electroporation theory alone.

Electroporation of Neuronal cells at non-uniform electric fields has been researched by Heida et al. [42]. Their results show that in general, cells located near an electrode collapse at lower field intensities in comparison to cells located further away. The position of the cell does not affect the area of membrane breakdown at low frequencies (10 and 100 kHz), but is affected at high frequencies (1 MHz to 30 MHz).

The mean value of the membrane rupture time strongly depends on the applied voltage. Chernomordik et al. [2] found that an increase of 100 *mV* in the voltage caused the mean membrane lifetime of the electroporated erythrocytes to decrease by nearly one order of magnitude.

The use of an extremely short pulse (60 *nsec*) of high strength field (95 *kV/cm*) to electroporate Jurkat cells was studied by Frey et al. [12]. They found, with a voltage sensitive dye that has sub-nanosecond voltage response time, only a small voltage drop (0.6-1.6 *V*) across the plasma membrane. During the pulse, almost all the voltage across the cell appears across the interior of the cell after the conduction pores through the membrane have formed. This means that nearly the entire imposed electric field penetrates into the interior of the cell.

In an earlier study, by Deng et al. [43], the temporal development of long versus short pulses of electric field on the permeability of the membrane was studied. Jurkat cells were exposed to fields from *3kV/cm* to *75kV/cm* with pulse durations ranging from 60 *nsec* to 100 *μsec*. Increasing the pulse amplitude was associated with increasing percentage of cells taking up PI (propidium iodide). As the pulse duration decreased, the median onset of PI uptake became progressively delayed.

For long pulses, the onset of fluorescence uptake followed shortly after pulse application, and fluorescence increased more rapidly in the part of the nucleus oriented towards the anode. For short pulses, onset of fluorescence was delayed, and the spatial development of the fluorescence occurred without obvious orientation. They interpret these results as proof that for short electrical pulses, the effects on the cell membrane cannot be described in terms of conventional electroporation. They believe that the temporal separation of the uptake from the pulse application (of short pulses) is more likely due to alterations of cellular function rather than direct effects on surface membrane structure. As to the observed distribution of dye uptake location following short pulses, it shows that the breaching of the membrane at these locations is a secondary effect, possibly induced by electric field interactions with sub-cellular structures that lead to secondary cellular events, including apoptosis induction.

- **Temperature**

There are two major concerns regarding temperature rise during the process of electroporation. The electric field pulse may heat the solution in which the cells reside, and cause unintentional thermal damage to the cells. Another concern is related to non-ohmic sources of heating such as exothermic electrochemical reactions at the electrodes. Pliquett et al. [44] addressed these concerns in a study of the heating done by an electric field in an electrolyte with aluminum electrodes. They have found that in some cases, depending on the specific electrolyte, the temperature rise is significantly larger than expected from ohmic dissipation with the electrolyte. The time-varying spatial gradients in temperature are most likely to affect recovery and survival than uptake since the onset phase of electroporation is relatively small.

Gallo et al. [45] found that the stratum corneum (the outer layer of the skin) is more susceptible to permeabilization at higher temperatures. The temperature significantly affects the resistance drop during the pulse and the critical voltage decreases with increased temperature. They claim that joule heating by the pulse is not likely to have much effect unless it is localized. Although the elevated temperature facilitates the formation of electropores in the stratum corneum, the high temperature samples had a more prolonged recovery time.

Murthy et al. [46] investigated the influence of temperature during and after electroporation of porcine epidermis. They witnessed a jump in transport in temperatures above 40°C that cannot be explained by electrophoresis induced by the pulse or by the increased diffusion kinesis of the molecules. They concluded that the enhanced transport is most likely due to the prolonged postpulse permeable state of the skin. The high temperature (above 40°C) increases the lifetime of the electropores thus causing enhanced transport of molecules. In an earlier study by Benz and Zimmermann [33], the long-lived pores due to an elevation in temperature were also found in lipid bilayers.

Another interesting temperature related phenomena has been studied by Díaz-Rivera and Rubinsy [47]. They applied voltage pulses (below critical electroporation threshold) on single cells in a planar micro-pore chip. A small gap between the individual cell and the micro-pore wall forms. Their results revealed that the size of the channels at the interface between the cell and the micro-pore is temperature dependent. The cell to pore wall distance can increase by as much as 60% when the temperature is lowered from 35°C to 0°C. The membrane surface tension becomes stronger as the system is cooled, which means that the membrane has a greater ability to oppose a deformation of the cell membrane. This means that when using devices *in-vitro*, that require higher seal quality (between cell and device) for achieving better sensitivity, working close to physiological temperatures is preferable.

The temperature effect during electroporation of V-79 and B16F-1 cells on the membrane fluidity and permeabilization has recently been examined by Kandušer et al. [48]. Their results show that cell membrane permeabilization depends on the temperature, which has a significant effect on cell membrane fluidity and domain structure. By lowering the temperature from physiological 37°C to 4°C cell membrane fluidity significantly decreases. At the same time, cells exposed to 4°C need higher voltage for permeabilization of their membranes, and even at the highest pulse amplitudes tested (900 *V/cm*), only about 40% of cells in suspensions are permeabilized. On the other hand, the permeabilization of the more fluid membranes of cells exposed to 37°C is significantly higher, reaching around 80%. The cell membrane fluidity and membrane domain structure play an important role in electro-permeabilization, however other factors may be involved.

- **Shape – of Cell, of Electric Field, of Surroundings (Channel, Patch-Clamp, etc.)**

Geometry also plays an important role in electroporation. The shape of the cell, for instance, has to be taken into account when calculating the induced transmembrane potential. A uniform electric field, at a steady state induces upon a spherical cell a transmembrane potential that is given by the Schwan equation [49]:

$$\Delta\Phi = \frac{3}{2} Er\cos\theta$$

Where $\Delta\Phi$ is the induced transmembrane potential, $\frac{3}{2}$ is the shape factor for a sphere, E is the external electric field, r is the cell radius and θ is the angle measured from the center of the cell with respect to the direction of the field. This equation has been generalized for cases of non-spherical cells, such as prolate and oblate spheroids [50-52].

The orientation of elongated cells, such as Chinese hamster ovary (CHO) cells, relative to the electric field has been examined by Valič et al. [51]. At 350 *V/cm* and 400 *V/cm* plated cells (low density of cells, therefore no contact between them) that were parallel to the field were permeabilized, and those perpendicular were not. At 600 *V/cm* orientation no long effects permeabilization.

The question of the effect of cell size and shape on single-cell electroporation was recently addressed by Agarwal et al. [53]. A549 cells (Human lung cancer cells) were electroporated using electrolyte filled capillaries. The experimental variables were the pulse duration and cell-capillary tip distance. A logistic regression was carried out in order to determine whether the probabilities of cell survival and electroporation depend on experimental conditions and cell properties. The outcome of this analysis was the conclusion that cells are more likely to survive and are more readily permeabilized if they are large and hemispherical, as opposed to small and ellipsoidal with high aspect ratio. They observed that the survival probabilities are related to the calculated fraction of the cell's surface area that is electroporated, while the success of electroporation is related to the maximum transmembrane achieved.

Electroporation may take place in a specifically designed chip, and its geometry may have an effect on electroporation. For example, Shin et al. [54] used a microchannel type electroporation chip to electroporate mammalian cells (SK-OV-3). They have found that the uptake of the cells in a narrow channel (100 μm) is higher than in a wide channel (500μm). This can be explained by the difference in the resistance of the media between the electrodes due to the change in the channel width. After the pulsation the diameter of the cells in a 150 μm channel increased by 23% and in the 500 μm wide the increase was only of 10%. This is further evidence for the difference in the degree of electroporation as a function of the geometry of the microchannel.

Changes in the Cell Caused by Electroporation

- **Conductivity**

As a result of applying high voltage pulses in a cell suspension, the permeability of the cells increases which is accompanied by an increase in membrane conductivity. Pavlin et al. [55] used B16F1 cells (Mouse melanoma cell line) in a cell suspension to measure the current-voltage during the pulse and the impedance before and after the pulse. They performed these measurements in low and high conductivity media. In analyzing their results they concluded that the change in conductivity consists of several contributions: (1) a transient increase during each pulse, which is observed above the threshold electric field and drops back to the initial level in less than a second; (2) a decrease in conductivity caused by colloid-osmotic swelling due to osmotic imbalance when the cell membrane is permeabilized (above the threshold field). The swelling decreases the conductivity by a few percent in seconds after the application of the pulses, and up to 80% on the order of minutes; (3) increase in conductivity due to Joule heating by a few percent in high-conductivity medium; and (4) efflux of ions from the cell interior, which is observed in low-conductive medium above the threshold electric field.

Hibino et al. [56] imaged with a pulsed-laser fluorescence microscope at sub-microsecond time resolution the transmembrane potential induced by an electric pulse in a sea urchin egg. The cell membrane was stained with the voltage sensitive fluorescent dye RH292. Assuming proportionality between the fluorescence signal and the induced potential, they interpret the saturation of the signal as representing saturation of the membrane potential due to electroporation. They estimated the membrane conductance to be ~4.3 S/cm^2 at E=400 V/cm, and ~1.8 S/cm^2 at E=186 V/cm.

One of the first comprehensive studies on the phenomena of electroporation was done by Zimmermann et al. [57] in 1974. The breakdown of the membranes was studied with human and bovine red blood cells and *Escherichia coli* **B**, using a hydrodynamic focusing orifice. They found that the conductance of the membrane increases at a certain critical electric field strength in the orifice ($4 \cdot 10^6\ V/cm$ which yields about 1.6 V for the membrane potential).

The influence of the conductivity of the medium in a wide range was checked by Pucihar et al. [58]. They electroporated DC3F cells (spontaneously transformed Chinese hamster fibroblasts suspended in mediums with a conductivity of – 1.6 S/m, 0.14 S/m, 0.005 S/m and 0.001 S/m. Their results indicated that decreased conductivity of the media causes increase in the survival percentage of the cells, with no effect on permeabilization. They measured the pH, osmolarity, temperature and cell diameter and did not witness any side effect of the media on the cells. The conductivity of the media does have an effect on the resting transmembrane voltage. Different values of the induced and resting transmembrane voltage in media with different conductivities are the main reason for their experimental results. These results do not agree with the results obtained by Djuzenova et al. [59]. Their results show that by reducing the conductivity of the media a lower viability of the cells is obtained. They investigated a range of 0.08 to 0.37 S/m, which is a much narrower range than the one studied in [58], that might explain the different results.

- **Mechanical Stress, Swelling**

A high-intensity electric field pulse induces an electric dipole within the cell by the free charges that accumulate at the inner and outer boundaries of the plasma membrane (Maxwell-Wagner polarization mechanism). A transient deformation force is exerted on the interfaces by the Maxwell stress. The polarity of this force depends on the orientation of the induced dipole which, in turn, is determined by the effective polarizability of the cells relative to the suspending media [60]. Measurements of electroporated erythrocytes by Sukhorukov et al. [60] show that under low-conductivity conditions, the electric stretching force contributes significantly to the enlargement of 'electroleaks' in the plasma membrane generated by electric breakdown.

The notion of electromechanical deformation of vesicles and cells in electoroporation has been studied by Neumann et al. [61]. They show that unilamellar lipid vesicles subjected to high-voltage rectangular electric field pulses undergo elongation due to Maxwell stress. The shape elongation under the Maxwell stress is caused by and rapidly coupled to the electroporative increase in the membrane area. This fundamental result of electroporative shape deformation also applies to biological cell membranes, although one has to take into account the complex structure of the biological membrane. In Chang and Reese [62] the results support the notion that elecrto-pores are limited by the membrane-cytoskeletal interactions.

In a patch-clamp experiment using human embryonic kidney cells (HEK 293) Akinlaja and Sachs [63] tried to determine whether mechanical tension and electrical stress couple to cause membrane breakdown. In the membrane there is a far-field mechanical stretch, thinning the membrane, and tension caused by electrical stress that works as a stretching force (energy/area). They found that short pulses (50*μsec*) cause breakdown at a threshold of 0.5 *V*, and are tension dependent. Long pulses (50-100*msec*) require 0.2-0.4 *V* to cause breakdown, and are independent of tension. They conclude, in contradiction with Teissie and Rols [38] that the mechanism of high field/short pulse and low field/long pulse breakdown are fundamentally different. The interpretation they give to the results in this study is that in short pulses we witness a lipid breakdown whereas in long pulses the breakdown is a result of protein denaturation.

Griese et al. [64] applied a rectangular pulse on lipid unilamellar vesicles. Structural changes of membrane electroporation have no threshold, they are continuous. They estimate that Maxwell pressure has a very minor importance, if any at all, for the electrolyte transport. The ion-pore wall interaction modulates electrolyte release through the pores and thus modulates the pore conductivity. The pore conductivity is dependent on the electric field because of a reduction of the repulsive energy barrier of ions in the pore by the electric field.

In the study of Deng et al. [43], previously discussed in the 'voltage' section, they have witnessed morphological changes in the Jurkat cells. The morphological changes induced by long pulses were most striking: cells went from showing somewhat refractile edges with little intracellular detail visible before pulse application, to diminished refractile edges and appearance of intracellular detail. Similar changes occurred with short pulses, although they became less striking as the pulse duration went from 300 *nsec* to 60 *nsec*. Cells that underwent multimicrosecond and 300 *nsec* pulse exposures all increased their two-dimensional cell areas while cells exposed to 60 *nsec* pulses had no significant increase in their cell area. This increase (for pulses longer than 60 *nsec*) likely represents colloid osmotic swelling of cells in physiological buffer, and as such, would be typical of electroporation with delayed resealing of the pores.

Irreversible Electroporation Studies in Mammalian Cells

Research on irreversible electroporation of cells was done primarily with microorganisms, in support of the use of IRE in the food industry. Electroporation in mammalian cells was studied primarily in the context of reversible electroporation. The study of IRE of mammalian cells in the context of tissue ablation is a new area of research of importance. Two recent studies of this type suggest interesting new findings. Leor et al. [65] have studied the parameters required for complete non-thermal IRE ablation of human hepatocarcinoma HepG2 cells. They have found that the application of 1500 V/cm in three sets separated by minutes of ten pulses of 300 microseconds each can produce complete cancer cell ablation. It is quite possible that shorter pulses could have produced the same effect. The same study has found that the use of multiple pulses appears to be more effective for cancer cell ablation than the application of the same energy in one single pulse. The importance of a properly designed sequence for cancer cell ablation and the value of protocols using multiple pulses were also found in the subsequent work of Rubinsky et al. [66]. Working with prostate PC3 adenocarcinoma cells, in a range of fields from 2000 to 250 V/cm it was found that the number of repetitions of the pulses are important in producing the IRE. A total of 90 pulses of a relatively low field of 250 V/cm delivered in 100 microsecond pulses was able to produce complete cancer cell destruction.

Discussion

We have seen in this chapter that there is great diversity in the exact parameters that suit various types of cells in the process of electroporation. To understand the mechanism of electroporation one has to extract the basic physical properties that are common phenomena. Looking at the process of pore formation, we may conclude that from the information that is available to us now there are the following key steps:

(1) Pore formation, induced by an electric field that drives the transmembrane potential above a critical threshold (typically 0.2 to 1 *V*). This step has stochastic features and an asymmetric nature (in a unipolar field).
(2) Transport across the cell membrane (mainly post-pulse). The type of the molecules can indicate the size of the pores and their lipid reorganization.
(3) Resealing of the pores, which can be divided into at least two main stages: (a) fast (*msec*) process in which the conductance returns to its pre-pulse level, and the size of the pores diminishes; (b) long (*sec-min*) process in which the pores disappear and the cell is fully recovered.

Much of the fundamental research on mammalian cells has been done in the context of reversible electroporation. However, with the advent of non-thermal irreversible electroporation there is a need and an opportunity to explore the broad range of irreversible electrical parameters and their effect on cells.

References

[1] Stämpfli, R.: Reversible electrical breakdown of the excitable membrane of a Ranvier Node. An. da Acad. Brasileira de Ciências 30, 57–63 (1957)
[2] Chernomordik, L.V., Sukharev, S.I., Popov, S.V., Pastushenko, V.F., Sokirko, A.V., Abidor, I.G., Chizmadzhev, Y.A.: The electrical breakdown of cell and lipid membranes: the similarity of phenomenologies. Biochim. Biophys. Acta 902, 360–373 (1987)
[3] Abidor, I.G., Arakelyan, V.B., et al.: Electric Breakdown of Bilayer Lipid-Membranes.1. Main Experimental Facts and Their Qualitative Discussion. Bioelectrochemistry and Bioenergetics 6, 37–52 (1979)

[4] Weaver, J.C., Chizmadzhev, Y.A.: Theory of electroporation: A review. Bioelectrochemistry and Bioenergetics 41, 135–160 (1996)
[5] Kotulska, M., Koronkiewicz, S., Kalinowski, S.: Self-similar processes and flicker noise from a fluctuating nanopore in a lipid membrane. Phys. Rev. E. Stat. Nonlin. Soft. Matter Phys. 69, 031920 (2004)
[6] Teruel, M.N., Meyer, T.: Electroporation-induced formation of individual calcium entry sites in the cell body and processes of adherent cells. Biophys. J. 73, 1785–1796 (1997)
[7] Neumann, E.: Membrane Electroporation and Direct Gene-Transfer. Bioelectrochemistry and Bioenergetics 28, 247–267 (1992)
[8] Weaver, J.C.: Electroporation of biological membranes from multicellular to nano scales. IEEE Transactions on Dielectrics and Electrical Insulation 10, 754–768 (2003)
[9] Glaser, R.W., Leikin, S.L., Chernomordik, L.V., Pastushenko, V.F., Sokirko, A.I.: Reversible electrical breakdown of lipid bilayers: formation and evolution of pores. Biochim. Biophys. Acta 940, 275–287 (1988)
[10] Kinosita, K., Tsong, T.Y.: Formation and Resealing of Pores of Controlled Sizes in Human Erythrocyte-Membrane. Nature 268, 438–441 (1977)
[11] Kinosita Jr., K., Tsong, T.Y.: Voltage-induced conductance in human erythrocyte membranes. Biochimica et Biophysica Acta (BBA) - Biomembranes 554, 479–497 (1979)
[12] Frey, W., White, J.A., Price, R.O., Blackmore, P.F., Joshi, R.P., Nuccitelli, R., Beebe, S.J., Schoenbach, K.H., Kolb, J.F.: Plasma membrane voltage changes during nanosecond pulsed electric field exposure. Biophys. J. 90, 3608–3615 (2006)
[13] Ryttsen, F., Farre, C., Brennan, C., Weber, S.G., Nolkrantz, K., Jardemark, K., Chiu, D.T., Orwar, O.: Characterization of single-cell electroporation by using patch-clamp and fluorescence microscopy. Biophys. J. 79, 1993–2001 (2000)
[14] Gehl, J.: Electroporation: theory and methods, perspectives for drug delivery, gene therapy and research. Acta Physiol. Scand. 177, 437–447 (2003)
[15] Gothelf, A., Mir, L.M., Gehl, J.: Electrochemotherapy: results of cancer treatment using enhanced delivery of bleomycin by electroporation. Cancer Treat. Rev. 29, 371–387 (2003)
[16] Mir, L.M., Glass, L.F., Sersa, G., Teissie, J., Domenge, C., Miklavcic, D., Jaroszeski, M.J., Orlowski, S., Reintgen, D.S., Rudolf, Z., Belehradek, M., Gilbert, R., Rols, M.P., Belehradek Jr., J., Bachaud, J.M., DeConti, R., Stabuc, B., Cemazar, M., Coninx, P., Heller, R.: Effective treatment of cutaneous and subcutaneous malignant tumours by electrochemotherapy. Br. J. Cancer 77, 2336–2342 (1998)
[17] Golzio, M., Rols, M.P., Teissie, J.: In vitro and in vivo electric field-mediated permeabilization, gene transfer, and expression. Methods 33, 126–135 (2004)
[18] Tekle, E., Astumian, R.D., Chock, P.B.: Selective and asymmetric molecular transport across electroporated cell membranes. Proc. Natl. Acad. Sci. U. S. A. 91, 11512–11516 (1994)
[19] Tekle, E., Astumian, R.D., Chock, P.B.: Electro-permeabilization of cell membranes: effect of the resting membrane potential. Biochem. Biophys. Res. Commun. 172, 282–287 (1990)
[20] Neumann, E., Toensing, K., Kakorin, S., Budde, P., Frey, J.: Mechanism of electroporative dye uptake by mouse B cells. Biophys. J. 74, 98–108 (1998)
[21] Phez, E., Faurie, C., Golzio, M., Teissie, J., Rols, M.P.: New insights in the visualization of membrane permeabilization and DNA/membrane interaction of cells submitted to electric pulses. Biochim. Biophys. Acta 1724, 248–254 (2005)

[22] Gabriel, B., Teissie, J.: Direct observation in the millisecond time range of fluorescent molecule asymmetrical interaction with the electropermeabilized cell membrane. Biophys. J. 73, 2630–2637 (1997)
[23] Sukharev, S.I., Klenchin, V.A., Serov, S.M., Chernomordik, L.V., Chizmadzhev Yu, A.: Electroporation and electrophoretic DNA transfer into cells. The effect of DNA interaction with electropores. Biophys. J. 63, 1320–1327 (1992)
[24] Pavlin, M., Leben, V., Miklavcic, D.: Electroporation in dense cell suspension–theoretical and experimental analysis of ion diffusion and cell permeabilization. Biochim. Biophys. Acta 1770, 12–23 (2007)
[25] Hibino, M., Itoh, H., Kinosita Jr., K.: Time course of cell electroporation as revealed by submicrosecond imaging of transmembrane potential. Biophysical Journal 64, 1789–1800 (1993)
[26] Eppich, H.M., Foxall, R., Gaynor, K., Dombkowski, D., Miura, N., Cheng, T., Silva-Arrieta, S., Evans, R.H., Mangano, J.A., Preffer, F.I., Scadden, D.T.: Pulsed electric fields for selection of hematopoietic cells and depletion of tumor cell contaminants. Nat. Biotechnol. 18, 882–887 (2000)
[27] Onik, G., Mikus, P., Rubinsky, B.: Irreversible electroporation: implications for prostate ablation. Technol. Cancer Res. Treat. 6, 295–300 (2007)
[28] Lavee, J., Onik, G., Mikus, P., Rubinsky, B.: A novel nonthermal energy source for surgical epicardial atrial ablation: irreversible electroporation. Heart Surg Forum 10, E162–E167 (2007)
[29] Rubinsky, B.: Irreversible electroporation in medicine. Technol. Cancer Res. Treat. 6, 255–260 (2007)
[30] Saulis, G., Venslauskas, M.S., et al.: Kinetics of Pore Resealing in Cell- Membranes after Electroporation. Bioelectrochemistry and Bioenergetics 26, 1–13 (1991)
[31] Bier, M., Hammer, S.M., Canaday, D.J., Lee, R.C.: Kinetics of sealing for transient electropores in isolated mammalian skeletal muscle cells. Bioelectromagnetics 20, 194–201 (1999)
[32] Schmeer, M., Seipp, T., Pliquett, U., Kakorin, S., Neumann, E.: Mechanism for the conductivity changes caused by membrane electroporation of CHO cell-pellets. Phys. Chem. Chem. Phys. 6, 5564–5574 (2004)
[33] Benz, R., Zimmermann, U.: The resealing process of lipid bilayers after reversible electrical breakdown. Biochim. Biophys. Acta 640, 169–178 (1981)
[34] Kanduser, M., Sentjurc, M., Miklavcic, D.: Cell membrane fluidity related to electroporation and resealing. Eur. Biophys. J. 35, 196–204 (2006)
[35] He, H., Chang, D.C., Lee, Y.K.: Nonlinear current response of micro electroporation and resealing dynamics for human cancer cells. Bioelectrochemistry 72, 161–168 (2008)
[36] Huang, Y., Rubinsky, B.: Micro-Electroporation: Improving the Efficiency and Understanding of Electrical Permeabilization of Cells. Biomedical Microdevices 2, 145–150 (1999)
[37] Kramar, P., Miklavcic, D., Lebar, A.M.: Determination of the lipid bilayer breakdown voltage by means of linear rising signal. Bioelectrochemistry 70, 23–27 (2007)
[38] Teissie, J., Rols, M.P.: An experimental evaluation of the critical potential difference inducing cell membrane electropermeabilization. Biophys. J. 65, 409–413 (1993)
[39] Kotnik, T., Pucihar, G., Rebersek, M., Miklavcic, D., Mir, L.M.: Role of pulse shape in cell membrane electropermeabilization. Biochim. Biophys. Acta 1614, 193–200 (2003)

[40] Pliquett, U., Joshi, R.P., Sridhara, V., Schoenbach, K.H.: High electrical field effects on cell membranes. Bioelectrochemistry 70, 275–282 (2007)

[41] Kennedy, S.M., Ji, Z., Hedstrom, J.C., Booske, J.H., Hagness, S.C.: Quantification of electroporative uptake kinetics and electric field heterogeneity effects in cells. Biophys. J. 94, 5018–5027 (2008)

[42] Heida, T., Wagenaar, J.B., Rutten, W.L., Marani, E.: Investigating membrane breakdown of neuronal cells exposed to nonuniform electric fields by finite-element modeling and experiments. IEEE Trans. Biomed. Eng. 49, 1195–1203 (2002)

[43] Deng, J., Schoenbach, K.H., Buescher, E.S., Hair, P.S., Fox, P.M., Beebe, S.J.: The effects of intense submicrosecond electrical pulses on cells. Biophys. J. 84, 2709–2714 (2003)

[44] Pliquett, U., Gift, E.A., Weaver, J.C.: Determination of the electric field and anomalous heating caused by exponential pulses with aluminum electrodes in electroporation experiments. Bioelectrochemistry and Bioenergetics 39, 39–53 (1996)

[45] Gallo, S.A., Sen, A., Hensen, M.L., Hui, S.W.: Temperature-dependent electrical and ultrastructural characterizations of porcine skin upon electroporation. Biophys. J. 82, 109–119 (2002)

[46] Murthy, S.N., Sen, A., Zhao, Y.L., Hui, S.W.: Temperature influences the postelectroporation permeability state of the skin. J. Pharm. Sci. 93, 908–915 (2004)

[47] Diaz-Rivera, R.E., Rubinsky, B.: Electrical and thermal characterization of nanochannels between a cell and a silicon based micro-pore. Biomed. Microdevices 8, 25–34 (2006)

[48] Kanduser, M., Sentjurc, M., Miklavcic, D.: The temperature effect during pulse application on cell membrane fluidity and permeabilization. Bioelectrochemistry 74, 52–57 (2008)

[49] Schwan, H.P.: Electrical properties of tissue and cell suspensions. Adv. Biol. Med. Phys. 5, 147–209 (1957)

[50] Kotnik, T., Miklavcic, D.: Analytical description of transmembrane voltage induced by electric fields on spheroidal cells. Biophys. J. 79, 670–679 (2000)

[51] Valic, B., Golzio, M., Pavlin, M., Schatz, A., Faurie, C., Gabriel, B., Teissie, J., Rols, M.P., Miklavcic, D.: Effect of electric field induced transmembrane potential on spheroidal cells: theory and experiment. Eur. Biophys. J. 32, 519–528 (2003)

[52] Gimsa, J., Wachner, D.: Analytical description of the transmembrane voltage induced on arbitrarily oriented ellipsoidal and cylindrical cells. Biophys. J. 81, 1888–1896 (2001)

[53] Agarwal, A., Zudans, I., Weber, E.A., Olofsson, J., Orwar, O., Weber, S.G.: Effect of cell size and shape on single-cell electroporation. Anal. Chem. 79, 3589–3596 (2007)

[54] Shin, Y.S., Cho, K., Kim, J.K., Lim, S.H., Park, C.H., Lee, K.B., Park, Y., Chung, C., Han, D.C., Chang, J.K.: Electrotransfection of mammalian cells using microchannel-type electroporation chip. Anal. Chem. 76, 7045–7052 (2004)

[55] Pavlin, M., Kanduser, M., Rebersek, M., Pucihar, G., Hart, F.X., Magjarevic, R., Miklavcic, D.: Effect of cell electroporation on the conductivity of a cell suspension. Biophys. J. 88, 4378–4390 (2005)

[56] Hibino, M., Shigemori, M., Itoh, H., Nagayama, K., Jr. Kinosita, K.: Membrane conductance of an electroporated cell analyzed by submicrosecond imaging of transmembrane potential. Biophys. J. 59, 209–220 (1991)

[57] Zimmermann, U., Pilwat, G., Riemann, F.: Dielectric breakdown of cell membranes. Biophys. J. 14, 881–899 (1974)

[58] Pucihar, G., Kotnik, T., Kanduser, M., Miklavcic, D.: The influence of medium conductivity on electropermeabilization and survival of cells in vitro. Bioelectrochemistry 54, 107–115 (2001)
[59] Djuzenova, C.S., Zimmermann, U., Frank, H., Sukhorukov, V.L., Richter, E., Fuhr, G.: Effect of medium conductivity and composition on the uptake of propidium iodide into electropermeabilized myeloma cells. Biochim. Biophys. Acta 1284, 143–152 (1996)
[60] Sukhorukov, V.L., Mussauer, H., Zimmermann, U.: The effect of electrical deformation forces on the electropermeabilization of erythrocyte membranes in low- and high-conductivity media. J. Membr. Biol. 163, 235–245 (1998)
[61] Neumann, E., Kakorin, S., Toensing, K.: Membrane electroporation and electromechanical deformation of vesicles and cells. Faraday Discuss, 111–125 (1998); discussion 137-57
[62] Chang, D.C., Reese, T.S.: Changes in membrane structure induced by electroporation as revealed by rapid-freezing electron microscopy. Biophys. J. 58, 1–12 (1990)
[63] Akinlaja, J., Sachs, F.: The breakdown of cell membranes by electrical and mechanical stress. Biophys. J. 75, 247–254 (1998)
[64] Griese, T., Kakorin, S., Neumann, E.: Conductometric and electrooptic relaxation spectrometry of lipid vesicle electroporation at high fields. Phys. Chem. Chem. Phys. 4, 1217–1227 (2002)
[65] Miller, L., Leor, J., Rubinsky, B.: Cancer cells ablation with irreversible electroporation. Technology in Cancer Research and Treatment 4(6), 699–706 (2005)
[66] Rubinsky, J., Onik, G., Mikus, P., Rubinsky, B.: Optimal Parameters for the Destruction of Prostate Cancer Using Irreversible Electroporation. J. Urol. (2008)

Mechanism of Irreversible Electroporation in Cells: Insight from the Models*

Wanda Krassowska Neu[1,**] and John C. Neu[2]

[1] Department of Biomedical Engineering, Duke University, Durham, NC 27708, USA
[2] Department of Mathematics, University of California, Berkeley, CA 94720, USA
** Corresponding author E-mail: wanda@eel-mail.bme.duke.edu

1 Introduction

In the last 20 years, electroporation studies have focused primarily on reversible electroporation because of its importance for drug and gene delivery. Irreversible electroporation (IRE) was mostly studied in the context of the delayed tissue damage in high-voltage accidents [70, 69], the postshock arrhythmias during defibrillation [51, 50], and biofouling control [46, 95]. Very recently, IRE has shown great promise as a nonthermal technique for the ablation of tumors and arrhythmogenic regions in the heart [20, 27, 91, 85, 4, 68]. IRE's ability to create a complete and predictable cell ablation with a sharp transition between normal and necrotic tissue, while sparing neighboring blood vessels, connective tissue, and nerves, has great advantages in a variety of medical applications.

However, relatively little is known about the mechanism by which IRE causes cell death. There have been numerous experimental studies on cell viability following the delivery of an electric pulse [46, 47, 73, 35, 66]. Some of these studies proposed empirical relationships connecting the probability of cell death to pulse strength, duration, or number of pulses [47, 43, 58, 66]. Yet, there are disagreements within the literature. Some studies find a correlation between cell death and the total energy delivered by the pulse [84, 58] while others do not [95, 107], and yet others correlate cell death with the total pulse charge [66].

Currently, we have a fairly good understanding of the mechanism of IRE in two special cases:

1. *IRE in planar lipid bilayer membranes:* Early experimental work [1, 8, 16] resulted in the development of the theory of the bilayer energetics, which offers an elegant explanation of IRE. Namely, there exists an energy maximum at a certain "critical radius" of a pore. When a pore with a supercritical radius is created, it will expand spontaneously leading to mechanical rupture of the membrane [1, 15, 6, 111, 25].

* Supported in part by the National Science Foundation Grant DMS-0515616.

B. Rubinsky (Ed.): Irreversible Electroporation, BIOMED, pp. 85–122.
springerlink.com

2. *IRE in cells and tissues subject to ultrashort (up to 100s of nanoseconds) and high-intensity (hundreds of kV/cm) pulses:* Pulses of this nature, termed supraelectroporation, are believed to directly affect intracellular organelles without irreversibly damaging the cell membrane [94, 86]. Cell death is attributed to mitochondria-induced apoptosis, release of calcium from endoplasmic reticulum, and damage to DNA [51, 50, 75, 93, 7].

This chapter will not address cell death caused by supraelectroporation. Extensive modeling of IRE under supraelectroporation conditions has been conducted by Joshi et al. [54, 55, 44, 53] and Weaver et al. [100, 38, 39, 106, 29]. Instead, it will address cell death caused by conventional electroporation pulses (micro- to milliseconds duration, up to a few kV/cm strength). Here, the mechanism of cell death is different from that in supraelectroporation because pulses of a few kV/cm are not sufficiently strong to affect small intracellular organelles. However, micro- to milliseconds pulses are sufficiently long to engage several electrical, mechanical, thermal, and mass transport processes. The existence of multiple competing processes makes it difficult to pinpoint the exact mechanism of IRE in cells subject to conventional electroporation pulses.

First, to give some background, we will review the theory of IRE in planar membranes and in cells under voltage-clamp conditions. Then we will present modeling studies of the processes taking place during IRE and evaluate their potential role in cell death. Finally, based on the current understanding of cell electroporation, we will summarize the potential mechanisms of cell death during IRE.

2 Background: IRE in Planar Membranes

2.1 Experimental Observations

Planar lipid bilayer membranes are formed by pipetting solutions of phospholipids upon a 1-mm opening in the partition of a two-compartment Teflon chamber [77]. As an example, we will summarize the voltage clamp experiment of Chernomordik et al. [16] on oxidized cholesterol and UO_2^{2+}-modified azolectin membranes. In this study, Ag/AgCl electrodes immersed in the two compartments delivered rectangular voltage-clamp pulses across the membrane. A current through the membrane was measured, which allowed one to continuously monitor changes in the membrane conductance for as long as it remained well below the chamber's conductance.

This experiment has shown that IRE consists of two stages. First, the conductance increases gradually by a factor of $10^5 - 10^6$. This increase is reversible if the pulse is terminated sufficiently early (e.g., at 20 μs for 700 mV pulses applied to oxidized cholesterol membrane): the same increase in conductance is reproduced during the repeated pulses. If, however, the pulse duration is increased to 40 μs, the second stage occurs, in which the conductance has an abrupt jump indicating irreversible membrane rupture. The

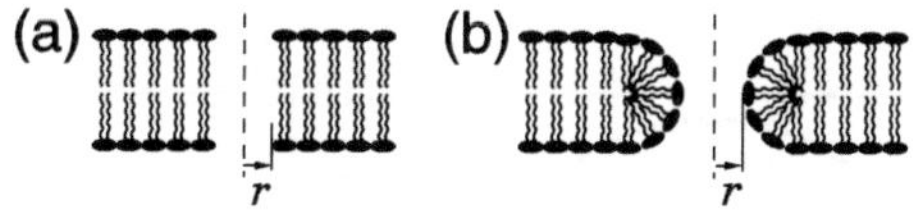

Fig. 1. The structure of (a) hydrophobic and (b) hydrophilic pores (based on Refs. [1, 108, 36]). Pore radius is denoted by r.

lifetime of the membrane, measured from the onset of the pulse to the abrupt conductance jump, is a random quantity but its mean value decreases with voltage.

These results show that under constant voltage, IRE starts with reversible increase in conductance and, after sufficient time, membrane rupture occurs and the process becomes irreversible. Thus, the same theoretical framework should describe both reversible and irreversible electroporation.

2.2 Theory of Electroporation in Planar Membranes: Pore Energy

The theory assumes the existence of two types of pores sketched in Fig. 1 [1, 108, 36]. *Hydrophobic* pores are simply gaps in the lipid bilayer of the membrane, formed as a result of its thermal fluctuations. *Hydrophilic* pores have their walls lined with the water-attracting heads of lipid molecules so their surface properties are identical with the membrane. Hence, hydrophilic pores allow the passage of water-soluble substances, such as ions, and they conduct electric current. We will focus on hydrophilic pores, as they are relevant to modeling IRE.

The initial reversible increase in conductance is due to the creation of hydrophilic pores. They appear with the initial radius $r_* \approx 0.5$ nm, which is the minimum radius of hydrophilic pores [36]. The pore creation rate is exponentially dependent on the square of the transmembrane potential V_m [36, 79]:

$$\frac{dN}{dt} = \alpha\, \mathrm{e}^{(V_m/V_{ep})^2} \left(1 - \frac{N}{N_{eq}(V_m)}\right), \tag{1}$$

where $N(t)$ is the density of hydrophilic pores, α is the creation rate coefficient, and V_{ep} is the characteristic voltage of electroporation. $N_{eq}(V_m)$ is the equilibrium pore density for a given voltage V_m,

$$N_{eq}(V_m) = N_0\, e^{q\,(V_m/V_{ep})^2}, \tag{2}$$

where N_0 and q are constants.

Once created, hydrophilic pores expand or shrink in response to two factors: diffusion and drift [1, 88]. Diffusion refers to random increases and decreases of pore radius induced by thermal fluctuations. Drift refers to definite

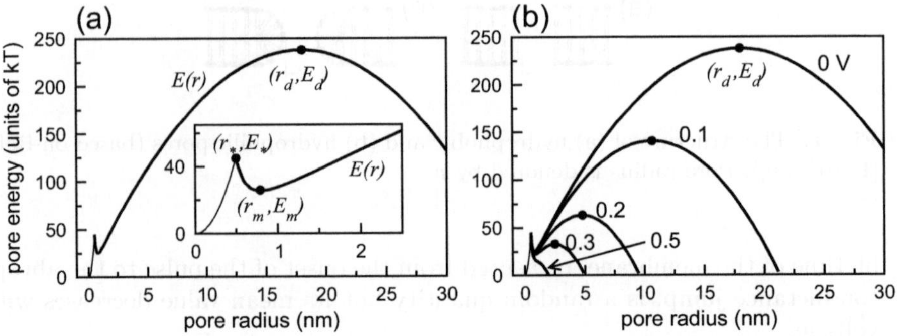

Fig. 2. (a) Energy of the hydrophobic pores ($r < r_*$, thin line) and the hydrophilic pores ($r \geq r_*$) at $V_m = 0$. Inset shows the pore energy for small radii. Filled circles mark positions of the energy minimum (r_m, E_m), and of the energy barriers for creation of hydrophilic pores (r_*, E_*) and for rupture (r_d, E_d). (b) Dependence of pore energy on the transmembrane potential V_m computed from the formula of Abidor et al. (5). Labels specify values of V_m, filled circles mark positions of the energy barrier for rupture (r_d, E_d).

time-rate of change of pore radius leading to a decrease of the energy of the system. To determine the drift velocity, we introduce the *pore energy*, E, which is the energy cost of introducing a single pore of radius r (Fig. 2):

$$E(r, V_m) = \beta \left(\frac{r_*}{r}\right)^4 + 2\pi\gamma r - \pi\sigma r^2 + E_{el}(r, V_m) \quad \text{in } r \geq r_*. \qquad (3)$$

The first term represents the steric repulsion between lipid heads lining the pore and is responsible for the increase in pore energy as r shrinks to r_* [79]; β is the steric repulsion energy. The second term represents the energy required to bend the bilayer in order to form the pore perimeter; γ is the linear tension of the pore perimeter. The third term represents the decrease in the energy due to the effect of a pore on the tension of the membrane; σ is the membrane surface tension. Finally, the fourth term is the electrical energy $E_{el}(r, V_m)$, which represents the effect of the external electric pulse (to be described in Sec. 2.3 and 2.4).

If we ignore thermal fluctuations, the pore radius r will evolve with the drift velocity u:

$$\frac{dr}{dt} = u(r, V_m) = -\frac{D}{kT}\frac{dE}{dr} = \frac{D}{kT}\left\{4\beta\left(\frac{r_*}{r}\right)^4 \frac{1}{r} - 2\pi\gamma + 2\pi\sigma r + F_{el}(r, V_m)\right\} \text{ in } r \geq r_*, (4)$$

where D is the diffusion coefficient associated with random fluctuation of pore radii, k is the Boltzman constant, T is the absolute temperature, and $F_{el} \equiv -dE_{el}/dr$ is the electric force expanding the pore.

Figure 2a shows that for the case of no imposed voltage ($V_m = 0$), $E(r)$ has two local maxima separated by a local minimum. The maximum E_* at

r_* is the energy barrier for creation of hydrophilic pores. The maximum E_d at r_d is the energy barrier for spontaneous expansion of pores. The drift velocity (4) implies that a pore tries to assume the radius corresponding to the lowest energy. Thus, the pores created with the initial radius r_* expand spontaneously to the minimum-energy radius r_m. Thermal fluctuations cause random expansion and shrinkage of pore radii. If a pore expands to a radius $r_m < r < r_d$, it will shrink to r_m and the membrane is stable. However, if any pore radius exceeds the critical value r_d, this pore will begin spontaneously expanding. This unbounded expansion is responsible for the jump in membrane conductance and the rupture seen in Chernomordik et al. experiments [16].

2.3 Mechanism of Membrane Rupture

Using the values of parameters from Appendix I, we see that for $V_m = 0$, the critical radius r_d is 18 nm and the corresponding energy E_d is 238 kT. With so high an energy barrier, the probability a pore reaching a supercritical radius is small and the membrane is stable. However, an applied voltage V_m has dramatic effect on E_d and r_d. To account for V_m, Abidor et al. approximated the electrical component E_{el} of pore energy by the dielectric energy that arises due to the change of the specific capacitance of the membrane when lipid molecules are replaced by water in the region of the pore [1],

$$E_{el}(r, V_m) = -\frac{\pi}{2h}\left(\epsilon_w - \epsilon_m\right) V_m^2 \, r^2, \tag{5}$$

where ϵ_w, ϵ_m are permittivities of the water solution and the membrane, respectively, and h is the membrane thickness.

Introducing (5) to (3), we obtain the formulas for the critical radius r_d,

$$r_d = \frac{\gamma}{\sigma + \frac{1}{2h}(\epsilon_w - \epsilon_m) V_m^2} \tag{6}$$

and for the height of the barrier for rupture E_d,

$$E_d = \frac{\pi\gamma^2}{\sigma + \frac{1}{2h}(\epsilon_w - \epsilon_m) V_m^2} \tag{7}$$

(Formulas (6-7) ignore the first term in (3), because steric repulsion of lipid heads decreases very quickly for $r > r_m$.) For $V_m = 0.3$ V, $r_d \approx 2.5$ nm and $E_d \approx 33$ kT. With the difference $E_d - E_m \approx 7$ kT, the appearance of a supercritical pore due to thermal fluctuation becomes quite likely.

Pastushenko et al. have estimated the mean lifetime of the membrane as a function of V_m by evaluating the time when the critical radius is first reached [88]. For a membrane having a population of K noninteracting pores, the expected membrane lifetime $\bar{t}$ is

$$\bar{t} = \frac{1}{4\pi D K}\left(\frac{kT}{\gamma}\right)^{3/2} \sqrt{r_d}\, \exp\left(\frac{E_d}{kT}\right). \tag{8}$$

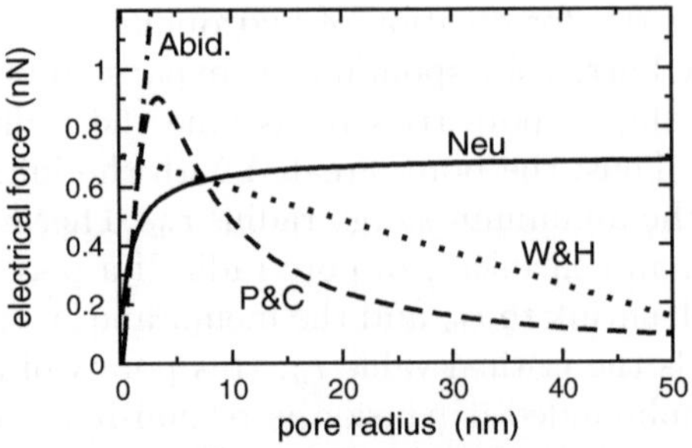

Fig. 3. Electrical force expanding the pore as a function of the pore radius at the transmembrane potential $V_m = 1$ V. The labels mark force predictions computed from: 'Abidor' – formula (5) of Abidor et al. [1], 'P&C' – Pastushenko and Chizmadzhev [87], 'W&H' – Winterhalter and Helfrich [113], 'Neu' – formula (9) of Neu et al. [81].

This lifetime $\bar{t}$ decreases exponentially with the imposed voltage V_m: from 4.7×10^{85} s at 0 V, to 30 s at 0.25 V, and 0.2 μs at 0.3 V.

2.4 Revised Effect of the Transmembrane Potential on Pore Energy

The above estimates, originally proposed by Abidor et al. in 1979, must be revised in view of recent theoretical results on pore energetics. Specifically, formula (5) stipulates that both the membrane and the water filling the pore are purely dielectric, and that the upper and lower surfaces of the membrane and the pore are kept at constant potentials that differ by V_m. These assumptions apply only to very small pores of negligible conductance. However, pores that lead to rupture are large and highly conductive, causing V_m to collapse as the pore radius increases, which reduces the electric pressure on the pore walls [10, 28]. For large pores, the purely capacitive energy (5) must be replaced the mechanical work required to deform a dielectric body in an ionic solution with steady-state electric current. Three different estimates of E_{el} for large pores have been developed by Pastushenko and Chizmadzhev [87], Winterhalter and Helfrich [113], and Neu et al. [81].

To compare these estimates, Fig. 3 plots the electric force expanding the pore, F_{el}. We see that the original Abidor's formula (5) has the force increasing linearly with pore radius. Interestingly, the three more recent estimates predict qualitatively different behavior of the force for large pores: the force decays to zero as $1/r$ in [87], decreases linearly in [113], and increases to a constant value in [81].

The discrepancy between these three estimates is surprising because all three studies started from the same assumption, a dielectric membrane in a dielectric/conductive fluid, and used Maxwell stress tensors to evaluate the force expanding the pore. However, Pastushenko and Chizmadzhev [87], as well as Winterhalter and Helfrich [113], replaced the original boundary value problem by an approximate one that admits an analytical solution. As shown in [81], the solution to the approximate problem underestimates the magnitude of the electric field next to the pore wall, resulting in smaller electrical force on a pore. Neu at al. avoided this pitfall by using the numerical solution of the full problem to compute the force. Thus, unless stated otherwise, the remainder of this chapter will use the heuristic formula proposed by Neu et al. for pores with toroidal inner surface [81]:

$$F_{el}(r, V_m) = \frac{F_{max}}{1 + \frac{r_h}{r+r_t}} V_m^2, \tag{9}$$

where F_{max}, r_h, and r_t are constants. The electrical component of the pore energy is evaluated from (9) as

$$E_{el}(r, V_m) = -\int_0^r F_{el}(r', V_m)\, dr'. \tag{10}$$

With (10), the energy barrier for rupture E_d and the lifetime of pores $\bar{t}$ are less sensitive to V_m than predicted originally. For example, with $V_m = 0.3$ V, r_d and E_d are 9 nm and 75 kT, rather then 2.5 nm and 33 kT. It takes larger V_m (≈ 0.38 V) to reduce the difference $E_d - E_m$ to 7 kT.

2.5 Dynamics of Membrane Rupture

Finally, we briefly review studies that investigated the actual kinetics of the rupture, not only the membrane lifetime. Wilhelm et al. [111], Winterhalter et al. [115, 112], and Diederich et al. [23, 24, 25] measured the growth rate of pores in planar bilayers that leads to irreversible breakdown. To model the dynamic behavior of a pore prior to breakdown, the authors assumed the pore opening is driven by the elastic force of the membrane tension, $2\pi r\sigma$ (the electric, edge, and steric repulsion contributions to pore energetics were neglected).

If the elastic force is balanced by the viscosity of the lipid,

$$2\pi r\sigma = 8\pi\eta\dot{r}, \tag{11}$$

it yields an exponential increase of the pore radii with time,

$$r(t) = r(0)\, e^{\frac{\sigma}{4\eta}t}. \tag{12}$$

In (11-12), η is the surface viscosity of the lipid bilayer. With $\eta = 10^{-9}$ Ns/m and $\sigma = 10^{-3}$ N/m, the time constant of the exponential rise $4\eta/\sigma = 4$ μs. Such an exponential expansion of pores on the microsecond time scale was observed experimentally by Sukharev et al. [101].

As shown by Diederich et al. [25], exponential growth forms only an initial transient of the approximately 0.3 ms time course; during the remaining time, the pore growth is linear in time. To explain this observation, the authors assumed that the elastic force is balanced by the inertial force,

$$2\pi r\sigma = \frac{d}{dt}\left(\pi r^2 h\rho\right)\dot{r} \tag{13}$$

where ρ is the membrane density. Assuming that the lipid material of the pore moves with uniform velocity, Wilhelm at al. obtained the formula that predicts the linear pore growth:

$$r(t) = r(0) + \sqrt{\frac{\Phi\sigma}{h\rho}}\, t \tag{14}$$

where Φ is a parameter that takes into account effects of viscosity and the unknown flow profile.

These modeling results imply that the tension-driven opening of the pore has two phases. The initial pore growth is exponential, due to balancing the surface tension by lipid viscosity. Later, the inertia of the lipid material takes over, and the pore growth rate becomes linear. Further understanding of the dynamics of rupture will come from molecular-level Monte Carlo and molecular dynamics simulations, which are carried out by several groups (e.g., [98, 30, 71]).

3 IRE in Cells under Voltage-Clamp Conditions

A question arises to what extent the theory of IRE in planar bilayers can be extended to cells. To answer it, Chernomordik et al. have used their voltage-clamp procedure to investigate electroporation in human erythrocytes and in L-cells (mice fibroblasts) [17]. Using glass microelectrodes, the cells were patch-clamped in the whole-cell recording mode. Rectangular voltage-clamp pulses were applied and the electrical breakdown was investigated by recording the current response.

This study has shown that IRE in cells under voltage clamp is qualitatively the same as in planar membranes. The same two-stage process was observed: a reversible increase in conductance followed by its abrupt jump signaling membrane rupture. The lifetime of the cell membrane showed a similar exponential dependence on the applied voltage. This qualitative similarity of the cellular and planar membrane breakdown suggests a common mechanism of these phenomena.

However, there were significant quantitative differences. For cells, the thresholds for IRE were higher and the time scale of the process was longer. Accordingly, cells had longer lifetimes: a 700 mV pulse required 40 μs to rupture an oxidized cholesterol membrane, but 10 ms to rupture an L-cell. These differences are due to fundamental differences between planar membranes and cells. First, in planar membranes, lipid molecules can easily move between the bilayer and solution, and they can accumulate on the perimeter of the Teflon apreture during the expansion of a supercritical pore. In contrast, cells are closed systems with a fixed number of molecules [17, 113, 116], which slows down the development of a supercritical pore. Second, planar membranes are under considerable tension (e.g., 10^{-3} N/m) while the tension of cell membranes is typically 2-3 orders of magnitude lower [17, 113, 116]. With $\sigma = 10^{-6}$ N/m, the critical radius r_d is 1800 nm and energy barrier for rupture E_d is $23,800\ kT$ when computed from formulas (6) and (7). Thus, for pore radii up to hundreds of nanometers, pore energy increases monotonically (see Fig. 6 in Sec. 4.1). With small membrane tension, larger voltage (0.6 rather than 0.3 V) is needed to sufficiently reduce the energy barrier for rupture, explaining the larger threshold for IRE observed in Chernomordik's experiments.

Note that the Chernomordik et al. study only demonstrated that cells *can* undergo rupture the same way planar membranes do. It did not prove that the rupture is the mechanism of IRE in practical applications. This is because their voltage-clamp conditions are very different from typical conditions of a cell exposed to an external electric field:

1. Under the whole-cell clamp, the transmembrane potential is independent of position on the cell membrane and the entire cell membrane electroporates. Under an external electric field, the cell is polarized, and V_m changes from positive on one pole to negative on the opposite pole of the cell [96]. Therefore, only polar caps of the cell membrane electroporate.
2. Under voltage clamp, V_m stays approximately constant, enabling continuing creation and expansion of pores. Under an external electric field, creation of pores decreases V_m in electroporated parts of the cell, interrupting further creation and growth of pores [42].
3. The curvature of cells makes them susceptible to deformation by an external electric field, which affects the membrane tension [114, 48, 103].

The next section will extend the theory developed for planar membranes to cells exposed to external electric fields.

4 IRE in Cells in External Electric Field

4.1 *Transmembrane Potential and Pore Creation*

Model of a Cell with Electroporating Membrane. Consider a spherical cell of radius a immersed in a conductive medium and exposed to an electric field **E**. At the time scales corresponding to conventional electroporation, the cytosol and the extracellular solution behave as purely conductive media. Thus, the intracellular and extracellular potentials, Φ_i and Φ_e, are governed by Laplace's equations:

$$\nabla^2 \Phi_i = 0 \text{ for } R < a \quad \text{and} \quad \nabla^2 \Phi_e = 0 \text{ for } R > a, \tag{15}$$

where R is the distance from the center of the cell. The external field $\mathbf{E}$ is included as a boundary condition on Φ_e:

$$\Phi_e(t, R, \theta) = -|\mathbf{E}| R \cos\theta \text{ as } R \to \infty, \tag{16}$$

where $|\mathbf{E}|$ is the field strength and θ is the polar angle measured with respect to the direction of the field. The current density is continuous across the cell membrane:

$$-\hat{n} \cdot (s_i \nabla \Phi_i) = -\hat{n} \cdot (s_e \nabla \Phi_e) = C_m \frac{\partial V_m}{\partial t} + g_l (V_m - V_{rest}) + I_p, \tag{17}$$

where $\hat{n}$ is the outward unit vector normal to the membrane surface, s_i and s_e are conductivities of the cytosol and extracellular medium, C_m is the surface capacitance of the membrane, V_{rest} is the rest potential of the cell, and the transmembrane potential is computed as $V_m(t,\theta) \equiv \Phi_i(t,a,\theta) - \Phi_e(t,a,\theta)$. The right-hand side of (17) shows that the transmembrane current consists of the capacitive current, current through protein channels (approximated by a constant surface conductance g_l), and current through electropores $I_p(t,\theta)$.

The creation and evolution of pores is described by the theory introduced in Sec. 2.2: hydrophilic pores appear with an initial radius r_* at a rate determined by equation (1) and subsequently evolve with the drift velocity (4), subject to thermal fluctuations. The difference is that (1) and (4) apply locally, and the value of V_m in (1) and (4) depends on the location θ on the cell. In the membrane segment $\theta, \theta + \Delta\theta$, which has K_θ pores of radii r_j, $j = 1, \ldots K_\theta$, the current density $I_p(t,\theta)$ is computed by summing currents through individual pores,

$$I_p(t,\theta) = \sum_{j=1}^{K_\theta} i_p(r_j, V_m)/\Delta A, \tag{18}$$

where i_p is the current-voltage relationship of an individual pore (to be discussed in Sec. 4.6) and ΔA is the surface area of the membrane segment $\theta, \theta + \Delta\theta$.

Evolution of the Transmembrane Potential and Membrane Conductance. Consider a cell that is initially at its rest potential V_{rest} and whose membrane contains no pores. Upon the application of an external field $\mathbf{E}$, the cell membrane charges like a parallel resistance and capacitance. The transmembrane potential increases in magnitude from its initial value of V_{rest} according to the formula [63, 96, 67]

$$V_m(t,\theta) = 1.5\, |\mathbf{E}| a \cos\theta \,(1 - e^{-t/\tau}) + V_{rest}, \text{ where } \tau = a\, C_m \left(\frac{1}{s_i} + \frac{1}{2 s_e} \right). \tag{19}$$

Time constant τ is 1.1 μs for the model parameters from Appendix I, although it is lower for smaller cells (e.g., 0.1-0.3 μs in red blood cells [63]).

Figure 4a shows the time evolution of V_m at the depolarized pole of the cell, $\theta = 0$. The passive charging is seen during the first 0.6 μs. When V_m at this location exceeds the threshold for electroporation (≈ 1 V in this model), there is a dramatic increase in the pore creation rate (Eq. 1), which leads to a jump in the local pore density N, illustrated in Fig. 4b. Figure 4c shows that pores also expand during this time. Both creation and expansion of pores lead to the increase in local membrane conductance, G (Fig. 4d). In consequence, V_m no longer follows the charging transient (19) that would be observed in a cell with a passive membrane. As seen in Fig. 4a, the magnitude of V_m peaks and then decreases below threshold. Consequently, the pore radii shrink, causing a decrease in membrane conductance (Fig. 4c-d).

Since different locations on the cell membrane reach the threshold at different times, the timing and degree of electroporation depend on position. Figure 5a shows how the transmembrane potential changes along the cell circumference. Up to 0.6 μs, V_m maintains a cosinusoidal profile predicted

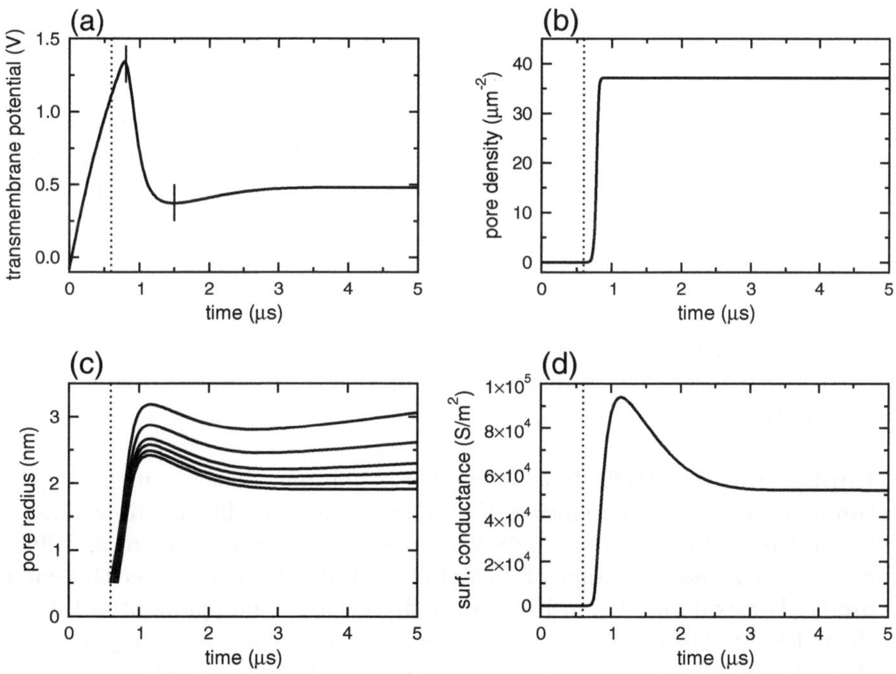

Fig. 4. Electroporation on the depolarized pole ($\theta = 0$) of a 50 μm-radius cell. (a) Transmembrane potential, (b) pore density, (c) radii of six representative pores, and (d) surface conductance of the membrane plotted as functions of time during the first 5 μs of a 400 V/cm pulse. Dotted vertical lines in all panels indicate the end of the passive charging phase. Short vertical lines in Panel (a) indicate time instants corresponding to pore energy plotted in Fig. 6. Simulation methods for the cell model are described in Krassowska and Filev [65].

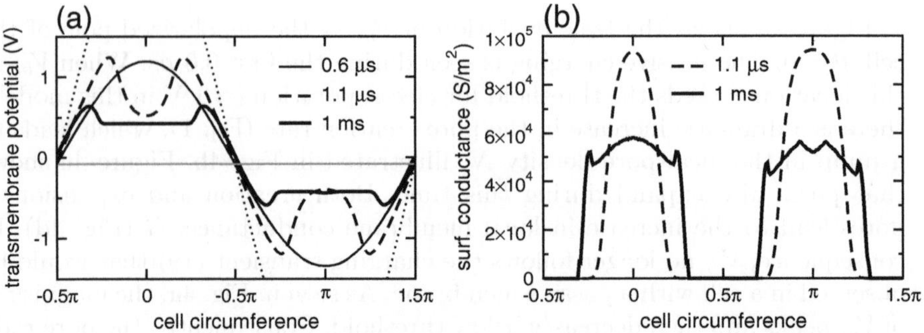

Fig. 5. Spatial distribution of electroporation in a cell. (a) Transmembrane potential $V_m(t,\theta)$ at times indicated in the legend. The dotted line is the steady-state potential that would have been achieved without electroporation. The time 0.6 μs is the end of the passive charging of the cell. (b) Membrane conductance $G(t,\theta)$ at times indicated in the legend. Small differences between depolarized $(-0.5\pi < \theta < 0.5\pi)$ and hyperpolarized $(0.5\pi < \theta < 1.5\pi)$ halves of the cell are caused by the negative rest potential.

by Eq. (19). By 1.1 μs, large dips develop in the V_m profile at the two polar regions of the cell. As seen in Fig. 5b, these dips are the direct consequence of the creation and growth of pores in these regions and the resulting increase in the membrane conductance. There is no conductance increase in the regions around the equator, where V_m has never reached the electroporation threshold. As pore radii evolve, $V_m(t,\theta)$ becomes nearly flat in the electroporated regions of the cell and the $G(t,\theta)$ distribution decreases in magnitude and broadens. The magnitude of V_m is larger near the border of the electroporated region, in which fewer pores provide fewer pathways for the transmembrane current. By 1 ms, spatial distributions of V_m and G are essentially at their steady states.

Implications for IRE. In cells, smaller membrane tension σ makes rupture through spontaneous expansion of a supercritical pore difficult to achieve. At the rest potential, $V_{rest} = -0.08$ V, pore energy increases monotonically for $r > r_m$ in the realistic range of radii (Fig. 6). Large V_m can change this energy profile. For example, V_m in Fig. 4a briefly reaches a maximum of ≈ 1.3 V on the poles. At that voltage, the pore energy decreases monotonically for $r > r_*$ (Fig. 6) and the pores expand spontaneously (Fig. 4c for $0.6 < t < 1$ μs). However, negative feedback takes over: the increased membrane conductance drops V_m to ≈ 0.37 V (Fig. 4a), for which the pore energy is again increasing monotonically for $r > r_m$ (Fig. 6). As seen in Fig. 4c, the pores stop growing and some even shrink. This way negative feedback between creation of pores and the transmembrane potential, as well as the smaller membrane tension in cells, prevent spontaneous expansion of pores. So far, the model suggests that IRE in cells cannot occur through the classical membrane rupture.

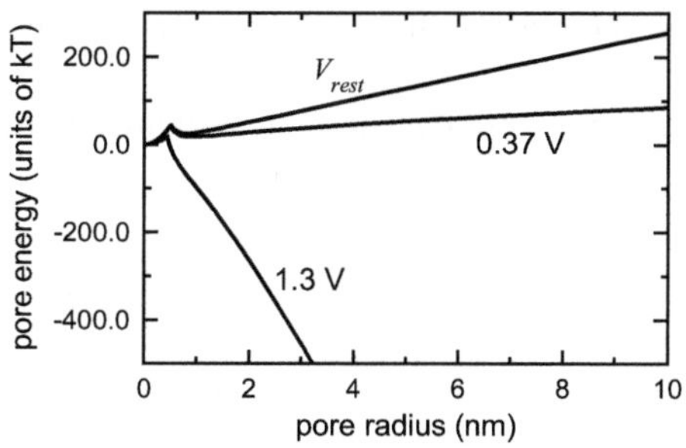

Fig. 6. Energy of a pore on a depolarized pole of a 50 μm-radius cell at transmembrane potentials indicated by labels. $V_m = 1.3$ and 0.37 V are indicated by short vertical lines in Fig. 4a. The difference between this figure and Fig. 2b is due to smaller membrane tension in cells ($\sigma = 10^{-6}$ N/m).

4.2 Cell Deformation by the Field

The basic scenario discussed in the previous section is modified by several physical processes. The first of those is the deformation of the cell shape by an electric field [96]. This field-induced deformation arises because of the closed, curved geometry of the cell; it does not arise in a flat membrane between identical electrolytes, because of the cancellation of the total electric stress [48].

Cell deformation is a fast process taking place immediately after the electric field is applied. At such short time scales, the membrane does not have enough time to exchange lipids with the natural reservoirs of the cell and there is no time to exchange solution between the cell and its exterior. Since the number of lipids forming the cell membrane and the cell volume are constant, the membrane stretches until a mechanical equilibrium is reached when the membrane tension balances the electric stress.

The above assumptions underly the analysis of Isambert, who estimated the increase in the membrane tension in a spherical cell in an external electric field [48]. Other studies of cell deformation (e.g., [114, 52]) use different assumptions that are less applicable to IRE. Isambert analyzes the electrical potential in the cell *before* the significant pore creation occurs, so that all three regions (cytosol, extracellular solution, and the membrane) have constant conductivites and their potentials are governed by Laplace's equations. Steady-state solutions for these potentials can be expressed in the following form:

$$\Phi_k(R,\theta) = \left(A_k\, R + \frac{B_k}{R^2}\right)\,\cos\theta, \tag{20}$$

where $k = i, m, e$ represents cytosol, membrane, and extracellular regions, respectively. Coefficients A_k and B_k are determined from the boundary conditions (16-17) using standard methods [26] and will not be repeated here.

From these potentials, Isambert determines the radial (E_r) and tangential (E_θ) components of the electric field at both membrane interfaces (formulas (3-8) in [48]).

The electric stress induced by these field components is calculated from the Maxwell stress tensor at each interface,

$$p_{ij}^k = \epsilon_k \left(E_i^k E_j^k - \frac{1}{2} |\mathbf{E}^k|^2 \delta_{ij} \right), \tag{21}$$

where i, j represent r or θ, and δ is Kronecker's delta. The net electric stress $\mathbf{p}^T$ applied to a membrane patch is

$$\mathbf{p}^T = \mathbf{p}^m - \mathbf{p}^{i/e}, \tag{22}$$

where $\mathbf{p}^m$ comes from the electric field in the membrane and $\mathbf{p}^{i/e}$ comes from the electric field in the cytosol or extracellular space. Of the two components, $\mathbf{p}^m$ dominates $\mathbf{p}^{i/e}$ in the typical case of a cell before electroporation. Using the computed electric field components, Isambert approximates the radial component of the total stress by

$$p_{rr}^T \sim p_{rr}^m \sim \frac{9}{4} \epsilon_m |\mathbf{E}|^2 \cos^2(\theta) \frac{a}{h}. \tag{23}$$

This formula indicates that radial stress tends to elongate the cell in the direction of the electric field, as observed experimentally by Kinosita at al. in giant liposomes and sea urchin eggs [61].

Next, the surface tension σ_{eq} at mechanical equilibrium is estimated from the Young-Laplace relation $p_{rr} = 2\sigma_{eq}/a$,

$$\sigma_{eq} = p_{rr}^T \frac{a}{2} = \frac{9\epsilon_m}{8h} (|\mathbf{E}|a)^2 \cos^2(\theta). \tag{24}$$

For a cell with a 50 μm radius in a 400 V/cm field, the maximum tension on the poles of the cell would reach 1.6×10^{-2} N/m, i.e., four orders of magnitude larger than the initial tension of 10^{-6} N/m. However, the formula (24) assumes that potentials and the cell deformation have reached steady state. We know from (19) that electric potentials and field components evolve with the time constant $\tau = 1.1$ μs for cell parameters from Appendix I. The time to deform that cell under stress p_{rr}^T is estimated by Isambert as $\tau_{eq} \sim \eta_i / p_{rr}^T$, where η_i is the viscosity of the cytosol [48]. Using the maximum value of p_{rr}^T from (23) and $\eta_i = 10^{-3}$ Ns/m, $\tau_{eq} \approx 1.6$ μs. Thus, the deformation process is indeed short compared with the time needed to change the cell volume or the number of the lipids comprising the membrane, justifying the assumptions of this analysis. However, it does compete with the creation and growth of pores.

Implications for IRE. Formula (24) indicates that stretching of the cell by the electric field can raise the maximum membrane tension by four orders of magnitude [103]. Such tension is above the experimentally-measured range of mechanical tensions leading to membrane rupture, $0.9 - 6.6 \times 10^{-3}$ N/m [116]. It is also comparable with that of planar lipid bilayers discussed in Sec. 2.2, so that the pore energy will have the energy barrier for rupture E_d (Fig. 2) rather than a monotonic increase (Fig. 6). Combined with the voltage-induced decrease of E_d, the deformation-induced rise in membrane tension may indeed lead to irreversible rupture of the membrane.

However, the time needed first to charge the membrane and then to deform the cell is on the order of a few microseconds, the same as the time scale for the creation of pores. As seen in Fig. 4a, V_m peaks at 1.3 V, less than a half of its steady-state value of 3 V, before leveling off at ≈ 0.5 V. Thus, the actual cell deformation and the resulting tension will be lower than the estimate from (24), which assumes that steady state has been reached. Clearly, we have a competition between cell deformation, which promotes rupture, and pore creation, which inhibits it. The final outcome will depend on the cell type and pulsing conditions.

4.3 Effect of Pores on Membrane Tension

An additional player in the competition described above is the direct effect of pores on membrane tension. As the pores are created and grow, they not only decrease transmembrane potential V_m but also relieve membrane tension, counteracting the increase in σ caused by cell deformation. This decrease in tension is not included in the original model of energetics for planar membranes (Sec. 2.2) because in planar membranes, lipid molecules accumulate on the boundary of the opening as the pores grow, and the membrane tension is essentially constant. In contrast, cells and vesicles have a fixed number of lipid molecules, so the membrane tension decreases with pore growth.

Mathematically, the effect of pores on membrane tension has been investigated by Isambert [48], Joshi and Schoenbach [54], and Neu and Krassowska [80]. We start with evaluating the membrane-tension component W_T of the energy of the *entire* lipid bilayer matrix of the cell membrane. Ignoring membrane proteins, W_T is proportional to the cell surface area A [80],

$$W_T = 2\sigma' A \left(1 + \frac{A_{opt}^2}{A^2}\right), \tag{25}$$

where σ' is the interfacial energy per area of the hydrocarbon–water interface and A_{opt} is the optimal surface area of the cell (determined by the minimum-energy distance between the lipid molecules [49]). The surface tension of this bilayer is computed as a derivative of W_T with respect to the area A [80],

$$\sigma_0 = 2\sigma' \left(1 - \frac{A_{opt}^2}{A^2}\right). \tag{26}$$

For cell membrane under tension, $A > A_{opt}$ and σ_0 is positive. As the pores are created and grow, the area subject to interfacial tension decreases from A to $(A - A_p)$, where A_p is the combined area of pores. Consequently, the bilayer energy W_T also decreases,

$$W_T(A_p) = 2\sigma' (A - A_p) \left(1 + \frac{A_{opt}^2}{(A - A_p)^2}\right). \tag{27}$$

We normalize (27) by subtracting the energy of the intact membrane, so that W_T is zero for $A_p = 0$,

$$W_T(A_p) = -2\sigma' \left(1 - \frac{A_{opt}^2}{A(A - A_p)}\right) A_p. \tag{28}$$

The derivative of (27) with respect to the decreased bilayer area $(A - A_p)$ determines the *effective tension* of a membrane with pores [80],

$$\sigma_{eff}(A_p) = 2\sigma' \left(1 - \frac{A_{opt}^2}{(A - A_p)^2}\right), \tag{29}$$

In (27-29), A_{opt} can be expressed in terms of the measurable tension σ_0, using (26).

The effect of pores on membrane tension modifies the drift velocity u of a pore: the formula (4) still applies but the constant membrane tension σ is replaced by σ_{eff} given by (29). Since now u depends not only on radius r of the pore and transmembrane potential V_m, but also on the total pore area A_p, the behavior of each pore is *coupled through membrane tension* to all other pores.

Figure 7 shows the combined the effect of the pore-induced decrease of both V_m and σ on pore energy. At each time instant, σ_{eff} is computed from the current ensemble of pores, and the tension σ in the individual pore energy (3) is set to the current value of σ_{eff}. We then plot E as a function of r, using the current value of V_m on the depolarized pole. Panel (a) shows that the creation and growth of pores lifts the right side of the energy curve, restoring the energy barrier at r_d (filled circles, not labeled in Fig. 7a) and producing a second energy minimum at $r_s \gg r_m$ (diamonds). Thus, there exist simultaneously two energy minima: at $r_m \approx 1$ nm and at $r_s \gg r_m$. We expect that in time, pores divide themselves into two populations: *small* pores with $r \approx r_m$ and *large* pores with $r \approx r_s$.

Evolution of the Pore Density Distribution. This prediction is confirmed by an example presented in Fig. 7b. As pores evolve in time, they indeed divide themselves into populations of small and large pores. Small pores greatly outnumber the large ones: At 1 ms, 97.8% of all pores on a cell are small, and the remaining 2.2% are large. Despite their much smaller number, large pores comprise 95.3% of the total pore area and they contribute 66% to the increased conductance of the cell [65].

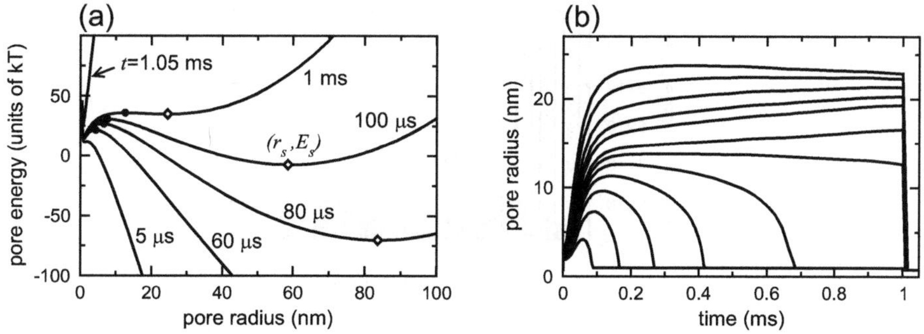

Fig. 7. Change in the pore energy as a result of the creation of pores, evolution of their radii, and accompanying changes in V_m and σ. Time instants are given by labels; filled circles in Panel (a) indicate the energy barrier at r_d and diamonds indicate the energy minimum at r_s, caused by the creation and growth of pores. The line labeled 't = 1.05 ms' corresponds to pore energy 50 μs after the end of the pulse. This figure shows the energy profile for a single pore added to the depolarized pole of the cell. Pore energy will be quantitatively different at other locations because of different value of V_m. (b) Evolution of pore radii during and after the pulse. The radii of 12 representative pores on the depolarized pole of the cell plotted as functions of time for a 1 ms, 400 V/cm pulse. After the end of the pulse, all pores shrink to r_m.

We now focus on the large pore population, since these pores may lead to cell death. As shown in Fig. 8, the number of large pores decreases with time while the distribution of their radii moves toward larger values. At 1 ms, there are 7,600 pores with radii of 22.8 ± 18.7 nm (mean and standard deviation). However, near the border of the electroporated region of the cell, pores grow more vigorously in order to compensate for their much smaller number and reach radii up to 419 nm [65].

Implications for IRE. The pore-induced decrease of membrane tension has two opposing effects on the irreversibility of electroporation. First, the presence of pores changes the energy profile: E_d is no longer a barrier for rupture but rather a barrier for creation of large but *stable* pores. Thus, the classical rupture through spontaneous expansion of a supercritical pore may not be possible in cells once the combined pore area is sufficiently large. Second, the evolution of pores results in an extremely heterogeneous population of pore radii, with some pores reaching radii of hundreds of nanometers. Such large pores, even though they are stable in the sense that they do not cause the mechanical rupture of the membrane, may still cause cell death by facilitating the exchange of the cell content with the exterior solution. Modeling these processes will be discussed in Sec. 4.5 and 4.6.

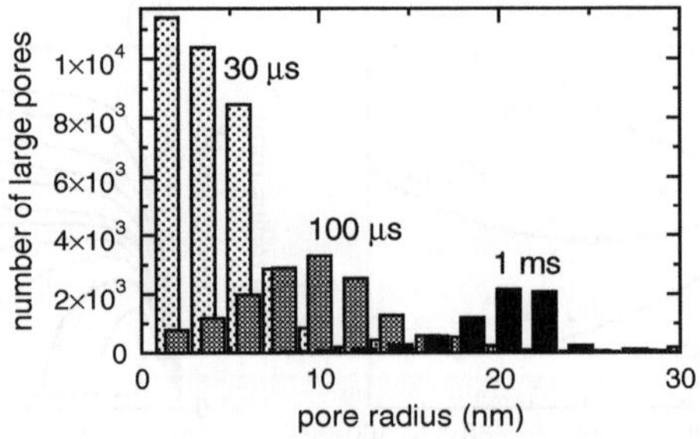

Fig. 8. Distribution of pore radii at 30 μs, 100 μs, and 1 ms. This distribution represents the large pore population from the entire cell. The bin width = 2 nm.

4.4 Postshock Pore Shrinkage and Coarsening

So far, we have concentrated on processes occurring during the electric pulse. Now we begin examining processes occurring after the pulse. In particular, this section will examine the behavior of pores after the pulse.

Once the electric field is turned off, the membrane discharges. Because of the presence of pores, the discharge rate is very fast and the transmembrane potential V_m drops to near zero, even if the cell has a nonzero rest potential. As we have seen in Sec. 4.3, the pulse has left a heterogeneous population of pores with radii ranging from r_m to above r_s. These pores can evolve according to two different scenarios: *shrinkage* or *coarsening* [48, 80, 99].

Shrinkage. In shrinkage, all large pores whose radii were close to r_s shrink to the minimum-energy radius r_m (Fig. 7b). Shrinkage relies on restoring the monotonic increase of the pore energy. As shown in Fig. 7a (plot labeled 't = 1.05 ms'), 50 μs after the pulse, when V_m has discharged to approximately zero, the pore energy has lost its second minimum at r_s and it now increases monotonically for $r > r_m$. Hence, all pores must shrink and we will have a large population of small, $\approx$ 1 nm radius pores. These pores eventually reseal. Mathematically, resealing is included in Eq. (1): after the pulse, pore density N is much larger than N_0, the equilibrium pore density for $V_m = 0$. Thus, the right-hand side of (1) is negative and the pore density decreases. In cells, resealing is a slow process, on the time scale of minutes [90, 37, 9]. Thus, the cell retains increased permeability to small molecules for several minutes after the pulse.

Coarsening. In coarsening, one pore expands to a giant radius, which can exceed 20% of the cell radius, while all other pores shrink to r_m. Coarsening is a general phenomenon that occurs in many physical systems such as solid crystals nucleating from liquid solution [72]. In electroporation, a single giant pore with a micrometer-size radius has been observed in liposomes by Kinosita et al. [61], Zhelev and Needham [116], and Brochard-Wyart et al. [12].

The mechanism of postshock coarsening relies heavily on tension coupling of all pores. When pores are coupled by tension, they evolve to minimize the energy of an *entire bilayer*. However, for a large number of relatively small pores, we can still use the concept of the single-pore energy to explain qualitatively the pore behavior (as we did in Fig. 7). This approximation breaks down for cases like coarsening, which involve a small number of very large pores. Here, we need to examine the bilayer energy W, which is computed as follows: steric repulsion and line energies are summed for all pores but the membrane tension energy is not summed; instead, we include formula (28) to obtain,

$$W = \sum_{j=1}^{K} \left(\beta \left(\frac{r_*}{r_j} \right)^4 + 2\pi r_j \gamma \right) - 2\sigma' \left(1 - \frac{A_{opt}^2}{A(A - A_p)} \right) A_p. \qquad (30)$$

(We ignore the electric energy because postshock $V_m \approx 0$.)

The formula (30) shows that W is a function of the *combined area* of all pores, so it cannot be plotted as a function of pore radius for an arbitrary distribution of pores. Thus, we explain the mechanism of coarsening for a special case of K pores whose radii are distributed uniformly between r_m and r_{max}.

Figure 9a shows the bilayer energy as a function of r_{max} for different values of pore density, $N = K/A$. If a strong pulse creates a sufficiently high pore density, the energy W that we see after the pulse will increase monotonically for $r_{max} > r_m$. Thus, the pore with the maximum radius r_{max} will decrease to r_m. The remaining pores will decrease as well because their radii $r_j < r_{max}$, and thus their drift velocity u computed from (4) is below that of r_{max}. Thus, we will have postshock shrinkage described above.

A weaker pulse, which creates pore density below a critical value N_{cr}, makes W nonmonotonic in r_{max} and sets the stage for coarsening. N_{cr} increases with the initial membrane tension σ_0. This is why coarsening is easiest to observe in vesicles that have been pressurized or stretched [116, 12]. In simulations (Figs. 9-11), we increase σ_0 to 10^{-3} N/m.

Coarsening proceeds as follows. A pulse creates pore density below N_{cr}, but it is sufficiently long to push r_{max} past the energy barrier at r_d. So, r_{max} and some number of the largest pores will continue increasing after the pulse while the remaining pores will shrink to r_m. Based on the bilayer energy plotted as a function of r_{max} (Fig. 9a), one would expect that the largest pores will eventually reach the stable radius r_s. However, the pores that shrink to r_m contribute very little to the total pore area A_p (Sec. 4.3), so the effective density of large pores decreases. Hence, the bilayer energy W must be examined as a function of both r_{max} and N. Figure 9b shows W as a surface in the three-dimensional space: note that the energy minimum at r_s, shown in Fig. 9a, is a part of a "valley" that becomes deeper as $N \to 0$ and $r_{max} \to \infty$. The trajectory overlaid on this surface shows that $r_{max}(t)$ eventually follows the bottom of this valley, in order to minimize the bilayer

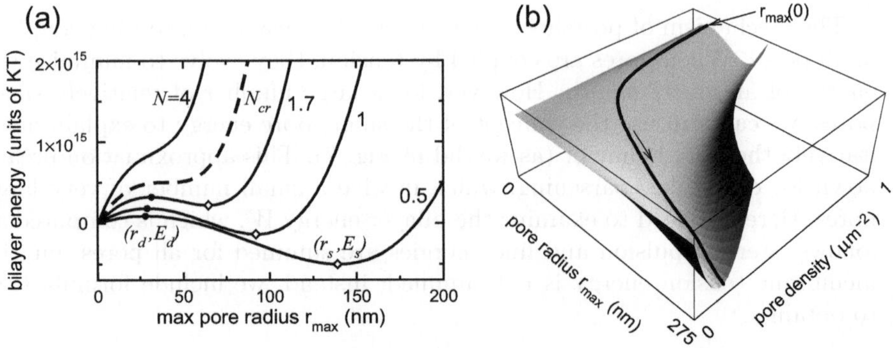

Fig. 9. Mechanisms of postshock shrinkage and coarsening. (a) Bilayer energy W plotted as a function of the maximum radius r_{max} of a uniform distribution of pores. Labels indicate the pore density N (μm^{-2}); dashed line is the plot corresponding to N_{cr}, the critical pore density separating shrinkage and coarsening. Filled circles and open diamonds indicate positions of the energy barrier and energy minimum on plots for $N < N_{cr}$. (b) Bilayer energy plotted as a function of both the maximum radius r_{max} and the pore density N. The solid line shows the trajectory $r_{max}(t)$ that starts at 25 nm. Initial pore density N(0) = 1 μm^{-2}. In this figure, the initial membrane tension σ_0 was increased to 10^{-3} N/m.

energy. As r_{max} grows, membrane tension decreases, forcing even the largest pores, which initially grow, to turn around and start shrinking. Ultimately, all pores except r_{max} shrink, and r_{max} continues to grow until it relieves all membrane tension except a small amount needed to balance the force exerted by its line energy.

Even though the above explanation assumes a uniform distribution of pores, the same mechanism of coarsening applies also to realistic pore distributions. An example is shown in Fig. 10, which illustrates the postshock evolution of the six largest pores created on a 50-μm cell. Five of these pores, with radii $r < r_{max}$, first grow, then shrink to r_m. Smaller pores (not shown) shrink even faster, and the coarsening process leads to only one giant pore. Note the time scale of the process: it takes approximately 30 ms for r_{max} to reach its final radius. The final value of r_{max} is determined by the minimum of the bilayer energy W (30), evaluated for one pore. For this example, the r_{max} grows to 11.2 μm, which is 22.4% of the cell radius.

Implications for IRE. From the viewpoint of cell survival, shrinkage is preferable, as pores with radii near r_m readily reseal. Thus, postshock shrinkage would be a prerequisite for *reversible* electroporation. Nevertheless, complete resealing takes several minutes in cells, so there will be a population of small, approximately 1-nm pores, which is responsible for increased cell permeability to ions, water, and other small molecules. That can lead to irreversibility by effects discussed in Sec. 4.6.

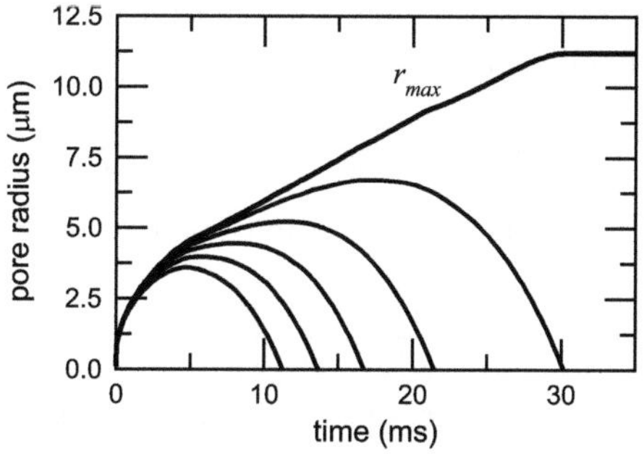

Fig. 10. Postshock evolution of pore radii illustrating the coarsening process. A 50-μm cell with the initial tension σ_0 increased to 10^{-3} N/m was subject to a 10-μs, 160 V/cm pulse. The radii of the six largest pores created by this pulse are plotted as functions of time. The heavy line shows the evolution of the largest radius, r_{max}.

Coarsening is most likely to happen in the parts of cell membrane where voltage V_m is just above the threshold for electroporation. We have already seen that this border zone has a small number of pores which tend to be very large, hundreds of micrometer in radius (Sec. 4.3). That sets the conditions for coarsening: after the end of the pulse, one of these pores would expand to micrometers-size radius. Such a giant pore reseals very slowly [61], leaving the cell vulnerable to leakage and death.

4.5 Membrane Tension-Induced Leakage through Pores

The resealing of a giant pore that formed as a result of coarsening is possible but it involves the change of cell volume resulting from leakage of the solution out of the cell. This process has been studied experimentally in vesicles by Zhelev and Needham [116] and Brochard-Wyart et al. [12].

The model of resealing of a giant pore induced by the cell volume change, which we report below, is based on the analysis of Zhelev and Needham [116] and Moroz and Nelson [76]. Here, we have simplified their model by ignoring the parts that deal with aspiration of the vesicle into the micropipette. We consider only a spherical cell of radius a with a single pore of radius r (although the analysis can be extended to a population of pores). The cell is under tension σ, caused by the pressure difference Δp, which are related by the Young-Laplace relation $\Delta p = 2\sigma/a$. Volumetric flow Q through the pore is given by

$$Q = \frac{\Delta p\, r^3}{3\eta_i}, \tag{31}$$

where η_i is the viscosity of the cytosol. The flow Q can also be determined from the change of cell volume V_{cell} as

$$Q = -\frac{dV_{cell}}{dt} = -4\pi a^2 \frac{da}{dt}. \tag{32}$$

Combining (31-32), one obtains the equation governing the rate of change of cell radius:

$$\frac{da}{dt} = \frac{-1}{6\pi} \frac{\sigma}{\eta_i} \frac{r^3}{a^3}. \tag{33}$$

In (33), two quantities are affected by the change of cell radius: membrane tension σ and pore radius r. For σ, recall from Sec. 4.3 the formula (29) for the effective tension of the membrane with pores of total area A_p. For a single pore of radius r, we substitute $A_p = \pi r^2$, $A = 4\pi a^2$, and $A_{opt} = 4\pi a_0^2 \sqrt{1 - \sigma_0/2\sigma'}$, where where a_0 and σ_0 are the initial values of cell radius and membrane tension. Thus, (29) becomes

$$\sigma_{eff}(r,a) = 2\sigma' - (2\sigma' - \sigma_0)\left[\frac{4a_0^2}{4a^2 - r^2}\right]^2. \tag{34}$$

In contrast to (29), here σ_{eff} depends on both the pore size and the cell size.

Next, to determine the dependence of the pore radius r on the cell size, Moroz and Nelson [76] assumed that the time scale of pore evolution is much shorter than the time scale of changes in cell radius. Thus, the pore radius instantaneously follows the second minimum, r_s, of the bilayer energy (30). Rewriting (30) for the case of a single pore we obtain the formula,

$$W(r,a) = \beta\left(\frac{r_*}{r}\right)^4 + 2\pi r\gamma - \left(2\sigma' - (2\sigma' - \sigma_0)\frac{(4a_0^2)^2}{4a^2\,(4a^2 - r^2)}\right)\pi r^2, \tag{35}$$

in which the bilayer energy depends also on the cell radius a. This dependence is illustrated in Fig. 11, which plots $W(r,a)$ for the cell radius decreasing from 50 to 49.5 μm. The decrease of the cell radius, governed by (33), makes the energy minimum at r_s shallower and moves it toward smaller values. At a critical cell size a_{crit}, energy minimum at r_s disappears because the membrane tension is too small to keep the pore open. Thus, our once giant pore quickly shrinks to the remaining minimum-energy radius, $r_m \approx 1$ nm (which is not visible on the micrometer scale used in this figure) and eventually reseals.

Implications for IRE. One needs to recognize that the resealing of a giant pore through tension-induced leakage and change of cell volume is a slow process. It took several seconds in experiments of Zhelev and Needham [116], which were performed on pressurized vesicles. Resealing may take substantially longer in cells, whose membrane tension is naturally small and has been further decreased by the growth of pores. Thus, there may be a long period of time during which a large-diameter pore is open. The flow of solution through this pore may, by convection, carry out macromolecules and ions important for cell function, and thus may stress the cell beyond recovery.

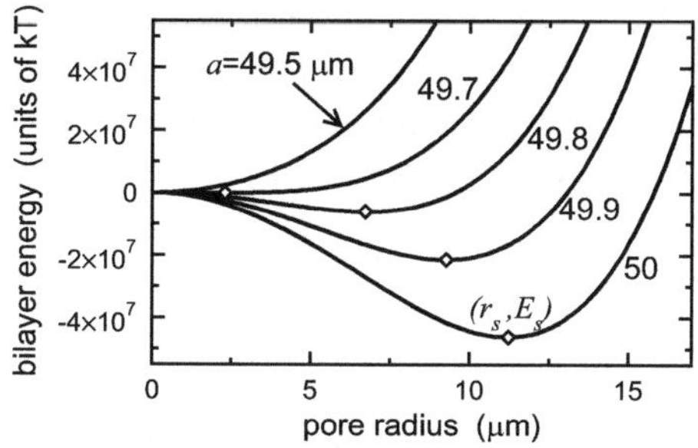

Fig. 11. Energy of a bilayer with a single, giant pore when the cell size decreases as a result of leakout. Labels indicate the cell radius a. Note the unit of the horizontal axis: pore radius is now measured in micrometers, not nanometers. Open diamonds indicate positions of the energy minimum. In this figure, initial membrane tension σ_0 was increased to 10^{-3} N/m.

4.6 Change in Ionic Concentrations and Colloidal-Osmotic Swelling

The membrane tension-driven leakage of cellular content, described in the previous section, may under some conditions coexist and compete with the flow of water driven by osmotic pressure. This process has been studied experimentally in L-cells [17] and especially in red blood cells, in which colloidal-osmotic swelling is the main cause of electric field-induced hemolysis [63, 97, 105, 37].

Colloidal-osmotic swelling involves both ion and water transport through the electropores [2, 105, 37]. Consider a red blood cell that is in osmotic equilibrium with extracellular solution: the total molar concentrations of ions plus hemoglobin inside and ions outside the cell are roughly equal. Thus, the *ion* concentration in cells is lower than in the extracellular solution. The pores created in cell membrane by an electric pulse are readily permeable to small ions, but much less to intracellular macromolecules such as hemoglobin. Thus, ions move easily down their concentration gradients, and the cell exchanges its internal K^+ ions for Na^+ and Cl^- ions from the solution. For example, Kinosita and Tsong have observed a complete reversal of intracellular K^+ and Na^+ concentrations in pulsed red blood cells, which correlated with the rate of hemolysis [63]. As a result, osmolarity inside the cell becomes higher than in the solution. To maintain osmotic balance, water moves in and the cell swells, which may lead to colloidal-osmotic lysis.

Ion Transport Through Pores. We first consider modeling the flow of ions through electropores and the resulting changes in intracellular concentrations. The first step is to choose the appropriate form for $i_p(r, V_m)$, the current-voltage relationship of an individual pore, which determines the total current through pores (18). In most models of electroporation, the formula for i_p

provides only the electric current though a pore, without specifying its ionic composition [36, 89, 5]. To develop an *ion-sensitive* formula for i_p, we adapt the independent electrodiffusion model used in the context of channel proteins [3]: the total current through a pore is computed by summing currents carried by all ions:

$$i_p(r, V_m) = \sum_X z_X F\, f_X(r, V_m), \tag{36}$$

where X stands for Na^+, K^+, Cl^-, Ca^{2+} and possibly other ions or charged molecules; z_X is the valence of ion X, and F is the Faraday constant. The flux of ion X through a pore is given by the Goldman constant-field model [3],

$$f_X(r, V_m) = z_X P_X\, \pi r^2\, v_m \frac{[X]_i - [X]_o \exp(-z_X v_m)}{1 - \exp(-z_X v_m)}. \tag{37}$$

Here, P_X is the permeability of ion X, $[X]_i$ and $[X]_o$ are its intra- and extracellular concentrations, $v_m \equiv V_m F/RT$ is the scaled transmembrane potential, R is the gas constant, and T is temperature.

This basic formula can be further refined. For small pores, the flux is reduced by an energy barrier inside the pore [36, 5, 22, 106] and by steric hinderance [6, 106]. For large pores, the transpore potential decreases (see Sec. 2.4), reducing the flux. This effect has been incorporated by assuming that the voltage drop V_m occurs not across the pore resistance but across the pore resistance and the "input resistance" connected in series [82, 87, 6, 111, 99, 106].

In a cell with K pores with radii r_j, $j = 1, ...K$, the change of the intracellular concentration $[X]_i$ is governed by the equation

$$\frac{d[X]_i}{dt} = \frac{1}{V_{cell}} \sum_{j=1}^{K} f_X(r_j, V_m). \tag{38}$$

For the case when the volume of the cell is simultaneously changing, Glaser et al. reformulated (38) in terms of the *amount* of ion X inside the cell [37],

$$\frac{d\left([X]_i V_{cell}\right)}{dt} = \sum_{j=1}^{K} f_X(r_j, V_m). \tag{39}$$

Combined Ion and Water Transport Through Pores. Using the above theory, Glaser et al. have developed a model that determines changes of the concentrations and the volume of red blood cells occurring on the time scale from seconds to a few hours after an electroporation pulse [37]. The state variables of the model are the amounts of Na^+, K^+, Cl^-, H^+ and of sugar inside and outside the cell. The fluxes of ions were determined from (37) while the sucrose flux was calculated from the Fick's law. In addition, the model included a neutral Cl^--H^+ cotransport to determine the pH of the cytosol and extracellular solution. Assuming that equilibrium holds on the long time scale of seconds to hours, the transmembrane potential V_m was computed from the Goldman voltage equation,

$$V_m = \frac{RT}{F} \ln \frac{P_K[\mathrm{K}]_o + P_{Na}[\mathrm{Na}]_o + P_{Cl}[\mathrm{Cl}]_i}{P_K[\mathrm{K}]_i + P_{Na}[\mathrm{Na}]_i + P_{Cl}[\mathrm{Cl}]_o}. \tag{40}$$

To compute the cell volume V_{cell} appearing in (39), a constraint of the osmotic balance between the cytosol and the extracellular solution was used:

$$\sum_X g_X[X]_i = \sum_X g_X[X]_o, \tag{41}$$

where g_X is the osmotic coefficient of substance X. In addition, Glaser et al. used an empirical formula for the decrease of permeabilities as a result of postpulse resealing of pores,

$$P_X(t) = P_X(0) \exp\left(\frac{t}{t_{init} + t/t_1}\right), \tag{42}$$

where $P_X(0)$ is the permeability of substance X measured immediately after the pulse, and t_{init} and t_1 are constants. This model successfully calculated experimentally measurable characteristics, such as volume changes of the red blood cells, increase in the conductivity of the suspension, and the hemolysis rate [37].

Implications for IRE. The exchange of molecules and water transport through pores can cause irreversibility in two ways. First, the change in ionic concentrations and loss of intracellular content can reach the point beyond which the cell cannot recover [43, 110, 109]. Second, colloidal-osmotic swelling can lead directly to membrane rupture [17]. The swelling occurs on the time scale of minutes [17, 105, 2], so it competes with resealing of pores, which proceeds with roughly the same rate. If the resealing is faster, swelling does not exceed 5-6% of cell volume before it reverses, and the cell recovers. If the swelling is faster, the cell enlarges 2-3 times, leading to membrane rupture [17]. This colloid-osmotic process, and thus the irreversibility, can be controlled by adjusting the composition of the external medium [37, 97].

4.7 Other Processes Affecting Irreversibility of Electroporation

There are other processes that affect the time course of electroporation and thus may either aid or prevent cell death. This section will shortly review a few most important ones, pointing the reader to relevant articles.

Membrane Composition. The dynamics of pore growth and membrane stability are affected by the type of lipids included in the membrane, and by inclusion of membrane proteins, surfactants, and lipid-attached molecules. Detailed investigations of the effect of membrane composition on the dynamics of electroporation was performed on planar membranes. For example, Winterhalter et al. [115, 112] and Diederich et al. [23, 24, 25] tested the effect of incorporating macromolecules into the lipid membrane on the velocity of pore opening that leads to a rupture. These studies have shown that the velocity is reduced by adding surfactants and lipid-attached macromolecules, while no such effect is observed for substances that have no interaction with the membrane. These results were explained theoretically in the framework of a model of pore opening kinetics described in Sec. 2.5 [111, 25]. Recall that in the final stage before rupture, pore widening is driven by the balance of surface tension and inertial forces (13). Incorporating macromolecules changes the effective mass of the membrane by modifying its thickness h and/or density ρ. As seen from Eq. (14), increases of h or ρ decrease the velocity of pore growth.

Similar experiments, although less comprehensive, were performed on cells [78, 57]. While most of the work in this area did not involve modeling, Kandušer et al. [56] and Fošnarič et al. [31] developed a model for the effect of membrane inclusions on stability of pores. The experiments on Chinese hamster lung fibroblast cells have shown that in surfactant-treated cells, thresholds for reversible and irreversible electroporation are equal, indicating that electroporation becomes irreversible as soon as it occurs [56]. The model assumes that the protein/surfactant or lipid/surfactant complexes stabilize the pore by changing its edge energy. Specifically, the complexes change the local toroidal shape of the pore edge, which adds a curvature-related term to the pore energy. For some shapes of the complexes, the combined edge and curvature energy has a local minimum at a radius up to 100 nm. The existence of such a minimum would prevent the pore from shrinking to the minimum size of r_m, making it resistive to resealing.

Membrane Cytoskeleton. Experimental and theoretical studies have recognized that the cytoskeletal network has an important effect on the stability of the membrane. Rols and Teissié used colchicine to alter the cytoskeleton of Chinese hamster ovary cells and observed a three-fold increase of the resealing rate [90]. In red blood cells, the alteration of the spectrin-actin complex by temperature increased both the resealing rate and, to a lesser degree, the threshold for electroporation. In consequence, the hemoglobin leakage was detected for only a short period immediately following the pulse [90].

To model the effect the membrane cytoskeleton, Sung and Park [102] proposed to incorporate an additional positive term in the pore energy (3), $0.5K_{cyto}r^2$, where K_{cyto} is the spring constant of the cytoskeleton. Since this term depends on r^2, it can be thought of as decreasing membrane tension and raising the energy barrier E_d for the formation of supercritical pores that lead to rupture. However, this simple model does not explain the effect of the cytoskeleton on the resealing rate, observed in the experiments.

The presence of cytoskeleton may also modify the mechanism of rupture by growth of supercritical pores. Specifically, Weaver put forward a hypothesis that a portion of the membrane bounded by cytoskeletal elements can behave like a small planar membrane and can therefore exhibit prompt rupture [110, 109]. Thus, rupture in the cell would involve not a single pore growing without bounds, but instead multiple pores whose growth would be arrested mechanically by the membrane cytoskeleton. Extending Weaver's hypothesis, one can imagine that the cytoskeleton can also influence postshock coarsening: if the tension coupling between pores is limited only to pores in the same opening of the network, the coarsening process would lead to one pore per opening, instead of one pore per cell [80]. Such postshock growth of pores and stabilization of their radii at 20-60 nm has been observed by Chang and Reese, who used rapid-freezing electron microscopy to visualize the evolution of pores in red blood cells [14]. This effect of the cytoskeleton on coarsening can also explain why Kinosita et al. failed to see development of a giant pore in sea urchin eggs [61]. While pores bounded by the cytoskeletal elements would not rupture the entire cell, they would be stable and large enough to facilitate irreversible loss of cell content.

Joule Heating. Most researchers believe that cell death in IRE is non-thermal: several experimental and theoretical studies documented that lethal effects of the field occur with a temperature rise of less than 10°C [46, 66, 19, 27, 68]. However, Kekez et al. argue that nonuniformity of the electric field on the subcellular level may lead to large temperature increase within the membrane, especially in the proximity of the pore [58]. Using a model of a spherical cell with one pore on each of the poles, the authors compute the Joule energy density that is dissipated in the pore and the rise of the temperature inside the pore. Next, they assume that this temperature rise increases the pressure inside the cell, which leads to cell swelling. This model yields a criterion for cell death: the pulse energy must exceed the minimum value that results in critical cell swelling. Despite simplified assumptions of the model, the Joule-energy criterion for cell death explains some of the experimental results from the literature, e.g., Hülsheger et al. on bacteria [46, 47] and Kinosita and Tsong on red blood cells [63]. However, other studies found no correlation between pulse energy and cell viability [95, 107, 66].

Even if Joule heating is not the primary cause of cell death during IRE, the mild heating induced by the electric pulses may still affect irreversibility. This is because higher temperature speeds up many of the processes involved in IRE. In particular, the resealing is strongly temperature dependent [62, 37]. Thus, a non-lethal temperature increase may facilitate faster resealing of the pores, protecting the cell from colloidal-osmotic swelling and irreversible changes in ionic concentrations.

Electrochemical Processes. Finally, one must recognize that cell death may also be due to the production of toxic chemical substances during the pulse. These effects include electrolysis of the media, release of ions from electrodes, and generation of free radicals.

Hülsheger and Nieman [46] have demonstrated that pulse-treated NaCl solutions are remarkably toxic: E. coli K12 cells added to the solution right *after* the pulse show a hundred-fold decrease in survival. In contrast, no such decrease was observed when sulfate and phosphate solutions were used. The authors attribute this toxicity to the electrolytic production of free chlorine and oxygen, and possibly direct anodal oxidation of other substances. For HEp-2 cancer cells, Krassowska et al. [66] observed a difference between viability of cells pulsed in the solution versus cells introduced to the prepulsed solution, but it did not reach statistical significance.

Potential toxic effects of stainless steel electrodes are of concern because of such electrodes are widely used and usually considered inert. Tomov and Tsoneva reported that pulsing through stainless steel electrodes releases ferrous ions from them [104]. This study did not assess the effect of these ions on cell survival, although the authors hypothesized that the ferrous ions might potentiate the effect of the bleomycin, a drug used in electrochemotherapy. However, Kotnik et al. [64] have a demonstrated statistically significant loss of viability of DC-3F cells after 1 h of incubation in the medium containing above 1.5 mM ferrous ions. Likewise, Krassowska et al. [66] observed that survival of HEp-2 cancer cells was higher for gold electrodes than for stainless steel ones. Thus, the release of ferrous ions affects cell viability even in the absence of bleomycin. This effect can be diminished by using bipolar pulses [64].

Generation of reactive-oxygen species was studied in Chinese hamster ovary cells by Gabriel and Teissié [34] and by Bonnafous et al. [11]. These studies revealed a linear relationship between the oxidative jump intensity and long-term cell death when the cells were subject to milliseconds-long electric pulses. However, no such correlation was found for trains of short (microseconds) pulses of the same cumulative duration, even though such pulse trains were extremely lethal to cells.

The above observations imply that electric field produces toxic chemical substances that may contribute to the mechanisms of cell death during IRE. Thus, electrochemical processes taking place during electroporation merit further experimental and theoretical studies.

5 Possible Mechanisms of Cell Death during IRE

The above review of the modeling studies of electroporation in single cells shows that conventional IRE involves several processes, some of which aid cell death while others prevent it. Based on this review, we can identify the following possible mechanisms of irreversibility:

1. *Classical membrane rupture through creation of a supercritical pore that expands spontaneously.* This mechanism is aided by the deformation of

cells by an electric field, which increases membrane tension and helps drive opening of the pore. It is prevented by pore-induced increase of membrane conductance, which decreases V_m, which in turn slows down or reverses pore growth. On a longer time scale, the growth of pores releases membrane tension and stabilizes the pore size. Thus, classical rupture needs to happen in the very early stages of electroporation, before the protective mechanisms take over.

2. *Membrane rupture through colloidal-osmotic swelling.* This mechanism is facilitated by the creation of a large number of *small* pores, which are permeable to ions and water, but not to macromolecules. It is prevented by prompt resealing of these pores, which proceeds on the same time scale as swelling, seconds to minutes. Colloidal-osmotic swelling can be controlled by a proper choice of extracellular solution.
3. *Irreversible changes in ionic concentrations through a large number of long-living, small pores.* This mechanism is prevented by resealing of pores, which can be slowed down by decreasing the temperature. Also, increasing pulse duration or following the electroporating pulse by a smaller pulse aids ion transport by electrophoresis.
4. *Loss of cellular content through one (or a few) giant pores developing as a result of postshock coarsening.* This mechanism is prevented by leakout, which decreases the cell size and facilitates the shrinkage and resealing of giant pores. The leakout can be driven either by membrane tension or by osmotic pressure. Mild colloidal-osmotic swelling, which prevents decrease in cell size, aids irreversibility.

Each of these mechanisms involves a competition of two or more processes. As the rates of these processes depend on cell parameters and experimental conditions, the actual mechanism of cell death may be different for different cell lines and experiments. It may explain *disagreements* in the literature regarding empirical formulas that relate probability of cell death to pulse strength, duration, or number of pulses [47, 84, 58, 95, 107, 66]. It may also explain difficulty of developing *theoretical* guidelines for planning medical treatments: at present, the extent of tissue destroyed by the pulse is estimated from empirical criteria: typically, a threshold value of $\mathbf{E}$ or V_m that is taken from experimental measurements [20].

One way to determine the actual mechanism of IRE under given conditions is to develop a comprehensive model of electroporation in a cell that would allow one to evaluate the relative contributions of all the processes. There has been tremendous progress in this area in recent years, which advanced modeling of single-cell electroporation from the simple Schwan equation (which assumes intact membrane) to full spatiotemporal dynamics of pore creation and growth [21, 55, 99, 100, 39, 65]. While most of these models still leave out mechanical, thermal, and transport effects, which may be of primary importance for cell death, they have nevertheless prepared the ground for the development of a comprehensive model of conventional IRE in cells.

These theoretical studies should be supplemented by experiments. Here, recent advances resulted in the development of several single-cell electroporation technologies [74, 45, 18, 83, 60, 32], so it is now possible to directly measure the response of individual cells instead of inferring it from measurements on cell suspensions. Especially promising are the techniques that incorporate feedback control [59] or combine multiple types of measurements of the cell response, such as the study of Ryttsén et al. who studied single-cell electroporation using a combination of patch-clamp and fluorescence microscopy [92]. These novel technologies, used alongside theoretical models, hold great promise for pinpointing the specific mechanism of IRE in single cells.

Appendix: Parameters of the Model

Pore creation and resealing

r_*	0.5×10^{-9} m	minimum radius of hydrophilic pores at $V_m=0$ [36]
α	$1 \times 10^9\ \mathrm{m}^{-2}\,\mathrm{s}^{-1}$	creation rate coefficient [21]
V_{ep}	0.258 V	characteristic voltage of electroporation [21]
N_0	$1.5 \times 10^9\ \mathrm{m}^{-2}$	equilibrium pore density at $V_m = 0$ [21]
q	2.56	constant in Eq. (1) for pore creation rate [21]

Pore energy and drift velocity

β	1.4×10^{-19} J	steric repulsion energy [79]
γ	1.8×10^{-11} N	edge energy coefficient [36, 33]
σ_0	$10^{-6}\ \mathrm{N\,m}^{-1}$	membrane tension in cells [40]
σ_0	$10^{-3}\ \mathrm{N\,m}^{-1}$	membrane tension in planar membranes [33]
σ'	$2 \times 10^{-2}\ \mathrm{N\,m}^{-1}$	tension of hydrocarbon-water interface [49]
D	$5 \times 10^{-14}\ \mathrm{m}^2\,\mathrm{s}^{-1}$	diffusion coefficient for pore radius [33]

Effect of $\mathbf{V}_m$

ϵ_0	$8.85 \cdot 10^{-12}\ \mathrm{Fm}^{-1}$	permittivity of vacuum
ϵ_m	$2\epsilon_0$	permittivity of the lipid bilayer [36, 33]
ϵ_w	$80\epsilon_0$	permittivity of the water filling the pores [36, 33]
F_{max}	$0.70 \times 10^{-9}\ \mathrm{N\,V}^{-2}$	maximum electric force for $V_m = 1$ V [81]
r_h	0.97×10^{-9} m	characteristic length for electric force [81]
r_t	0.31×10^{-9} m	correction for toroidal pores [81]

Miscellaneous

h	5×10^{-9} m	membrane thickness [36]
T	310 K	absolute temperature (37°C)

Geometry and parameters of the cell

a	50×10^{-6} m	cell radius [41]
s_i	$0.455\ \mathrm{S\,m}^{-1}$	conductivity of the cytosol [21]
s_e	$5\ \mathrm{S\,m}^{-1}$	conductivity of the extracellular solution [21]
C_m	$10^{-2}\ \mathrm{F\,m}^{-2}$	surface capacitance of the membrane [41]
g_l	$2\ \mathrm{S\,m}^{-2}$	surface conductance of the membrane [41]
V_{rest}	-0.08 V	rest potential [13]

References

[1] Abidor, I.G., Arakelyan, V.B., Chernomordik, L.V., Chizmadzhev, Y.A., Pastushenko, V., Tarasevich, M.R.: Electric breakdown of bilayer lipid membranes: I. Main experimental facts and their qualitative discussion. Bioelectrochem. Bioenerg. 6, 37–52 (1979)

[2] Abidor, I.G., Li, L.-H., Hui, S.W.: Studies of cell pellets: II. Osmotic properties, electroporation, and related phenomena: Membrane interactions. Biophys. J. 67, 427–435 (1994)

[3] Aidley, D.J., Stanfield, P.R.: Ion Channels. Molecules in Action. Cambridge University Press, Cambridge (1996)

[4] Al-Sakere, B., Bernat, C., Andr, F., Connault, E., Opolon, P., Davalos, R.V., Mir, L.M.: A study of the immunological response to tumor ablation with irreversible electroporation. Technol. Cancer Res. Treat. 6, 301–305 (2007)

[5] Barnett, A.: The current-voltage relation of an aqueous pore in a lipid bilayer membrane. Biochim. Biophys. Acta 1025, 10–14 (1990)

[6] Barnett, A., Weaver, J.C.: Electroporation: a unified, quantitative theory of reversible electrical breakdown and mechanical rupture in artificial planar bilayer membranes. Bioelectrochem. Bioenerg. 25, 163–182 (1991)

[7] Beebe, S.J., Blackmore, P.F., White, J., Joshi, R.P., Schoenbach, K.H.: Nanosecond pulsed electric fields modulate cell function through intracellular signal transduction mechanisms. Physiol. Meas. 25, 1077–1093 (2004)

[8] Benz, R., Beckers, F., Zimmermann, U.: Reversible electrical breakdown of lipid bilayer membranes: A charge-pulse relaxation study. Membrane Biol. 48, 181–204 (1979)

[9] Bier, M., Hammer, S.M., Canaday, D.J., Lee, R.C.: Kinetics of sealing for transient electropores in isolated mammalian skeletal muscle cells. Bioelectromagnetics 20, 194–201 (1999)

[10] Bivas, I., Danelon, C.: Fields and forces acting on a planar membrane with a conducting channel. Phys. Rev. E 69(041901), 1–12 (2004)

[11] Bonnafous, P., Vernhes, M., Teissié, J., Gabriel, B.: The generation of reactive-oxygen species associated with long-lasting pulse-induced electropermeabilisation of mammalian cells is based on a non-destructive alteration of the plasma membrane. Biochim. Biophys. Acta 1461, 123–134 (1999)

[12] Brochard-Wyart, F., de Gennes, P.G., Sandre, O.: Transient pores in stretched vesicles: role of leak-out. Physica A 278, 32–51 (2000)

[13] Chambers, E.L., de Armendi, J.: Membrane potential, action potential and activation potential of eggs of the sea urchin, lytechinus variegatus. Exp. Cell. Res. 122, 203–218 (1979)

[14] Chang, D.C., Reese, T.S.: Changes in membrane structure induced by electroporation as revealed by rapid-freezing electron microscopy. Biophys. J. 58, 1–12 (1990)

[15] Chernomordik, L.V., Chizmadzhev, Y.A.: Electrical breakdown of lipid bilayer membranes. Phenomenology and mechanism. In: Neumann, E., Sowers, A.E., Jordan, C.A. (eds.) Electroporation and Electrofusion in Cell Biology, pp. 83–95. Plenum Press, New York (1989)

[16] Chernomordik, L.V., Sukharev, S.I., Abidor, I.G., Chizmadzhev, Y.A.: Breakdown of lipid bilayer membranes in an electric field. Biochim. Biophys. Acta 736, 203–213 (1983)

[17] Chernomordik, L.V., Sukharev, S.I., Popov, S.V., Pastushenko, V.F., Sokirko, A.V., Abidor, I.G., Chizmadzhev, Y.A.: The electrical breakdown of cells and lipid membranes: The similarity of phenomenologies. Biochim. Biophys. Acta 902, 360–373 (1987)
[18] Davalos, R., Huang, Y., Rubinsky, B.: Electroporation: bio-electrochemical mass transfer at the nano scale. Microscale Therm. Eng. 4, 147–159 (2000)
[19] Davalos, R.V., Mir, L.M., Rubinsky, B.: Theoretical analysis of the thermal effects during in vivo tissue electroporation. Bioelectrochemistry 61, 99–107 (2003)
[20] Davalos, R.V., Mir, L.M., Rubinsky, B.: Tissue ablation with irreversible electroporation. Ann. Biomed. Eng. 33, 223–231 (2005)
[21] DeBruin, K.A., Krassowska, W.: Modeling electroporation in a single cell. I: Effects of field strength and rest potential. Biophys. J. 77, 1213–1224 (1999)
[22] DeBruin, K.A., Krassowska, W.: Modeling electroporation in a single cell. II: Effects of ionic concentrations. Biophys. J. 77, 1225–1233 (1999)
[23] Diederich, A., Bähr, G., Winterhalter, M.: Influence of polylysine on the rupture of negatively charged membranes. Langmuir 14, 4597–4605 (1998)
[24] Diederich, A., Bahr, G., Winterhalter, M.: Influence of surface charges on the rupture of black lipid membranes. Phys. Rev. E 58, 4883–4889 (1998)
[25] Diederich, A., Strobel, M., Meier, W., Winterhalter, M.: Viscosity- and inertia-limited rupture of dextran-supported black lipid membranes. J. Phys. Chem. B 103, 1402–1407 (1999)
[26] Dixon, C.: Applied Mathematics of Science and Engineering. John Wiley & Sons, New York (1971)
[27] Edd, J.F., Horowitz, L., Davalos, R.V., Mir, L.M., Rubinsky, B.: In-vivo results in rats of a new focal ablation technique: Irreversible electroporation. IEEE Trans. Biomed. Eng. 53, 1409–1415 (2006)
[28] El-Hag, A.H., Zheng, Z., Boggs, S.A., Jayaram, S.H.: Effect of pore size on the calculated pressure at biological cells pore wall. IEEE Trans. Nanobiosc. 5, 157–163 (2006)
[29] Esser, A.T., Smith, K.C., Gowrishankar, T.R., Weaver, J.C.: Toward solid tumor treatment by irreversible electroporation: Intrinsic redistribution of fields and currents in tissue. Technol. Cancer Res. Treat. 6, 261–273 (2007)
[30] Fournier, L., Joós, B.: Lattice model for the kinetics of rupture of fluid bilayer membranes. Phys. Rev. E 67(051908), 1–11 (2003)
[31] Fošnarič, M., Kralj-Iglič, V., Bohinc, K., Iglič, A., May, S.: Stabilization of pores in lipid bilayers by anisotropic inclusions. J. Phys. Chem. B 107, 12519–12526 (2003)
[32] Fox, M.B., Esveld, D.C., Valero, A., Luttge, R., Mastwijk, H.C., Bartels, P.V., van den Berg, A., Boom, R.B.: Electroporation of cells in microfluidic devices: a review. Anal. Bioanal. Chem. 385, 474–485 (2006)
[33] Freeman, S.A., Wang, M.A., Weaver, J.C.: Theory of electroporation of planar bilayer membranes: Predictions of the aqueous area, change in capacitance, and pore-pore separation. Biophys. J. 67, 42–56 (1994)
[34] Gabriel, B., Teissié, J.: Generation of reactive-oxygen species induced by electropermeabilization of chinese hamster ovary cells and their consequence on cell viability. Eur. J. Biochem. 223, 25–33 (1994)
[35] Gabriel, B., Teissié, J.: Control by electrical parameters of short- and long-term cell death resulting from electropermeabilization of chinese hamster ovary cells. Biochim. Biophys. Acta 1266, 171–178 (1995)

[36] Glaser, R.W., Leikin, S.L., Chernomordik, L.V., Pastushenko, V.F., Sokirko, A.I.: Reversible electrical breakdown of lipid bilayers: formation and evolution of pores. Biochim. Biophys. Acta 940, 275–287 (1988)

[37] Glaser, R.W., Wagner, A., Donath, E.: Volume and ionic composition changes in erythrocytes after electric breakdown. Bioelectrochem. Bioenerg. 16, 455–467 (1986)

[38] Gowrishankar, T., Weaver, J.C.: Electrical behavior and pore accumulation in a multicellular model for conventional and supra-electroporation. Biochem. Biophys. Res. Comm. 349, 643–653 (2006)

[39] Gowrishankar, T.R., Esser, A.T., Vasilkoski, Z., Smith, K.C., Weaver, J.C.: Microdosimetry for conventional and supra-electroporation in cells with organelles. Biochem. Biophys. Res. Comm. 341, 1266–1276 (2006)

[40] Hénon, S., Lenormand, G., Richert, A., Gallet, F.: A new determination of the shear modulus of the human erythrocyte membrane using optical tweezers. Biophys. J. 76, 1145–1151 (1999)

[41] Hibino, M., Itoh, H., Kinosita Jr., K.: Time courses of cell electroporation as revealed by submicroscopic imaging of transmembrane potential. Biophys. J. 64, 1789–1800 (1993)

[42] Hibino, M., Shigemori, M., Itoh, H., Nagayama, K., Kinosita Jr., K.: Membrane conductance of an electroporated cell analyzed by submicrosecond imaging of transmembrane potential. Biophys. J. 59, 209–220 (1991)

[43] Hofmann, G.A.: Cells in electric fields. Physical and practical electronic aspects of electro cell fusion and electroporation. In: Neumann, E., Sowers, A.E., Jordan, C.A. (eds.) Electroporation and Electrofusion in Cell Biology, pp. 389–407. Plenum Press, New York (1989)

[44] Hu, Q., Viswanadham, S., Joshi, R.P., Schoenbach, K.H., Beebe, S.J., Blackmore, P.F.: Simulations of transient membrane behavior in cells subjected to a high-intensity ultrashort electric pulse. Phys. Rev. E 71(031914), 1–9 (2005)

[45] Huang, Y., Rubinsky, B.: Micro-electroporation: improving the efficiency and understanding of electrical permeabilization of cells. Biomed. Microdev. 2, 145–150 (1999)

[46] Hülsheger, H., Niemann, E.-G.: Lethal effects of high voltage pulses on E. coli K12. Radiat. Environ. Biophys. 18, 281–288 (1980)

[47] Hülsheger, H., Potel, J., Niemann, E.-G.: Killing of bacteria with electric pulses of high field strength. Radiat. Environ. Biophys. 20, 53–65 (1981)

[48] Isambert, H.: Understanding the electroporation of cells and artificial bilayer membranes. Phys. Rev. Lett. 80, 3404–3407 (1998)

[49] Israelachvili, J.: Intermolecular and Surface Forces, 2nd edn. Academic Press, London (1992)

[50] Jones, J.L., Jones, R.E., Balasky, G.: Microlesion formation in myocardial cells by high-intensity electric field stimulation. Am. J. Physiol. 253, H480–H486 (1987)

[51] Jones, J.L., Proskauer, C.C., Paull, W.K., Lepeschkin, E., Jones, R.E.: Ultrastructural injury to chick myocardial cells in vitro following “electric countershock”. Circ. Res. 46, 387–394 (1980)

[52] Joshi, R., Hu, Q., Schoenbach, K., Hjalmarson, H.: Theoretical predictions of electromechanical deformation of cells subjected to high voltages for membrane electroporation. Phys. Rev. E 65(021913), 1–10 (2002)

[53] Joshi, R., Nguyen, A., Sridhara, V., Hu, Q., Nuccitelli, R., Beebe, S., Kolb, J., Schoenbach, K.: Simulations of intracellular calcium release dynamics in response to a high-intensity, ultrashort electric pulse. Phys. Rev. E 75(041920), 1–10 (2007)
[54] Joshi, R., Schoenbach, K.: Mechanism for membrane electroporation irreversibility under high-intensity, ultrashort electrical pulse conditions. Phys. Rev. E 66(052901), 1–4 (2002)
[55] Joshi, R.P., Hu, Q., Schoenbach, K.H.: Modeling studies of cell response to ultrashort, high-intensity electric fields—implications for intracellular manipulation. IEEE Trans. Plasma Sci. 32, 1677–1686 (2004)
[56] Kandušer, M., Fošnarič, M., Šentjurc, M., Kralj-Iglič, V., Hägerstrand, H., Iglič, A., Miklavčič, D.: Effect of surfactant polyoxyethylene glycol ($C_{12}E_8$) on electroporation of cell line DC3F. Colloid. Surface A 214, 205–217 (2003)
[57] Kandušer, M., Šentjurc, M., Miklavčič, D.: Cell membrane fluidity related to electroporation and resealing. Eur. Biophys. J. 35, 196–204 (2006)
[58] Kekez, M.M., Savic, P., Johnson, B.F.: Contribution to the biophysics of the lethal effects of electric field on microorganisms. Biochim. Biophys. Acta 1278, 79–88 (1996)
[59] Khine, M., Ionescu-Zanetti, C., Blatz, A., Wang, L.-P., Lee, L.P.: Single-cell electroporation arrays with real-time monitoring and feedback control. Lab Chip 7, 457–462 (2007)
[60] Khine, M., Lau, A., Ionescu-Zanetti, C., Seo, J., Lee, L.P.: A single cell electroporation chip. Lab Chip 5, 38–43 (2005)
[61] Kinosita, K., Hibino, M., Itoh, H., Shigemori, M., Hirano, K., Kirino, Y., Hayakawa, T.: Events of membrane electroporation visualized on a time scale from microseconds to seconds. In: Chang, D.C., Chassy, B.M., Saunders, J.A., Sowers, A.E. (eds.) Guide to Electroporation and Electrofusion, pp. 29–46. Academic Press, San Diego (1992)
[62] Kinosita, K., Tsong, T.Y.: Formation and resealing of pores of controlled sizes in human erythrocyte membrane. Nature 268, 438–441 (1977)
[63] Kinosita, K., Tsong, T.Y.: Voltage-induced pore formation and hemolysis of human erythrocytes. Biochim. Biophys. Acta 471, 227–242 (1977)
[64] Kotnik, T., Miklavcic, D., Mir, L.: Cell membrane electropermeabilization by symmetrical bipolar rectangular pulses—Part II. Reduced electrolytic contamination. Bioelectrochemistry 54, 91–95 (2001)
[65] Krassowska, W., Filev, P.D.: Modeling electroporation in a single cell. Biophys. J. 92, 404–417 (2007)
[66] Krassowska, W., Nanda, G.S., Austin, M.B., Dev, S.B., Rabussay, D.P.: Viability of cancer cells exposed to pulsed electric fields: The role of pulse charge. Ann. Biomed. Eng. 31, 80–90 (2003)
[67] Krassowska, W., Stone, B.A., Neu, J.C.: Electrical stimulation of cardiac cells. In: Cabo, C., Rosenbaum, D.S. (eds.) Quantitative Cardiac Electrophysiology, pp. 139–197. Marcel Decker, Inc., New York (2002)
[68] Lavee, J., Onik, G., Mikus, P., Rubinsky, B.: A novel nonthermal energy source for surgical epicardial atrial ablation: Irreversible electroporation. Heart Surg. Forum 10, E162–E167 (2007)
[69] Lee, R.C.: Cell injury by electric forces. Ann. NY Acad. Sci. 1066, 85–91 (2005)
[70] Lee, R.C., Kolodney, M.S.: Electrical injury mechanisms: Electrical breakdown of cell membranes. Plast. Reconst. Surg. 80(5), 672–679 (1987)

[71] Leontiadou, H., Mark, A.E., Marrink, S.J.: Molecular dynamics simulations of hydrophilic pores in lipid bilayers. Biophys. J. 86, 2156–2164 (2004)
[72] Lifshitz, E.M., Pitaevskii, L.P.: Physical Kinetics. Pergamon Press, Oxford (1981)
[73] Lubicki, P., Jarayam, S.: High voltage pulse application for the destruction of the Gram-negative bacterium Yersinia enterocolitica. Bioelectrochem. Bioenerg. 43, 135–141 (1997)
[74] Lundqvist, J., Sahlin, F., Aberg, M., Stromberg, A., Eriksson, P., Orwar, O.: Altering the biochemical state of individual cultured cells and organelles with ultramicroelectrodes. Proc. Natl. Acad. Sci. 95, 10356–10360 (1998)
[75] Meaking, W.S., Edgerton, J., Wharton, C.W., Meldrum, R.A.: Electroporation-induced damage in mammalian cell DNA. Biochim. Biophys. Acta 1264, 357–362 (1995)
[76] Moroz, J.D., Nelson, P.: Dynamically stabilized pores in bilayer membranes. Biophys. J. 72, 2211–2216 (1997)
[77] Mueller, P., Rudin, D.O., Tien, H.T., Wescott, W.C.: Methods for the formation of single bimolecular lipid membranes in aqueous solution. J. Phys. Chem. 67, 534–535 (1963)
[78] Mussauer, H., Sukhorukov, V.L., Haase, A., Zimmermann, U.: Resistivity of red blood cells against high-intensity, short-duration electric field pulses induced by chelating agents. Membrane Biol. 170, 121–133 (1999)
[79] Neu, J.C., Krassowska, W.: Asymptotic model of electroporation. Phys. Rev. E 59, 3471–3482 (1999)
[80] Neu, J.C., Krassowska, W.: Modeling postshock evolution of large electropores. Phys. Rev. E 67(021915), 1–12 (2003)
[81] Neu, J.C., Smith, K.C., Krassowska, W.: Electrical energy required to form large conducting pores. Bioelectrochemistry 60, 107–114 (2003)
[82] Newman, J.: Resistance for flow of current to a disk. J. Electrochem. Soc. 113, 501–502 (1966)
[83] Nolkrantz, K., Farre, C., Brederlau, A., Karlsson, R., Brennan, C., Eriksson, P., Weber, S., Sandberg, M., Orwar, O.: Electroporation of single cells and tissues with an electrolyte filled capillary. Anal. Chem. 73, 4469–4477 (2001)
[84] Okino, M., Tomie, H., Kanesada, H., Marumoto, M., Esato, K., Suzuki, H.: Optimal electric conditions in electrical impulse chemotherapy. Jpn. J. Cancer Res. 83, 1095–1101 (1992)
[85] Onik, G., Mikus, P., Rubinsky, B.: Irreversible electroporation: Implications for prostate ablation. Technol. Cancer Res. Treat. 6, 295–300 (2007)
[86] Pakhomov, A., Kolb, J., White, J., Joshi, R., Xiao, S., Schoenbach, K.: Long-lasting plasma membrane permeabilization in mammalian cells by nanosecond pulsed electric field (nspef). Bioelectromagnetics 28, 655–663 (2007)
[87] Pastushenko, V., Chizmadzhev, Y.A.: Stabilization of conducting pores in BLM by electric current. Gen. Physiol. Biophys. 1, 43–52 (1982)
[88] Pastushenko, V., Chizmadzhev, Y.A., Arakelyan, V.B.: Electric breakdown of bilayer lipid membranes: II. Calculations of the membrane lifetime in the steady-state diffusion approximation. Bioelectrochem. Bioenerg. 6, 53–62 (1979)
[89] Powell, K.T., Derrick, E.G., Weaver, J.C.: A quantitative theory of reversible electrical breakdown of bilayer membranes. Bioelectrochem. Bioenerg. 15, 243–255 (1986)

[90] Rols, M.-P., Teissié, J.: Experimental-evidence for the involvement of the cytoskeleton in mammalian-cell electropermeabilization. Biochim. Biophys. Acta 1111, 45–50 (1992)
[91] Rubinsky, B., Onik, G., Mikus, P.: Irreversible electroporation: A new ablation modality–clinical implications. Technol. Cancer Res. Treat. 6, 37–48 (2007)
[92] Ryttsén, F., Farre, C., Brennan, C., Weber, S.G., Nolkrantz, K., Jardemark, K., Chiu, D.T., Orwar, O.: Characterization of single-cell electroporation by using patch-clamp and fluorescence microscopy. Biophys. J. 79, 1993–2001 (2000)
[93] Schoenbach, K., Joshi, R., Kolb, J., Chen, N., Stacey, M., Blackmore, P., Buescher, E., Beebe, S.: Ultrashort electrical pulses open a new gateway into biological cells. Proc. IEEE 92, 1122–1137 (2004)
[94] Schoenbach, K.H., Beebe, S.J., Buescher, E.S.: Intracellular effect of ultrashort electrical pulses. Bioelectromagnetics 22, 440–448 (2001)
[95] Schoenbach, K.H., Peterkin, F.E., Aldeen III, R.W., Beebe, S.J.: The effect of pulsed fields on biological cells: Experiments and applications. IEEE Trans. Biomed. Eng. 25, 284–292 (1997)
[96] Schwan, H.P.: Dielectrophoresis and rotation of cells. In: Neumann, E., Sowers, A.E., Jordan, C.A. (eds.) Electroporation and Electrofusion in Cell Biology, pp. 3–21. Plenum Press, New York (1989)
[97] Schwister, K., Deuticke, B.: Formation and properties of aqueous leaks induced in human erythrocytes by electrical breakdown. Biochim. Biophys. Acta 816, 332–348 (1985)
[98] Shillcock, J.C., Boal, D.H.: Entropy-driven instability and rupture of fluid membranes. Biophys. J. 71, 317–326 (1996)
[99] Smith, K.C., Neu, J.C., Krassowska, W.: Model of creation and evolution of stable macropores for DNA delivery. Biophys. J. 86, 2813–2826 (2004)
[100] Stewart, D.A., Gowrishankar, T.R., Weaver, J.C.: Transport lattice approach to describing cell electroporation: Use of a local asymptotic model. IEEE Trans. Plasma Sci. 32, 1696–1708 (2004)
[101] Sukharev, S.I., Arakelyan, V.V., Abidor, I.G., Chernomordik, L.V., Pastushenko, V.F.: BLM destruction as a result of electrical breakdown. Biofizika 28, 756–760 (1982)
[102] Sung, W., Park, P.J.: Dynamics of pore growth in membranes and membrane stability. Biophys. J. 73, 1797–1804 (1997)
[103] Timoshkin, I.V., MacGregor, S.J., Fouracre, R.A., Given, M.J., Anderson, J.G.: Forces acting on biological cells in external electrical fields. In: 2006 Annual Report Conference on Electrical Insulation and Dielectric Phenomena, pp. 676–679 (2006)
[104] Tomov, T., Tsoneva, I.: Are the stainless steel electrodes inert? Bioelectrochemistry 51, 207–209 (2000)
[105] Tsong, T.Y.: Time sequence of molecular events in electroporation. In: Chang, D.C., Chassy, B.M., Saunders, J.A., Sowers, A.E. (eds.) Guide to Electroporation and Electrofusion, pp. 47–61. Academic Press, San Diego (1992)
[106] Vasilkoski, Z., Esser, A.T., Gowrishankar, T.R., Weaver, J.C.: Membrane electroporation: The absolute rate equation and nanosecond time scale pore creation. Phys. Rev. E 74(021904), 1–12 (2006)

[107] Vernhes, M.-C., Cabanes, P.-A., Teissié, J.: Chinese hamster ovary cells sensitivity to localized electrical stresses. Bioelectrochem. Bioenerg. 48, 17–25 (1999)

[108] Weaver, J.C.: Molecular basis for cell membrane electroporation. In: Lee, R.C., Capelli-Schellpfeffer, M., Kelly, K.M. (eds.) Electrical Injury. A Multidisciplinary Approach to Therapy, Prevention, and Rehabilitation. Ann. N. Y. Acad. Sci., vol. 720, pp. 141–152. New York Academy of Sciences, New York (1994)

[109] Weaver, J.C.: Electroporation in cells and tissues: A biophysical phenomenon due to electromagnetic fields. Radio Sci. 30, 205–221 (1995)

[110] Weaver, J.C.: Electroporation theory. Concepts and mechanisms. In: Nickoloff, J.A. (ed.) Animal Cell Electroporation and Electrofusion Protocols. Methods in Molecular Biology, vol. 48, pp. 3–28. Humana Press Inc., Totowa (1995)

[111] Wilhelm, C., Winterhalter, M., Zimmermann, U., Benz, R.: Kinetics of pore size during irreversible electrical breakdown of lipid bilayer membranes. Biophys. J. 64, 121–128 (1993)

[112] Winterhalter, M.: On the defect growth after short electric field pulses in planar lipid bilayers. Colloids and Surfaces A 149, 161–169 (1999)

[113] Winterhalter, M., Helfrich, W.: Effect of voltage on pores in membranes. Phys. Rev. A 36, 5874–5876 (1987)

[114] Winterhalter, M., Helfrich, W.: Deformation of spherical vesicles by electric-fields. J. Colloid Interface Sci. 122, 583–586 (1988)

[115] Winterhalter, M., Klotz, K.-H., Benz, R., Arnold, W.M.: On the dynamics of the electric field induced breakdown in lipid membranes. IEEE Trans. Ind. Appl. 32, 125–130 (1996)

[116] Zhelev, D.V., Needham, D.: Tension-stabilized pores in giant vesicles: determination of pore size and pore line tension. Biochim. Biophys. Acta 1147, 89–104 (1993)

Thermal Aspects of Irreversible Electroporation

Rafael V. Davalos[1,*], Paulo A. Garcia[1], and Jon F. Edd[2]

[1] School of Biomedical Engineering and Sciences, Virginia Tech – Wake Forest University, Blacksburg, VA 24061, USA
[2] Center for Engineering in Medicine, Massachusetts General Hospital, Harvard Medical School and Shriners Hospital for Children, Boston, MA 02114, USA
and
Department of Mechanical Engineering, Vanderbilt University, Nashville, TN 37212, USA
* Corresponding author E-mail: davalos@vt.edu

1 Introduction

Irreversible electroporation (IRE) is a promising new technique for the ablation of tumors and arrhythmogenic regions in the heart (Davalos, Mir et al. 2005; Edd, Horowitz et al. 2006; Al-Sakere, Andre et al. 2007; Edd and Davalos 2007; Onik, Mikus et al. 2007; Rubinsky 2007). One of its primary advantages over other ablation techniques lies in its mechanism to kill undesirable cells by affecting the cell membrane without thermally damaging major blood vessels, connective tissue, nerves and the surrounding tissue. IRE's ability to create complete and predictable tissue ablation with sharp transition between normal and necrotic tissue will have great advantages in a variety of medical applications (Rubinsky 2007).

Electrical fields, including that for IRE, produce the Joule effect, as charge carriers transfer thermal energy to surrounding chemical species during their E-field induced migration. Davalos, Mir and Rubinsky postulated and demonstrated that irreversible electroporation can be isolated from traditional thermal damage and used to destroy substantial volumes of tissue *in vivo* (Davalos, Mir et al. 2005). Their finding that irreversible electroporation can be used as an independent modality for tissue ablation was subsequently confirmed in studies on cells (Miller, Leor et al. 2005), small animal models in the liver (Edd, Horowitz et al. 2006), and on tumors (Al-Sakere, Andre et al. 2007) as well as in large animal models in the liver (Rubinsky, Onik et al. 2007) and the heart (Lavee, Onik et al. 2007).

Perhaps one of the most important findings of these IRE animal experiments was that the special mode of non-thermal cell ablation has many beneficial effects. For instance, it allows extremely rapid replacement of ablated tissue with healthy tissue (Rubinsky, Onik et al. 2007) without scar formation, and it induces a beneficial immune response (Rubinsky, Onik et al. 2007). It also allows treatment in the heart (Lavee, Onik et al. 2007) and blood vessels (Maor, Ivorra et al. 2007) without the danger of coagulation in the blood stream and subsequent emboli. Therefore, the study of how electrical fields can be applied to living tissue to induce IRE damage to the cell membrane without causing thermal damage is an emerging area of research and it is critical to capitalize on all of the benefits of this new technique.

B. Rubinsky (Ed.): Irreversible Electroporation, BIOMED, pp. 123–154.
springerlink.com

To aid in the planning of IRE treatments, several theoretical studies have focused on computing the macroscopic electrical and thermal conditions within the treated tissue (Edd and Davalos 2007; Ivorra and Rubinsky 2007; Rubinsky 2007). For given electrode configurations and tissue properties, models can predict the electric field (E), pulse-induced transmembrane potential (V_m), and rise in temperature within the tissue (ΔT). The extent of tissue destroyed by the pulse is estimated from empirical criteria: typically, a threshold value of E is assumed based on experimental measurements (Davalos, Mir et al. 2005).

In order to design protocols for an IRE procedure, the electric field distribution, which is dependent on the procedure's specific electrode-tissue geometry, pulse amplitude, and tissue impedance distribution, must be determined. Furthermore, to verify that a specific protocol does not induce thermal effects, the temperature distribution can be calculated from the electric field distribution, the electric pulse parameters, and tissue electrical and thermal properties. Knowledge of the electric field and temperature distribution enables surgeons and researchers to reliably predict the results of an IRE procedure. This insight enables surgeons to plan and optimize the electrode geometry and voltage parameters for varying types of tissue and heterogeneities to:

- Ensure that the entire treatment region undergoes IRE, especially when multiple applications are necessary
- Minimize applied voltages and visualize where potential damage may occur in order to reduce thermal damage to surrounding tissues
- Minimize treatment time, invasiveness, and number of procedures to prevent unnecessary damage to sensitive tissues
- Superimpose medical images such as MRI and CT to plan treatment of the appropriate region

The primary goal of this chapter is to supply readers with the tools and understanding necessary to design appropriate protocols for use of IRE for non-thermal tissue ablation with particular application in cancer therapy. To this end, we first present the fundamental theory that determines how electric fields will result from a chosen electrode configuration, pulse characteristics, and the electrical properties of the tissue. We then proceed to describe the other important theoretical paradigm for modeling IRE, heat diffusion via the Pennes bioheat equation, and we detail the connection between them, namely Joule heating.

2 Effects of Electroporation Fields

2.1 Transmembrane Potential

Electroporation, which results in an increase in the permeabilization of the cell membrane, is initiated by exposing cells or tissues to electric pulses (Weaver and Chizmadzhev 1996; Weaver 2003). As a function of the resulting transmembrane potential (the electric potential difference across the plasma membrane), the electroporation pulse can either: have no effect on the cell membrane, reversibly permeabilize the cell membrane after which cells can survive, or irreversibly permeabilize the cell membrane, which leads to cell death presumably through a loss of homeostasis. The natural transmembrane potential is on the order of 70mV in healthy cells, resulting primarily from ion channels and pumps embedded within the lipid bilayer. If the potential drop across the membrane is made to exceed approximately 1V by the action of an applied electric field, structural rearrangement of the lipid bilayer occurs, creating permanent aqueous pathways or pores for ions and macromolecules to pass through, i.e. irreversible electroporation (Sale and Hamilton 1967).

The typical formula to approximate the induced transmembrane potential (V_m) resulting from an applied electric field for a cell in suspension is:

$$V_m = \lambda r E_a \cos(\theta) \cdot \left[1 + (f/f_s)^2\right]^{-0.5}, \tag{1}$$

where λ is the shape factor of the cell (1.5 for spherical cells), r is the radius of the cell, E_a is the applied electric field, θ is the angle between electric field and the vector from the cell center to any point on its surface, f_s is approximately equal to the frequency where the beta dielectric dispersion occurs (below which the cell membrane charge is in step with the electric field) and f is the frequency of the assumed sinusoidal E_a (Lee, Zhang et al. 2000). This results from the simplified model of a cell as a resistor (intra- and extra-cellular path-resistance) in series with a capacitor (membrane capacitance). Besides the attenuation of the magnitude of V_m in Eqn. 1 by $[1+(f/f_s)^2]^{-0.5}$, the phase shift of transmembrane potential with respect to the applied electric will be $tan^{-1}(-f/f_s)$. Since even a DC pulse can be decomposed into a superposition of various sine waves at different frequencies, this equation enables the transmembrane potential resulting from a short pulse to be computed in an approximate sense. For most cases, the transient terms can be neglected because the electroporation pulse (20μs-50ms) is much larger than the membrane charging time (about 1μs for spherical cells about 10 μm in diameter) (Weaver 2000).

Example

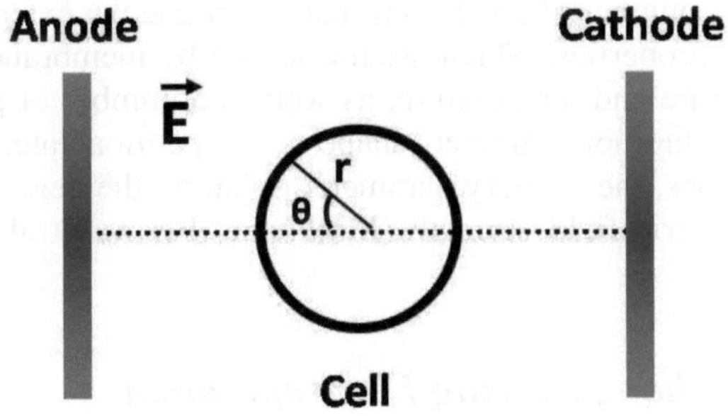

For example problems a-d, refer to Eqn. 1 and assume that the electrical pulse is longer than the typical cell membrane charging time.

Q(a) What is the electric field (V/cm) necessary to irreversibly electroporate a 10 μm (diameter) cell assuming a 1 V transmembrane potential is sufficient to kill the cell?

A(a) The typical formula to approximate the induced transmembrane potential (V_m) resulting from an applied electric field is given by:

$$V_m = \lambda r E_a \cos(\theta),$$

where λ is the shape factor of the cell (1.5 for spherical cells), r is the radius of the cell, E_a is the applied electric field and θ is the angle between the electric field and the vector from the cell center to any point on its surface.

$$V_m = \lambda r E_a \cos(\theta)$$

$$E_a = V_m / \lambda r \cos(\theta) \text{ **Assume } \cos(\theta) = 1 \text{ **}$$

$$E_a[V/cm] = 1[V]/[(1.5)(5x10^{-4})[cm]] = 1333V/cm$$

Q(b) What is the electric field (V/cm) necessary to irreversibly electroporate a 25 μm (radius) cell through a 1 V transmembrane potential?

A(b) $E_a[V/cm] = 1[V]/[(1.5)(2.5x10^{-3})[cm]] = 266.7V/cm$

Q(c) For what value of θ is the transmembrane potential at its maximum?

A(c) The value for θ is equal to 0 or π radians, where $\cos(\theta) = 1$.

Q(d) From the results found in a) and b), what can you conclude about the relationship between cell size and the applied electric field? Describe a situation in which this application would be useful.

A(d) As the size of the cell increases the applied electric field needed for irreversible electroporation decreases. Therefore, it can be used as a selectivity factor if the tumor cells are larger than the native cells.

2.2 *Electric Field Threshold*

In tissue, there are a number of conditions that determine the extent of electroporation, such as tissue properties, which are influenced by membrane properties, cell size, vascular structure and temperature, as well as a number of pulse parameters such as amplitude, duration, number, shape and repetition rate. However, for a given set of conditions, the primary parameter affecting the degree of electroporation is the local electric field strength (Miklavcic, Beravs et al. 1998; Davalos, Otten et al. 2002).

2.3 *Properties Change during Electroporation*

Researchers have shown that there is a change in tissue impedance during and after pulsing, as result of electroporation (Bhatt, Gaylor et al. 1990; Pavlin and

Miklavcic 2003; Davalos, Otten et al. 2004; Miklavcic, Sel et al. 2004). More specifically, the magnitude of specific electrical conductivity of tissue during electroporation increases as a result of the opening of the previously membrane-blocked intracellular spaces to electrical conduction (Lee, Zhang et al. 2000; Pavlin and Miklavcic 2003; Davalos, Otten et al. 2004; Miklavcic, Sel et al. 2004). As described in the following example problem, these changes can be readily incorporated into the user's numerical models but can be ignored to approximate the area treated (Davalos, Otten et al. 2002; Davalos, Otten et al. 2004).

Since the magnitude of specific impedance of tissue changes during IRE, it provides an active means for the physician to monitor the procedure by measuring the change in current. It allows for imaging the tissue region that has been irreversibly electroporated using electrical impedance tomography to verify that the targeted region has been successfully treated (Davalos, Otten et al. 2002; Davalos, Otten et al. 2004; Lee, Loh et al. 2007).

Examples

The change in electrical conductivity that results from local permeation of cell membranes affects what happens during subsequent electroporation pulses. To see how this happens, consider the simple one-dimensional case where two infinite area plate electrodes are applied across a similarly infinite sheet of tissue with thickness of 6 mm, comprised of two distinct tissue types of equal thickness that are in series with one another between the electrodes. To simplify the situation, we assume that the electrical conductivity of each section of tissue is independent of frequency. The first tissue has a normal electrical conductivity of 0.3 S/m where tissue type II is less conductive at 0.05 S/m. Further, we assume that full electroporation will occur when the electric field strength is at least 1000 V/cm in either tissue, and that electroporation will lead to a tissue conductivity of 1 S/m for either tissue.

Q(a): If the tissues had identical electrical conductivity, a pulse of 600 V or more would be required for electroporation. What is the voltage required to electroporate one section of tissue? What is required to electroporate both in the same pulse?

A(a): In this case, we can treat the two sections of tissue as a pair of resistors in series. The current density (amperes per cross electrode-normal area) will then be equal to the pulse voltage (V_p) divided by (3 mm / 0.3 S/m + 3 mm / 0.05 S/m), or V_p / (0.07 Ωm^2). The voltage drop across the first tissue will then be V_p / (0.07 Ωm^2) multiplied by the effective tissue resistance (3 mm / 0.3 S/m), so that the electric field in this region will be (V_p / 7) divided by 3 mm. So, to generate 1000 V/cm, we need to apply 2100 V across the electrodes. Since the second tissue has a lower conductivity, similar analysis reveals that a pulse of only 350 V will cause its electroporation. So, we could electroporate tissue two without electroporating tissue one if the applied voltage was at least 350 V but less than 2100 V. A pulse of 2100 V or greater would result in electroporation of both tissues.

Q(b): To minimize the deposition of Joule heat due to large pulse magnitudes, it is beneficial to apply multiple pulses so that electroporation in more resistive areas will allow more conductive regions to feel the required electric field during later pulses. To understand how this occurs, find the smallest voltage needed to cause electroporation in both tissues after the second pulse?

A(b): In this case, it is easier to work backwards from the second pulse to the first. Since tissue one is more conductive, it will be harder to electroporate. So, we will assume that the second tissue must be electroporated after the first pulse, leading to a modified conductivity of 1 S/m in that region. So, the voltage needed to electroporate the unaffected first tissue during the second pulse will be 390 V, found as in part (a). As we already computed, this voltage will be sufficient to electroporate the more resistive second tissue, but if this was not the case, we would need to increase the pulse voltage to that value instead. Note also that the current density will be larger during the second pulse as the total path resistance has fallen due to electroporation of the second tissue. Specifically, the current density increases from 5.57×10^3 to 3×10^4 A/m^2.

2.4 Joule Heating

Joule heating is the heat generation rate per unit volume caused by an electric field. The Joule heating source term is evaluated by solving the Laplace equation for the potential distribution associated with an electrical pulse: $\nabla \cdot (\sigma \nabla \phi) = 0$, where ϕ is the electrical potential and σ is the electrical conductivity of the tissue. The associated Joule heating rate per unit volume, $\dot{q}$, from an electric field, is the square of the local electric field magnitude, $|-\nabla \phi|$, times the electrical conductivity of the tissue.

$$\dot{q} = \sigma |-\nabla \phi|^2 . \tag{2}$$

A convenient equation to estimate the increase in temperature (ΔT) of homogeneous tissue for the parallel plate configuration from the Joule heating is:

$$\Delta T = \frac{\sigma}{\rho c_p} |\nabla \phi|^2 \Delta t , \tag{3}$$

where Δt is the total duration of the pulses, ρ is mass density and c_p is specific heat (Krassowska, Nanda et al. 2003). This equation assumes no heat dissipation between the pulses, and no fringe effects at the electrode edge. Furthermore, this equation assumes that the biological properties are uniform and the contributions from blood flow, metabolic heat and electrode heat dissipation are negligible.

Even though electroporation pulses are usually applied using a DC field, it should be noted that the electrical conductivity is frequency dependent as shown for select tissues in Table 1.

One final note on the electrical conduction equation is that if the tissue has directional orientation (e.g., muscle tissue), its properties are anisotropic. For example, muscle tissue conducts electricity more easily along the fibers than across them. For such a case, σ must be replaced with a 2nd order tensor (σ) that is symmetric ($\sigma_{ij}=\sigma_{ji}$) and positive definite.

Table 1. Tissue electrical conductivity at various frequencies. (Adapted from (Gabriel, Lau et al. 1996)).

σ (S/m)	*10² Hz*	*10³ Hz*	*10⁴ Hz*	*10⁵ Hz*	*10⁶ Hz*	*10⁷ Hz*
Blood	0.7000	0.7000	0.7000	0.7029	0.8221	1.0967
Heart	0.0936	0.1063	0.1542	0.2151	0.3275	0.5014
Liver	0.0381	0.0414	0.0535	0.0846	0.1866	0.3167
Muscle	0.2667	0.3212	0.3408	0.3618	0.5027	0.6168

2.5 Thermal Regimes

There have been numerous experimental studies on cell viability following the delivery of an electric pulse (Hulsheger and Niemann 1980; Gabriel and Teissie 1995; Lubicki and Jarayam 1997; Krassowska, Nanda et al. 2003). However, relatively little is known about the mechanism by which IRE causes cell death. There is debate whether IRE-induced cell death is caused by: (1) membrane rupture, (2) excessive leakage through pores, or (3) thermal damage to cells. The three proposed mechanisms have different scaling laws that relate the strength (E) and duration (d) of the threshold electric pulse for cell death. In particular, for rupture, $\ln(d) \sim 1/E^2$; for leakage, $d \sim 1/E$; and for thermal damage, $d \sim 1/E^2$. Some studies find a correlation between cell death and the total energy delivered by the pulse (Okino, Tomie et al. 1992; Kekez, Savic et al. 1996), while others do not (Schoenbach, Peterkin et al. 1997; Vernhes, Cabanes et al. 1999), and yet others correlate cell death with the total pulse charge (Krassowska, Nanda et al. 2003).

3 Determining the Temperature Distribution

3.1 Derivation of Heat Diffusion Equation from 1st Principles

In order to ensure that a particular IRE treatment does not produce thermal damage to the target tissue, it is necessary first to understand how heat is removed from the regions of high electric field strength where transient Joule heating may occur.

Conduction is a mode of heat transfer (as opposed to convection and radiation) in which thermal energy spreads from regions of higher temperature to those of lower temperature. Mechanistically, the macroscopic transfer of heat occurs through the integration of many atomic-scale collisions within the constituent matter in a process which tends to evenly distribute kinetic energy. In 1822, Jean

Baptiste Joseph Fourier proposed a simple law predicting the conduction of heat that was based on and has agreed with abundant experimental evidence: the heat flux (***q***, in W) from a region of higher temperature to that of a lower temperature is proportional to the temperature difference and to the area (in the plane orthogonal to ***q***) through which the heat is transferred. When one applies this law to the conduction of heat along the length (x-direction) of material of cross-sectional area *A*, the heat flux along this length becomes:

$$q_x = -kA\frac{\partial T}{\partial x}, \tag{4}$$

where *k* is the intrinsic thermal conductivity of the material through which heat flows, defined typically in W/m^2K and the minus sign ensures that heat flows from hot to cold material.

To derive the relation that allows temperature to be predicted as a function of space and time, we first need to write an appropriate energy balance for an infinitesimal control volume that represents any point within the object to be modeled, i.e. the tissue. Most simply, the rate of energy input into the control volume minus the rate of energy efflux, plus the rate of volumetric energy generation must equal the rate of energy storage at every point within the object. Energy enters and leaves the boundaries of the control volume via conductive heat fluxes, energy generation (or withdrawal) may come from many sources (metabolism, blood flow and Joule heating are most relevant here), and energy storage results in a temperature rise of the material. Here we take the control volume as a rectangular prism ($dx \times dy \times dz$) whose centroid is located at ($x+dx/2$, $y+dy/2$, $z+dz/2$). Grouping the heat fluxes according to the direction along which they enter or leave the control volume, the energy balance then becomes:

$$(q''_x - q''_{x+dx})dydz + (q''_y - q''_{y+dy})dxdz + (q''_z - q''_{z+dz})dxdy + \dot{q}dxdydz = \rho c_p \frac{\partial T}{\partial t}dxdydz, \tag{5}$$

where the double-prime marks heat flux per area, $\dot{q}$, is heat generated per unit volume, ρ is mass density and c_p is the specific heat at constant pressure. Recognizing that the limit of $(q''_{x+dx}-q''_x)/dx$ is by definition the derivative of q''_x with respect to *x* and dividing by volume $dx\,dy\,dz$, we rewrite as:

$$-\frac{\partial q''_x}{\partial x} - \frac{\partial q''_y}{\partial y} - \frac{\partial q''_z}{\partial z} + \dot{q} = \rho c_p \frac{\partial T}{\partial t}. \tag{6}$$

Substituting Fourier's law for conduction along each direction and simplifying results in:

$$\frac{\partial}{\partial x}\left(k\frac{\partial T}{\partial x}\right) + \frac{\partial}{\partial y}\left(k\frac{\partial T}{\partial y}\right) + \frac{\partial}{\partial z}\left(k\frac{\partial T}{\partial z}\right) + \dot{q} = \rho c_p \frac{\partial T}{\partial t}, \tag{7}$$

or in more compact form, $\nabla \cdot (k\nabla T) + \dot{q} = \rho c_p \frac{\partial T}{\partial t}$, which is the heat diffusion equation. In general, however, one can obtain reasonably accurate results by assuming isotropic thermal properties that are distributed homogeneously. With appropriate boundary conditions (heat flux and prescribed temperature are the most common, though convection cooling is also possible) and initial conditions (temperature at every point at time zero, or immediately after the pulse), one can make use of analytical or numerical techniques to predict the evolution of temperature.

3.2 Bioheat Equation

The Pennes Bioheat Equation is often used to assess the heating associated with a thermally relevant procedure because it can provide an estimate of important biological contributions. It takes into account heat conduction and also the dynamic processes that occur in tissues like blood perfusion and metabolism which modify the heat transfer process. Blood perfusion is an effective way to dissipate (take away) heat contrary to metabolic processes which generate heat in the tissue. Thus, the contributions of these two processes must be considered.

3.3 Bioheat Equation Transformation

The heating of the tissue resulting from electroporation pulses can be calculated by adding the Joule heating volumetric source term to the Pennes bioheat transfer equation (Pennes 1948). The modified Pennes bioheat equation has the following form:

$$\nabla \cdot (k\nabla T) - w_b c_b T + q''' + \sigma|\nabla \phi|^2 = \rho c_p \frac{\partial T}{\partial t} \tag{8}$$

where k is the thermal conductivity of the tissue, T is the tissue temperature minus the arterial temperature (typically 37°C), w_b (Table 2) is the blood perfusion per unit volume, c_b is the heat capacity of the blood, q''' is the metabolic heat generation (33,800 W/m^3) (Deng and Liu 2001), ρ is the tissue density, and c_p is the

Table 2. Tissue perfusion rates (w_b) for humans in resting state adapted from (Duck 1990)

Organ or Tissue	Blood Perfusion rate ml kg^{-1} min^{-1}
Brain	560
Fatty Tissue	28
Kidney (male)	4000
Liver (male)	1000
Lung	400
Pancreas	600
Skin	120

Table 3. Common thermal properties of tissue (Duck 1990)

Tissue	T, °C	k, W m^{-1} K^{-1}	c, J g^{-1} K^{-1}	ρ, kg m^{-3}
Artery, aorta	35	0.476 ± 0.041	-	-
Blood	36.6-39.6	0.507-0.513	3.84	1052 - 1064
Brain	37	0.503-0.576	3.6 (white), 3.68 (grey)	1030 - 1041
Breast	37	0.499±0.004	3.55	990 - 1060
Colon	37	0.556±0.009	-	-
Fat, Subcutaneous	*In-vivo*	0.23-0.27	-	916
Kidney	*37*	0.513-0.564	3.6, 3.89	1050
Liver	*37*	0.467-0.527	3.6	1050 - 1070
Lung	*37*	0.302-0.55	-	1040 - 1092
Muscle: cardiac	*37*	0.492-0.562	3.72	1060
Muscle: skeletal	*37*	0.449-0.546	-	1038 - 1056
Pancreas	*37*	0.294-0.588	-	1040 - 1050
Skin	In-vivo	0.293±0.016	-	1093 - 1190
Spleen	37	0.539	3.72	1054

heat capacity of the tissue (Table 3). It should be noted the term tissue refers to the aggregate of solid and blood and that q''' varies depending on the tissue type and state. Properties for metabolism and blood flow can be found in (Duck 1990; Deng and Liu 2001) even though it has been suggested that these factors have a negligible contribution to the overall temperature distribution as compared with Joule heating for time domains common of IRE procedures (Davalos, Rubinsky et al. 2003). Furthermore, it may be possible to neglect blood flow since it has been suggested that perfusion is interrupted or even stopped in the treated zone during IRE, which may assist in inducing total necrosis of the tissue (Gehl, Skovsgaard et al. 2002; Edd, Horowitz et al. 2006). Nevertheless, it may be necessary to consider when assessing heat transfer in nearby non-electroporated tissue that conduct heat from the pulse zone or in treatments that require a train of pulses.

3.4 Non-dimensionalized Form of the Bioheat Equation

Eqn. 8 can be non-dimensionalized in a similar fashion as described in (Foster, A.Lozano-Nieto et al. 1998), which uses the following assumptions: the domain is homogenous, the heat capacity of blood is equal to the tissue ($c_b = c_p = c$) and the metabolic heat generation is negligible. For irreversible electroporation, the joule heating term is non-dimensionalized by normalizing the electric field with the applied voltage-to-distance ratio:

$$\tilde{q}_o''' = \frac{\sigma E^2}{\sigma_o (V_a / L)^2} \tag{9}$$

where σ_o is the conductivity of the tissue before electroporation, V_a is the applied voltage and L is the distance between the electrodes. In the cases where cylindrical or spherical electrodes are used, L is the center-to-center distance between the electrodes. The following additional dimensionless terms are used to non-dimensionalize Eqn. 8:

$$\tilde{T} = \frac{Tk}{L^2 \sigma_o (V_a / L)^2} \tag{10}$$

$$\tilde{x} = \frac{x}{L} \tag{11}$$

$$\tilde{t} = \frac{tk}{\rho c L^2} \tag{12}$$

Expressing Eqn. 8 in terms of these non-dimensional quantities for temperature, length, and time reduces it to the following dimensionless form:

$$\nabla^2 \tilde{T} - w_b \frac{cL^2}{k} \tilde{T} + \tilde{q}_o''' = \frac{\partial \tilde{T}}{\partial \tilde{t}} \tag{13}$$

4 Modeling for Treatment Planning

4.1 Models

Numerical modeling is a technique to approximate the solutions to mathematical formulations that simulate physical processes in a system. The system is defined by the geometry and the physical properties of the material (i.e. tissue). Such simulations provide invaluable insight as to the potential response of the physical system to different stimuli. These stimuli are represented by a combination of boundary conditions, initial conditions or forcing functions that depend on the application of the numerical model.

In this section we show how numerical modeling can be used to predict the tissue area treated by IRE and to confirm that the procedure does not generate thermal tissue damage. The first set of numerical models explores the critical parameters of IRE treatment in two different electrode geometries (parallel plate and needle). The second set is an analysis of three different electrode geometries (spheres, cylinders and parallel plate) and their effect on the resulting electric field and the temperature distribution due to IRE. The computations described in this section were performed with a commercial finite element package (FEMLab, *Comsol AS*, Stockholm, Sweden).

The first two fundamental models are representative of common conditions associated with irreversible electroporation and are used to illustrate the critical parameters in designing treatments. These models are depicted in Fig. 1. The first configuration (Fig. 1a) involves two coaxial disk electrodes and the second (Fig. 1b) contains two parallel needle electrodes. These are common electrode geometries and many more geometries are certainly possible, yet through these two basic examples, we illustrate the trends that occur as a function of critical parameters such as electrode size and shape.

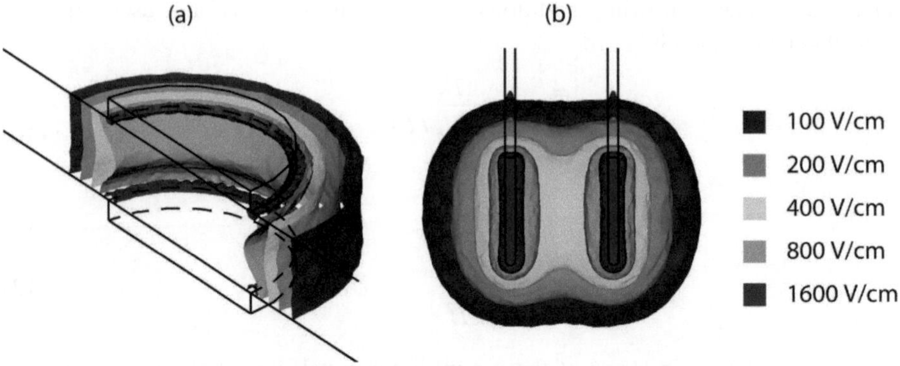

Fig. 1. Two commonly employed electrode designs – (**a**) coaxial disk electrodes and (**b**) parallel needle electrodes. Full 3D solution is depicted for each case, where the aspect ratio (d/D = interelectrode gap/electrode diameter) is 0.4 for (**a**) and 10 for (**b**), and where the applied voltage to distance ratio (V/d) is 1000 V/cm. Adapted from (Edd and Davalos 2007).

4.2 Accuracy of 2D versus 3D

The models used to generate the results in Figs. 3-5 are two-dimensional, as depicted in Fig. 2. For the disk electrodes, as long as conductivity is uniform, no loss of generality is incurred by using a cylindrically symmetric model. However, the 2D simplification for the needle electrodes implies infinitely long electrodes. As will be discussed in the following sections, this causes the cell constant, which can be used to predict power consumption or to estimate bulk tissue conductivity, to be overestimated, the treatment volume to be underestimated, and has minimal impact on the predicted maximal temperature rise due to the pulse. These errors become substantial when the ratio of electrode length (L) to electrode gap (d) is small, yet it will be shown that the 2D model is still a useful approximation for the purpose of IRE treatment planning.

4.3 Boundary Conditions

A model is fully defined and solvable using a numerical method once an appropriate set of boundary conditions and the properties of the tissue are defined. Boundary conditions most often include surfaces where electric potential is specified, as in the case of a source or sink electrode, or surfaces that are electrically insulating, as on the free surfaces of the tissue, for example.

For the case studies described, the surface of one electrode is assumed to have a prescribed voltage, and the other electrode is set to ground. Specifically, at the boundary where the tissue is in contact with one electrode:

$$\phi = V_o \tag{14}$$

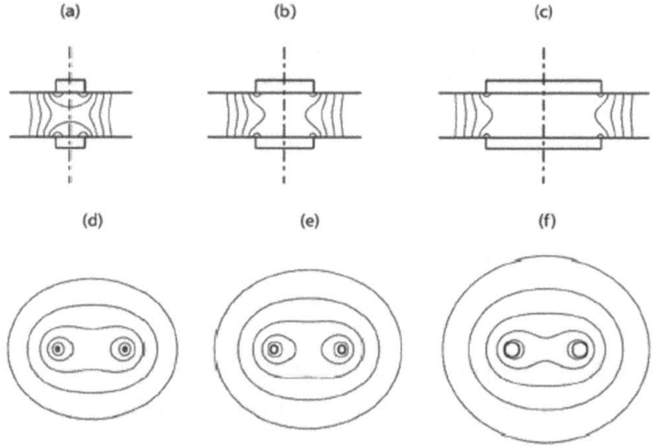

Fig. 2. Cross-sectional electric field distribution for some representative cases. Graphs show simplified 2D models where the same surfaces of constant electric field from Figure 1 are represented with contours. Electrode diameters are 2.5, 5, and 10 mm for (**a**)-(**c**) and 0.5, 1, and 2 mm for (**d**)-(**f**), respectively. All other parameters are identical with those listed for the corresponding graphs of Figure 1, so that (**c**) and (**e**) are for the same two cases as shown in Figure 1. For the disk electrodes in (**a**)-(**c**), cylindrical symmetry ensures the accuracy of the 2D model; however, the 2D simplification for the needle electrodes in (**d**)-(**f**), where the electrodes are perpendicular to the plane of the model, implies infinitely long electrodes. Adapted from (Edd and Davalos 2007).

where V_o is the applied voltage, and at the boundary where the tissue is in contact with the other electrode:

$$\phi = 0 \quad (15)$$

The remaining boundaries are treated as electrically insulating:

$$\frac{\partial \phi}{\partial n} = 0. \quad (16)$$

The analyzed domains extend far enough from the area of interest (i.e. the area near the electrodes) that the electrically and thermally insulating boundaries at the edges of the domain do not significantly influence the results in the treatment zone.

Several thermal boundary conditions can be employed to study their heat exchange between the electrodes and the tissue (Davalos, Rubinsky et al. 2003; Becker and Kuznetsov 2006; Becker and Kuznetsov 2007). However, in the examples in this chapter, the boundaries are taken to be adiabatic to the boundary of the analyzed domain to predict the maximum temperature rise in the tissue:

$$\frac{\partial T}{\partial n} = 0. \quad (17)$$

Other possibilities for electrode boundary conditions include those for passive electrodes, where a film conductance is specified, or those where the electrode interface impedance is included in an approximate manner, as in the complete electrode model (Somersalo, Cheney et al. 1992). In addition, one could incorporate a fin effect boundary condition in which there is thermal dissipation through the electrode itself (Davalos, Rubinsky et al. 2003). One could consider the electrodes to be infinite fins, thereby providing an upper limit for their ability to remove heat from the tissue using the following boundary condition at the electrode-tissue interface:

$$q' = \sqrt{A \cdot k_f \cdot h / P} \cdot (T_b - T_s) \tag{18}$$

where q' is the heat flux per unit length, h is heat transfer coefficient, T_b is the electrode-tissue interface temperature, T_s is the surrounding ambient temperature, P is the perimeter, A is the cross-sectional area, and k_f is the thermal conductivity of the electrode (Davalos, Rubinsky et al. 2003).

4.4 Cell Constants

The cell constant (K) is numerically equal to the product of the bulk conductivity (in S/m) of the tissue (σ) and the magnitude of electrical impedance (Z, in Ohms) between the electrodes, and has units of inverse length. They allow the surgeon to predict the amount of heat generated in total during an IRE procedure when the electrical conductivity is known. This heat would be heterogeneously deposited in general, but after it spreads during the inter-pulse time would give a rough estimate of the level of persistent heat deposited by IRE. Figure 3 presents the cell constants (K) computed for the two electrode configurations over a range of *d*/D values typically encountered. To obtain these data, the applied current was calculated by integrating the normal current density over the surface of one electrode. Most importantly, K provides: (i) an estimate of σ when Z has been measured and (ii) an estimate of the power consumption when σ is known ahead of time ($P = \sigma V^2/K$).

For example, consider the two sets of cases presented in Figure 1. For the disk electrode case where $d/D = 0.4$ (Fig. 2c), $KD \approx 0.37$ ($K = 37\ m^{-1}$) from Figure 1a. Similarly, for the needle electrode case where $d/D = 10$ (Fig. 2e), $KL \approx 0.96$ ($K = 96\ m^{-1}$ for $L = 10$ mm). If we assume $\sigma \approx 0.2$ S/m, then the power during the pulse will be around 0.86 kW ($P \approx 0.2{\cdot}4002/37$) for the disk electrode case in Figure 2c and about 2.1 kW ($P \approx 0.2{\cdot}10002/96$) for the needle electrode case in Figure 2e. For the needle electrode configuration, however, this 2D simplification has resulted in a lower than actual rate of power consumption during the pulse being predicted. The full 3D model predicts that the pulse will in fact consume 2.8 kW, so that K= 72 m-1. This discrepancy can be understood by considering that the impedance in the circuit falls when current can flow not only within the 10 mm thick section of tissue that bounds the electrodes, but also above and below it;

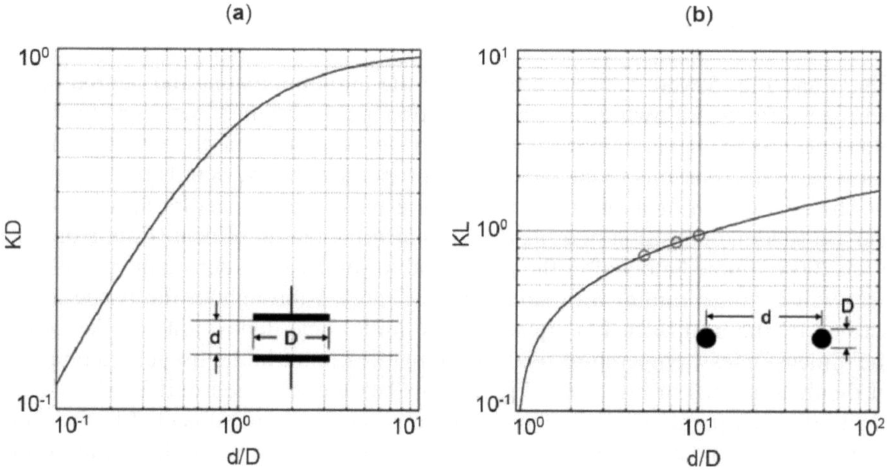

Fig. 3. Cell constant (K) as a function of aspect ratio (d/D) plotted for (**a**) coaxial disk electrodes and (**b**) parallel needle electrodes. Cell constant is non-dimensionalized in the first case by multiplying with the diameter (D) of the disk electrodes and in the second case by multiplying with the active electrode length normal to the page (L). Circles in (**b**) correspond with the three columns in Table 4 that report discrepancies between the 2D assumption (infinitely long parallel needle electrodes) and the fully 3D case. Adapted from (Edd and Davalos 2007).

since adding impedances in parallel will reduce the overall impedance (Z), the 3D model will always have higher power consumption and a lower cell constant than would be predicted from the 2D model. This effect becomes more pronounced as *L*/*d* is reduced.

Table 4 examines this discrepancy between the cell constant predicted for the case of 1 mm diameter parallel needle electrodes by the 2D model and that for the full 3D model. From these computed data, it can indeed be seen that as *L* increases for constant *d*, the 2D model becomes more accurate. Conversely, the 2D model becomes less accurate as *d* is increased across the three columns that correspond with the *d*/*D* points marked as circles in Figure 3b. By adjusting the data in Figure 3b according to the trends in Table 4, it is possible to generate reasonably accurate predictions of the cell constant with just the 2D model.

Table 4. Cell constant discrepancy (K_{3D}/K_{2D})

d =	5 mm	7.5 mm	10 mm
L = 2.5 mm	0.51	0.47	0.45
L = 5 mm	0.65	0.61	0.57
L = 10 mm	0.78	0.74	0.71
L = 20 mm	0.86	0.84	0.81

4.5 Heat Dissipation

Fig. 4 and Fig. 5 were generated to enable the duration, extent and degree of the transient effects following an electroporation pulse, applied between needle or plate electrodes in any uniform tissue, to be predicted based on approximate knowledge of tissue electrical and thermal properties and the details of the pulse.

The application of a short electroporation pulse across needle electrodes will cause a sudden deposition of a highly non-uniform amount of heat (Edd, Horowitz et al. 2006). Immediately after the pulse (and to some extent during the pulse), this heat will begin to diminish in strength as it spreads throughout the tissue as described by the heat diffusion equation. The initial heat deposition can be seen as the dimensionless temperature rise at each marked point in the tissue at the far left of the three graphs in Fig. 4 and Fig. 5. This heating is so non-uniform for needle electrodes that points far from the electrode surface are not subjected to the more intense heating near to the electrodes until some time has passed, as can be observed from the bump in temperature that occurs at progressively later times as one looks at points further from the electrode surface in Fig. 4 and from the axis of

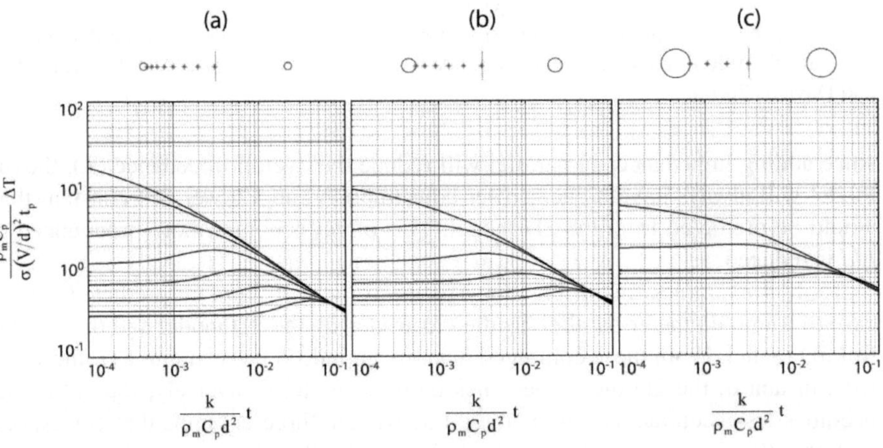

Fig. 4. Dissipation of heat generated by a single pulse between two parallel needle electrodes. Graphs show the evolution of temperature with time at various points along the centerline between electrode centers. Temperature rise (ΔT) and time (t) are given in a dimensionless form to allow general use. Asterisks in above electrode depictions mark exact locations from where the temperature is reported in the several plots in each pane. In each case, the highest and lowest curves correspond to the electrode inner surface and midpoint between electrodes, respectively. Eight curves in (a) are for d/D = 20, six curves in (b) are for d/D = 10, and four curves in (c) are for d/D = 5. Exact positions of marked points can be found by starting with the point on the inner electrode surface (D/2 from electrode center) and marking subsequent points at intervals of a^n (D/2), where this factor is 1.252 for (a), 1.203 for (b), and 1.151 for (c). Top horizontal line designates the maximal temperature rise within the object just at the end of the pulse. Adapted from (Edd and Davalos 2007).

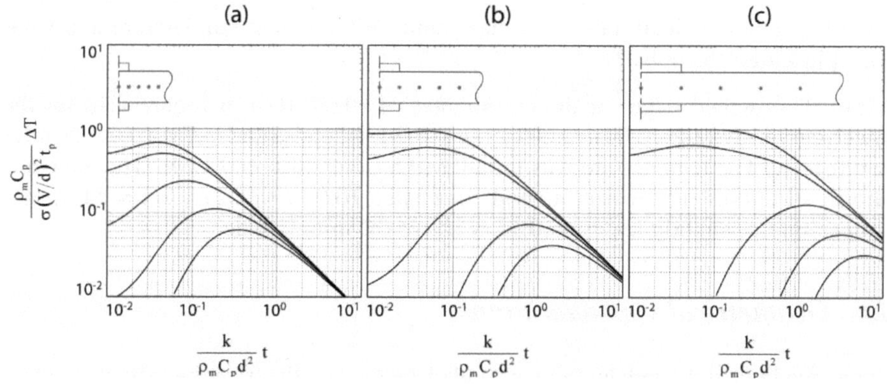

Fig 5. Dissipation of heat generated by a single pulse between two concentric disk electrodes. Graphs show the evolution of temperature with time along the middle plane between the two electrodes at radial positions of 0, 0.5, 1, 1.5 and 2 times D from the axis of cylindrical symmetry. Temperature (ΔT) and time (t) are given in dimensionless form. Asterisks in above electrode depictions mark exact locations from where ΔT is reported in the several plots in each pane. In each case, the highest and lowest curves correspond to the 0 D and 2 D positions respectively. Curves in (**a**)-(**c**) are for d/D = 1.6, 0.8 and 0.4 respectively. Top horizontal line designates the temperature rise at the end of the pulse expected for a local electric field of V/d.

symmetry in Fig. 5. By the time the temperatures have stabilized at the far right side of these plots; therefore, an additional pulse can be applied with minimal cumulative heating of the tissues.

Examples

Q(a): Find the maximum temperature rise during a 100 μs pulse of 1000 V that is applied across two 1 mm diameter parallel needle electrodes spaced 1 cm apart (center-to-center distance). Assume that ρ_m = 1000 kg/m^3, c_p = 4200 J/kg-K and k = 0.5 W/m-K. As disrupted cell membranes will cause the electrical conductivity to approach that for physiological saline, we can set an approximate upper bound on the Joule heating during the electroporation pulse with $\sigma \approx 1$ S/m.

A(a): According to Figure 4(b), $[\rho_m c_p / \sigma(V/d)^2 t_p]\ \Delta T$ is between 10 and 20 at the inner electrode surface at the end of the pulse. Taking the upper limit of 20, we rewrite to solve for ΔT as 20 (1 S/m) (1000 V / 0.01 m)^2 (100 μs) / (1000 kg/m^3) (4200 J/kg-K) = 4.8 K.

Q(b): How long should we wait until the maximum temperature rise drops below 1 K?

A(b): Finding the x-coordinate in Figure 4(b) where the y-coordinate drops to 20 (1 K / 4.8 K) = 4.2. This corresponds with $(k / \rho_m c_p d^2)\ t$ = 0.002, so t = 0.002*(1000 kg/m^3) (4200 J/kg-K) (0.01 m)^2 / (0.5 W/m-K), or about 1.7 seconds.

Q(c): How long will it take for heat diffusion to bring the temperature at the midpoint between the electrodes to its maximum? What is that temperature and what was it just after the pulse?

A(c): This would occur at the coordinates of (0.04, 0.6) in Figure 4(b) on the lower curve. So, t = (0.04 / 0.002) 1.7 s = 34 seconds, and ΔT = (0.6 / 20) 4.8 K = 0.14 K. The temperature was initially (0.5 / 20) 4.8 K = 0.12 K. The temperature in all curves intersect at a short time later, (0.06 / 0.002) 1.7 s = 51 s, at a temperature of slightly less than 0.14 K above the initial value.

4.6 Geometrical Considerations

Three fundamental models were provided below to illustrate the effect of electrode geometry using configurations commonly used for IRE on the electric field and the temperature distribution.

Case A: Two 1 mm diameter spheres separated by a distance of 1 cm.
Case B: Two 1 mm diameter cylinders separated by a distance of 1 cm.
Case C: Two infinite plates separated by a distance of 1 cm.

The models were chosen to provide insight, but other configurations can be used as well. It should be noted that these electrode configurations are similar to those commonly used in reversible electroporation. The models are depicted in Fig. 6. For the cylindrical and spherical cases, 1 cm is the center-to-center distance between the electrodes.

The spherical and cylindrical cases were solved numerically while the plate electrode configuration was solved analytically. The spherical model and the cylindrical model are 2D models. The spherical case is treated as axis-symmetric and

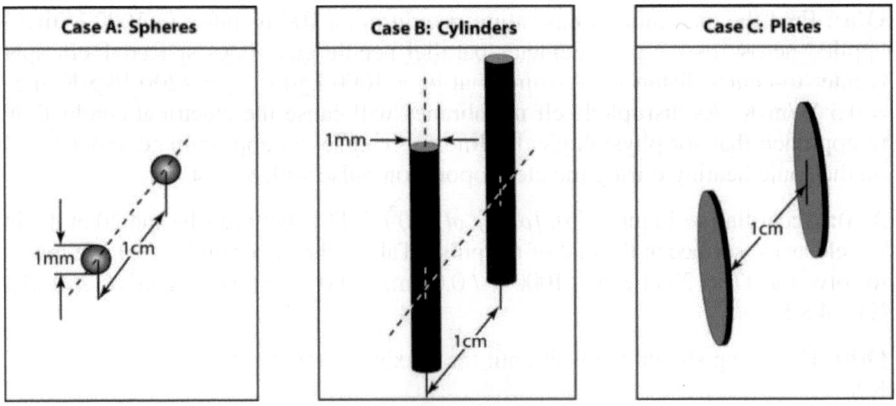

Fig. 6. Models used in study. **Case A:** Two 1 mm diameter spheres separated by a distance of 1 cm. The dashed line indicates the axis of symmetry. **Case B:** Two 1 mm diameter cylinders separated by a distance of 1 cm. The dashed lines indicate the plane of symmetry. **Case C:** Two plates separated by a distance of 1 cm. Adapted from (Davalos and Rubinsky 2008).

the cylindrical case is treated as symmetric with the plane of symmetry made by the axis of each electrode. As mentioned earlier, the 2D simplification for the needle electrodes implies infinitely long electrodes. This simplification does incur some error, particularly when the ratio of electrode length to electrode gap is small, but it is typically a good approximation for IRE treatment planning. Taking advantage of the geometric symmetry enables avoidance of computationally expensive 3D models.

The values of the tissue heat capacity ($c_p = 4$ kJ·kg^{-1}K^{-1}), electrical conductivity (σ =0.2 S·m^{-1}), thermal conductivity (k=0.5 W·m^{-1}K^{-1}), and density (ρ = 1000 kg·m^{-3}) used in the models are taken from the literature (Swarup, Stuchly et al. 1991; Deng and Liu 2001). The tissue temperature is assumed to be initially the same as the arterial temperature which is the physiological temperature (37 °C).

Fig. 7 shows the maximum temperature within the tissue as a function of pulse duration normalized for generality for the three cases (Fig. 6) described in this study. The temperature is normalized by the applied voltage-to-distance ratio as well as other properties as described in the methods but it should be noted that a number of other factors such as electrode diameter could have been used. These plots were obtained by calculating the pulse duration required to reach a maximal temperature of 40, 45, and 50 °C for an applied voltage of 500, 1000, or 2000 V for each case. At a specified voltage, the solutions are perfectly linear (i.e., $R^2 = 1$) for each of the three geometries. Therefore, these solutions can be extrapolated to determine the maximum pulse duration to stay below a specified temperature using the applied voltage for each electrode configuration. For example, pulse durations on the order of 100 μs are normally used for therapeutic electroporation, and such pulse lengths have been shown to be highly effective for tumor ablation using IRE (Al-Sakere, Andre et al. 2007). Fig. 7 can be used to obtain the maximum voltage allowable to stay below a certain temperature limit when a 100 μs pulse is applied.

Fig. 8 shows the electric field distribution and temperature distribution for the spherical case (Case A) and the cylindrical case (Case B) when 1000 V is applied across the electrodes. As described in the model section, only half of the domain is used for the solution to take advantage of the geometrical symmetry.

The surface plots are of the electric field distribution - the dominant parameter dictating which region of tissue has been electroporated (Miklavcic, Beravs et al. 1998). The magnitude of the electric field required to induce IRE is a function of a number of other parameters such as pulse number, pulse length, and tissue type and is an ongoing area of research (Al-Sakere, Andre et al. 2007; Al-Sakere, Bernat et al. 2007; Onik, Mikus et al. 2007). The surface plots reveal that the field distribution is strongly dependent on electrode shape. The results in Fig. 8 also show that the electric field distribution decays more gradually in the cylindrical case than in the spherical case. Since the associated Joule heating is strongly dependent on the electric field, the temperature distribution within the tissue is also more visually spread than for the spherical case.

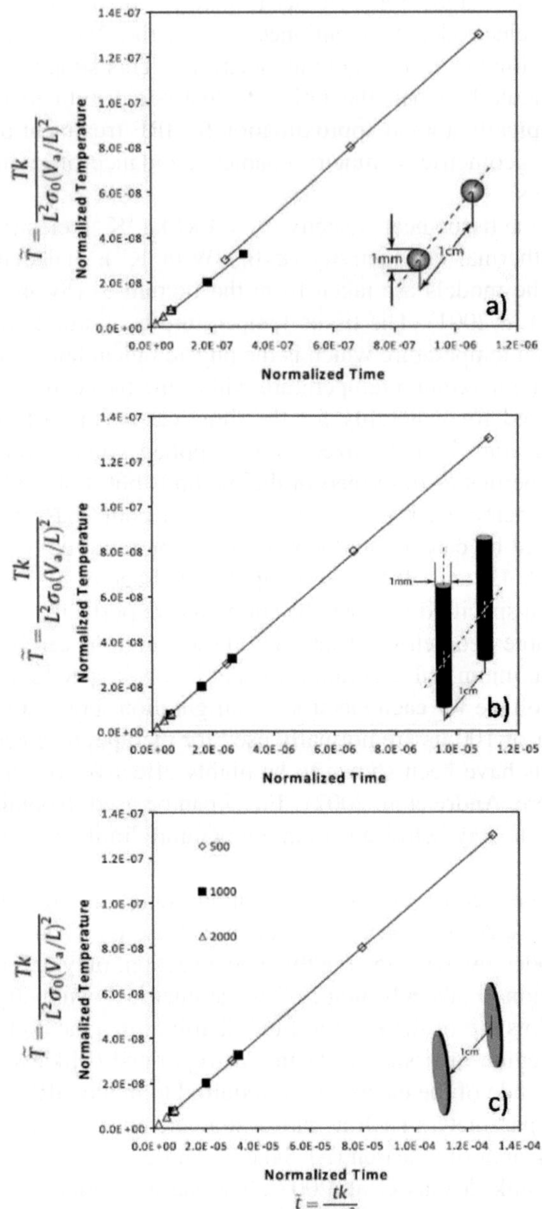

Fig. 7. Non-dimensionalized plot of temperature vs. time depicting the duration required to reach a maximal temperature of 40 °C, 45 °C and 50 °C when 500 V, 1000 V or 2000 V are applied for **(a)** two spheres, **(b)** two cylinders, and **(c)** two plates. Adapted from (Davalos and Rubinsky 2008).

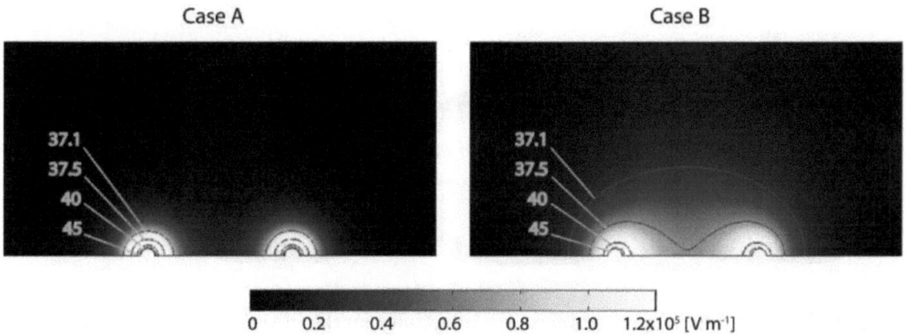

Fig. 8. The electric field and thermal distribution at 1000 V for **Case A:** Two spherical electrodes, and **Case B:** Two cylindrical electrodes. The surface plots show the electric field distribution and the superimposed isotherms show the ensuing temperature distribution when the maximal temperature reaches 50°C (at 242 μs for Case A and 2.11 ms for Case B). The contour lines are 37.1, 37.5, 40, and 45°C with the highest temperature curve being nearest to the electrode. Adapted from (Davalos and Rubinsky 2008).

5 Special Considerations When Modeling Tissues

5.1 Tissue Heterogeneity

In a treatment, if the impedance distribution in the targeted region is homogenous, the results in Fig. 1 can be applied directly to estimate the size of the treated region as a function of electrode geometry and voltage applied. However, there can be factors that would make the targeted domain heterogeneous, such as the presence of large blood vessels, multiple tissue types, or tissues with anisotropic properties, such as muscle. Fig. 9 is an example of how tissue heterogeneity can affect the electric field distribution. Under these circumstances, the user would need to use our guidelines to make their own model including this heterogeneous conductivity and tailor our guidelines to their specific procedure.

Fig. 9 and Fig 10 provide three examples of how local heterogeneities in tissue conductivity affect the electric fields that develop within the nearby tissue. The first graph (Fig. 9a and Fig. 10a) shows a 5-mm diameter sphere when one fifth background conductivity is present between the electrodes, where the electrodes are parallel needle electrodes in Fig. 9 and parallel plates in Fig. 10. As a result of the higher impedance path for current traveling through the inclusion along the electrode-electrode axis, the current density directly between the electrode surfaces and the nearest parts of the low conductivity sphere has fallen substantially. This has caused the 800 V/cm surface surrounding the electrodes to become locally smaller, falling to below 400 V/cm for parallel plates. The electric potential in this space is, therefore, much closer to the nearby electrode voltage and as a result will cause other regions of tissue to experience a higher than normal electric

s

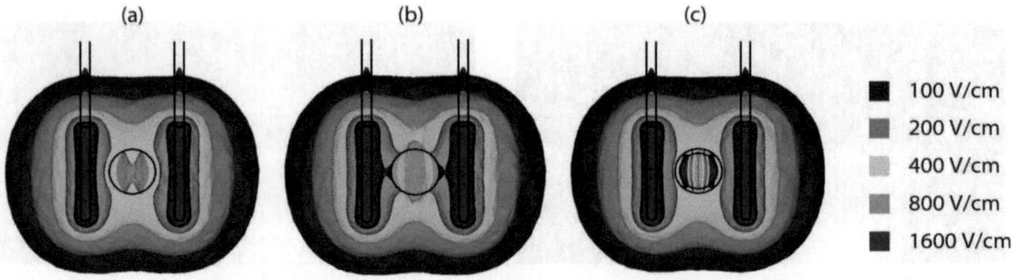

Fig. 9. Effect of a heterogeneous tissue conductivity on the electric field distribution. Each graph depicts the surfaces of constant electric field strength (for 100, 200, 400, 800, and 1600 V/cm) that would result from the presence of a 5 mm diameter spherical inclusion, located halfway between the electrodes. The inclusion is composed of tissue with one fifth or five times the background electrical conductivity in graphs (a) and (b) respectively, while (c) the third graph is similar to (a) but with a nested 4 mm diameter sphere of five times normal conductivity. Five surfaces of constant E-field strength are plotted for two needle electrodes with *d/D* = 10 and *V/d* = 1000 V/cm. Color map, electrode geometry, pulse amplitude, and background tissue properties are identical to the similar 3D model of needle electrodes in Fig. 1b. Adapted from (Edd and Davalos 2007).

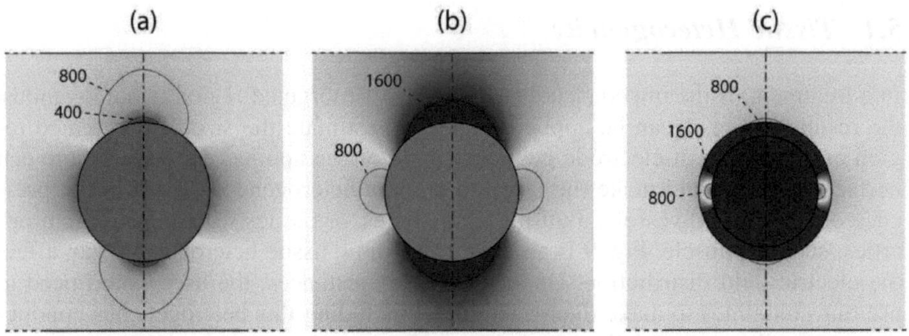

Fig. 10. Effect of various spherical inclusions on the electric field distribution for concentric disk electrodes within an otherwise homogeneous tissue. Electrodes are 4 mm apart and extend to 10 mm in diameter (horizontal dashed lines). Pulse amplitude was 400 V (V/d = 1000 V/cm). Conductivity of the 2 mm diameter spherical inclusion is five times lower and five times higher than the surrounding tissue in (**a**) and (**b**) respectively. Inclusion in (**c**) is identical to (**b**) except that a shell (0.8 mm thick) of five times lower conductivity surrounds the five times higher conductivity interior. Electric field is between 800 and 1600 V/cm in most of the tissue, though deviations occur near the inclusions (contours are labeled in V/cm). Electric field within each inclusion was highly uniform, at about 1320, 450 and 225 V/cm in (**a**)-(**c**) respectively, though in (**c**) it exceeds 5500 V/cm in the thin shell where it contacts the upper and lower extents of the enclosed sphere of high conductivity. E-field is plotted as contours and also with color in the same scaling as for Fig. 9 (blue to red corresponds to 0 to 1600 V/cm).

field as the electric potential establishes continuity. This is evident based on the appearance of a separate 800 V/cm surface within the inclusion with an hourglass shape in Figure 9a and a more uniform region in Figure 10a of much higher E-field strength. At this location, the electric potential will traverse the extra voltage necessary to connect the opposite regions that were varying more gradually than normal. In Fig. 9a, the higher electric field strength corresponds with a focusing conical flow of a relatively small amount of current into the surface of the inclusion from one side and leaving in a similar fashion from the opposite side, where the net current flow is in the direction of the electrode-electrode axis. In Fig. 10a, there is a more uniform distribution of currents within the inclusion.

The second graphs (Fig. 9b and Fig. 10b) reveal the effect when the inclusion is five times more conductive than the surrounding tissue. In these figures, the path impedance between the electrodes and through the inclusion is substantially less than normal, leading to elevated electric field strengths in the spaces between each electrode and the inclusion. This is evident from the extension of the 800 V/cm and 1600 V/cm surfaces to the inclusion surface. In these two regions, the effect is that the electric potential is further removed from the electrode voltages than normal. This will cause the electric potential to vary gradually over the high conductivity inclusion in order to monotonically connect the two regions at opposite ends. The result is that the electric field strength inside the inclusion (and extending laterally) will become smaller than normal, a fact which is illustrated by the appearance of a hole in the 400 V/cm region in Fig. 9b and nearly so in Fig. 10b that is centered within the high conductivity inclusion. The mechanism for the difference in this case is a shunting of extra current from surrounding normal tissue into one hemisphere and out of the other, where net current travels along the electrode-electrode axis. However, relatively little voltage drop will occur within the inclusion since it has a five-fold elevated conductivity and the extra current is not sufficient to offset this fact. Finally, the reason the hole extends into the normal tissue is that the high conductivity inclusion is providing a local short-cut for charges traveling near the inclusion surface but within the normal tissue.

The third graphs (Fig. 9c and Fig. 10c) show what will happen when a thin shell of low-conductivity tissue surrounds a core of high-conductivity tissue, as is the case within regions bounded by an endothelium, such as blood vessels. The path impedance from one electrode to the other and through the center of the inclusion will be between that for the two previous cases, but is approximately equal to the second cases, which can be seen by the nearly undisturbed shapes of the primary electric field surfaces. Though the electric potential and local current densities are relatively unchanged in the space between the electrodes and the inclusion boundaries, the core-shell structure has an interesting effect of focusing the electric field onto certain portions of the shell, while reducing it greatly within the core. If the two regions are quite different in terms of their conductivities, this can be understood by considering that the potential difference across the entire inclusion will be split almost entirely between the two primary locations where current either enters or exits the core through the shell. In Fig. 9c, this can be observed as the opposite halos of 1600 V/cm electric fields that correspond with conical influx or efflux of current from one electrode to the other and through the inclusion, while in Fig. 10c, this is present in the extreme E-field strengths in the shell sections closest to the electrodes. Notice also that the hole in the 400 V/cm surface in Fig. 9c is less pronounced than in Fig. 9b, since the concentrated current visible as the 800 V/cm and 1600 V/cm extended surfaces in Fig. 9b is spread and greatly diminished by the low conductivity shell, and the hole does not protrude into normal tissue since the shell prevents the local short-cut discussed for Fig. 9b. There is a uniformly low E-field strength region internally to the inclusion in Fig. 10c.

The lessons from these three cases also apply in a general sense to any local perturbation in conductivity within a section of tissue where a macroscopic electric field has been imposed. Analogous to the electrode voltages are the bounding electric potentials, set up across the local control volume that includes the perturbation, and induced by the more global variation of impedance and ultimately the electrode voltages. In fact, the line of reasoning from Fig. 9c and Fig. 10c can lead to an understanding of how local electric field strength translates into transmembrane potential. From these discussions an important phenomenon becomes apparent. The high electric fields that will develop across the shell of such core-shell objects will cause electroporation locally long before it occurs in the surrounding tissues. This will provide an easier path for current to reach the high-conductivity core, leading to a much increased flow of current across the newly electroporated shell regions, and increasing somewhat the total amount of Joule heating. Depending on the details of the case, this could involve a pattern of microscopically focused large increases in temperature, visible as small paths of thermal damage across strategic points in the endothelium of large blood vessels or other lumen-filled cavities, a phenomenon that has been observed in recent animal studies of IRE (Edd, Horowitz et al. 2006) and may be the cause of the vascular lock observed even with presumably reversible pulses (Ramirez, Orlowski et al. 1998; Sersa, Cemazar et al. 1999; Gehl, Skovsgaard et al. 2002). Though not primarily responsible for the effectiveness of IRE, this phenomenon will enhance its therapeutic effect by causing tissues that survive the pulse itself to die later of local vascular occlusion (Edd, Horowitz et al. 2006).

5.2 Temperature Dependent Properties

If it is necessary to take into consideration the thermal effects from a treatment, then other tissue properties such as the mass density, heat capacity and thermal conductivity are needed. If these properties cannot be directly measured, the properties of the tissue can be taken from the literature, for example from (Duck 1990). It should be noted that the thermal and electrical conductivities of biological tissues are dependent on temperature and their dependence can be found in literature and incorporated into the models if necessary. Since IRE produces negligible heating, the change in conductivity is not usually significant. For example, in liver near body temperature the thermal and electrical conductivities increase by about 0.25% and 1.5% per degree Celsius rise in temperature, respectively (Duck 1990).

6 Assessing Thermal Effects

6.1 Damage Equation

Since thermal damage is a function of temperature and duration at elevated temperatures, the negligible heating associated with these case studies is emphasized by the fact that an electroporation pulse is typically a fraction of a second long. The methods to calculate the thermal dose or damage associated with a procedure can be found in (Tropea and Lee 1992; Lee and Despa 2005; Al-Sakere, Andre et al. 2007; Becker and Kuznetsov 2007; Edd and Davalos 2007) and are briefly described here.

One of the distinguishing features of irreversible electroporation is that it does not induce thermal damage. To assess whether a particular set of voltage parameters will induce thermal damage in addition to irreversible electroporation, the thermal effects can be calculated. An example that determines the upper bound of irreversible electroporation (onset of thermal damage) through a theoretical calculation is provided as presented in (Shafiee, Garcia et al. 2009). The plots were developed using Wolfram Mathematica 6.0 for Students (Champaign, IL) and they take into account the physical properties of the cell sample (Table 5) and the electric pulse characteristics (Shafiee, Garcia et al. 2009).

Table 5. Physical properties used in the analysis of IRE and thermal damage

Property	Symbol	Value	Units	Reference
Electrical Conductivity	σ_{PBS}	1.4	$S{\cdot}m^{-1}$	-
	σ_{tissue}	0.2	$S{\cdot}m^{-1}$	(Davalos, Rubinsky et al. 2003)
Density	ρ	1000	$kg{\cdot}m^{-3}$	(Miller, Leor et al. 2005)
Heat Capacity	c_p	4200	$J{\cdot}Kg^{-1}{\cdot}K^{-1}$	(Miller, Leor et al. 2005)
Thermal Diffusivity	α	$1.34X10^{-7}$	m^2/s	(White 1991)
Frequency Factor	ζ	$1.19X10^{35}$	s^{-1}	(Feng, Tinsley Oden et al. 2008)
Activation Energy	E_a	$2.318X10^{5}$	$J{\cdot}mol^{-1}$	(Feng, Tinsley Oden et al. 2008)
Universal Gas Constant	R	8.314	$J{\cdot}K^{-1}{\cdot}mol^{-1}$	-

Thermal damage, Ω, is described and quantified by the Arrhenius rate equation proposed by Henriques (Henriques and Moritz 1947; Tropea and Lee 1992; Martin, Pliquett et al. 2002):

$$\Omega = \int \zeta \cdot e^{-E_a/RT}\, dt \tag{19}$$

where ζ is the frequency factor, E_a is the activation energy, R is the universal gas constant, and T is the absolute temperature in Kelvin. It has been shown that $\Omega = 0.53$ is the threshold for burn injuries in blood-perfused skin (Diller and Hayes 1983; Jiang, Ma et al. 2002). The Arrhenius equation, which traditionally has been used to study burn injuries in skin and transdermal drug delivery using electroporation, was adapted to investigate therapeutic IRE.

6.2 Analytical Model

The temperature rise due to joule heating, $\Delta T = T_i - T_0$, of cell solutions or homogeneous tissue for a parallel plate electrode configuration is given by Eqn. 3 during electroporation (Krassowska, Nanda et al. 2003). The heat transfer solution to transient conduction in a finite slab was used to determine the temperature profile versus position and time. This model is similar to the electrode geometry in which the electrodes serve as infinite fins that dissipate the heat away from the cell solution. The exact solution is given as a Fourier series in (Schneider 1955; White 1991)

$$\frac{T-T_0}{T_i-T_0} = \sum_{i=1}^{\infty} C_i e^{-\beta_i^2 \alpha t/L^2} \cos\left(\frac{\beta_i x}{L}\right) \tag{20}$$

where T is the temperature at time t, T_i is the initial temperature, T_0 is the room temperature, α is the thermal diffusivity, L is half thickness of the slab, and x is the distance from the centerline of the slab. C_i is calculated from:

$$C_i = \frac{4 \sin\beta_i}{2\beta_i + \sin(2\beta_i)} \tag{21}$$

and the constants β_i are calculated from the following equation:

$$\beta_i \tan(\beta_i) = h_0 L/k \tag{22}$$

where k is the thermal conductivity of the solution or tissue, and h_0 is the equivalent heat transfer coefficient assuming that the stainless steel electrodes act as infinite fins in free convection as previously described in (Davalos, Rubinsky et al. 2003). The temperature was determined from the first term of the Fourier series in the exact solution which is accurate within one percent since the dimensionless time $t^* = \alpha t/L^2$ is greater than 0.2 under the modeled conditions (Heisler 1947; White 1991). The temperature was assumed to be homogeneous throughout the domain equal to the temperature at the centerline of the sample to be conservative (x=0). Thus the exact solution (Eqn. 20) can be rewritten as

$$\frac{T-T_0}{T_i-T_0} = C_1 e^{-\beta_1^2 \alpha t/L^2} \tag{23}$$

By substituting Eqns. 3 and 23 into Eqn. 19, the thermal damage for cells suspended in Phosphate Buffered Saline (PBS) between two parallel plate electrodes can be estimated as

$$\Omega(t', t_p, E) = \int_0^{t'} \Psi(t_p, E)\,dt = \int_0^{t'} \zeta \cdot \exp\left(-\frac{E_a}{R\left[\frac{\sigma \cdot E^2}{\rho \cdot c_p} \cdot t_p C_1 \exp\left(-\frac{\beta_1^2 \alpha t}{L^2}\right) + T_0\right]}\right) dt \tag{24}$$

where t' is the heat dissipation time after electroporation, t_p is the pulse length and *E* is the applied electric field (Shafiee, Garcia et al. 2009). The heat dissipation is dependent on physical properties of the cell solution, cell type, pulse parameters and electrode configuration.

6.3 Analytical Results

The results are based on a highly conductive buffer solution (PBS). It can be infered that the electrical conductivity of the cell solution or tissue has great influence on the onset of thermal damage. For the same experimental conditions, buffers with lower electrical conductivities would generate lower joule heating in the sample. Consequently, IRE pulses with higher electric fields or longer durations could still be used prior to the onset of thermal damage.

Figure 11 shows the temperature profile and the function $\Psi(t_p, E)$ for the cases in which minimum ($\Omega = 0.53$) and third degree burns ($\Omega = 10{,}000$) due to joule heating occur. The temperature of the cell solution returns to room temperature independent of the initial temperature within 30 seconds. The function $\Psi(t_p, E)$ returns to zero within this time frame as well due to the effect of heat dissipation through the electrodes. Therefore, we used a time, t'=30s, to evaluate the thermal damage integral in Eqn. 24.

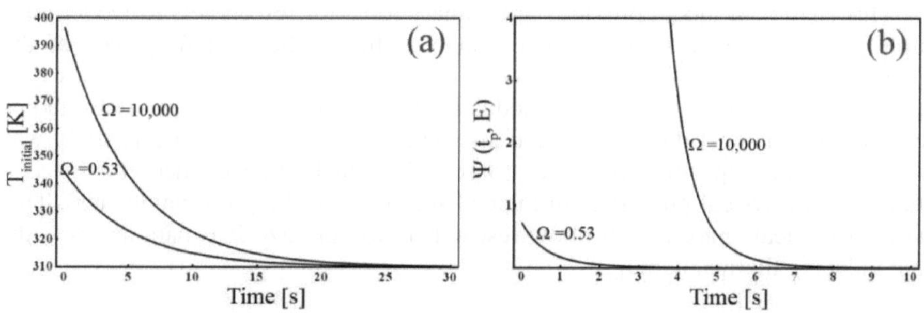

Fig. 11. a) Temperature profile and b) the function $\Psi(t_p, E)$ versus time for $\Omega = 0.53$ (onset of thermal damage) and $\Omega = 10{,}000$ (third-degree burn injury) (Diller and Hayes 1983; Jiang, Ma et al. 2002). The thermal damage magnitude, Ω, is the area under the $\Psi(t_p, E)$ curve (Eqn. 24). Adapted from (Shafiee, Garcia et al. 2009).

The thermal damage delineation from the potential IRE region for cells suspended in PBS (Figure 12a) and tissue (Figure 12b) is given as a function of the electrical pulse parameters (magnitude and duration). These plots can be used to select appropriate pulse parameters for non-thermal IRE treatments using parallel plate electrodes. The potential IRE regions for cells suspended in PBS is narrower than that for tissue due to its higher electrical conductivity since more joule heating is generated for the same pulse parameters. In tissue, blood perfusion and a larger volume are other factors that help dissipate the heat after IRE. The

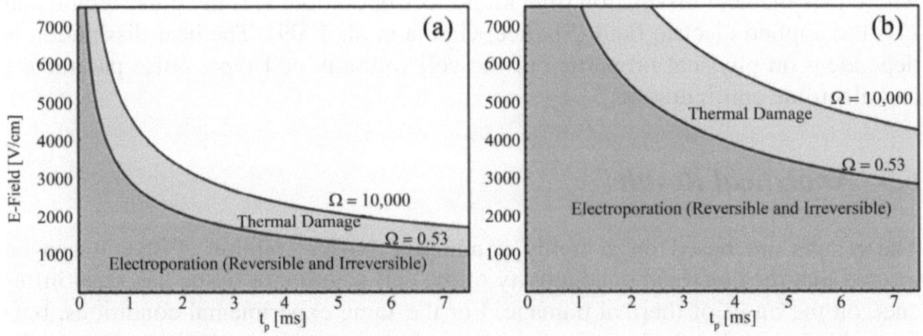

Fig. 12. Thermal damage delineation from the area of potential irreversible electroporation as a function of the IRE electrical pulse parameters (magnitude and duration) for a) cells suspended in PBS and b) tissue. $\Omega = 0.53$ represents the onset of thermal damage and $\Omega = 10{,}000$ represents a third-degree burn injury in skin (Diller and Hayes 1983; Jiang, Ma et al. 2002). Note: The onset of electroporation is not depicted. Adapted from (Shafiee, Garcia et al. 2009).

boundary between reversible electroporation and no effects on the cell membrane has been investigated and is not included in this study (Weaver and Powell 1989; Mir and Orlowski 1999).

This analytical model provides preliminary tools for researchers to model multiple pulses and assess the thermal damage. The number and frequency of the pulses are some other factors that need to be considered in subsequent studies. The proposed single pulse analytical model might be used for multiple IRE studies considering the heat dissipation time after each pulse. The goal of this study was to express the upper bound of IRE (onset of thermal damage) theoretically as a function of physical properties of the cell solution and the pulse parameters. This new theoretical analysis will allow researchers to optimize IRE parameters without inducing thermal damage.

6.4 Equivalent Thermal Dose Equation

For procedures involving time varying temperatures, thermal damage can be assessed by calculating the amount of time it would take to equivalently damage the tissue as if it was held at a constant temperature, typically 43°C (Becker and Kuznetsov 2006). The following expression is the duration necessary to hold the tissue at 43°C to result in an equivalent thermal dose:

$$t_{43} = \sum_{t=0}^{t=final} R^{(43-T_t)} \Delta t \tag{25}$$

where T_t is the average temperature during Δt with $R = 0.25$ when $T_t \leq 43°C$ and $R = 0.5$ when $T_t > 43°C$ (Sapareto and Dewey 1984; Damianou, Hynynen et al. 1993).

7 Conclusion

IRE has been studied in the context of the delayed cell damage in high-voltage accidents (Lee and Kolodney 1987; Lee 2005), the post-electric-shock arrhythmias during defibrillation (Jones, Proskauer et al. 1980), and biofouling control (Schoenbach, Peterkin et al. 1997). However, IRE had mostly been considered an undesirable outcome in medical applications and biotechnology until recently.

In order to design protocols for an IRE procedure, the electric field distribution must be determined, which is dependent on the procedure's specific electrode-tissue geometry and tissue impedance distribution. By predicting the electric field distribution for a specific scenario, the electrode geometry can be optimized to ablate the entire targeted region while minimally affecting the surrounding tissue. Furthermore, to verify that a specific protocol does not induce thermal effects, the temperature distribution can be calculated from the electric field distribution, the electric pulse parameters, and tissue properties.

To aid in the further development of IRE as a controlled non-thermal method for ablating undesirable tissues such as cancer, this chapter has been concerned with providing the tools necessary to model the electrical and thermal fields that result from the application of short high voltage pulses to biological tissues.

References

Al-Sakere, B., Andre, F., et al.: Tumor ablation with irreversible electroporation. PLoS ONE 2(11), e1135 (2007)

Al-Sakere, B., Bernat, C., et al.: A study of the immunological response to tumor ablation with irreversible electroporation. Technology in Cancer Research and Treatement 6, 301–306 (2007)

Becker, S.M., Kuznetsov, A.V.: Numerical modeling of in vivo plate electroporation thermal dose assessment. Journal of Biomechanical Engineering 128(1), 76–84 (2006)

Becker, S.M., Kuznetsov, A.V.: Thermal Damage Reduction Associated with in Vivo Skin Electroporation: A Numerical Investigation Justifying Aggressive Pre-Cooling. International Journal of Heat and Mass Transfer 50, 105–116 (2007)

Bhatt, D.L., Gaylor, D.C., et al.: Rhabdomyolysis due to pulses electric fields. Plast. Reconstr. Surg. 86(1), 1–11 (1990)

Damianou, C., Hynynen, K., et al.: Application of the Thermal Dose Concept for Predicting the Necrosed Tissue Volume During Ultrasound Surgery. In: Ultrasonics Symposium (1993)

Davalos, R.V., Mir, L.M., et al.: Tissue ablation with irreversible electroporation. Annals of Biomedical Engineering 33(2), 223–231 (2005)

Davalos, R.V., Otten, D.M., et al.: Electrical impedance tomography for imaging tissue electroporation. IEEE Transactions on Biomedical Engineering 51(5), 761–767 (2004)

Davalos, R.V., Otten, D.M., et al.: A feasibility study for electrical impedance tomography as a means to monitor tissue electroporation for molecular medicine. IEEE Transactions on Biomedical Engineering 49(4), 400–403 (2002)

Davalos, R.V., Rubinsky, B.: Temperature considerations during irreversible electroporation. Int. J. Heat. Mass. Tran. 51, 5617–5622 (2008)

Davalos, R.V., Rubinsky, B., et al.: Theoretical analysis of the thermal effects during in vivo tissue electroporation. Bioelectrochemistry 61(1-2), 99–107 (2003)

Deng, Z.S., Liu, J.: Blood perfusion-based model for characterizing the temperature fluctuations in living tissue. Phys. A STAT Mech. Appl. 300, 521–530 (2001)

Diller, K.R., Hayes, L.J.: A finite element model of burn injury in blood-perfused skin. J. Biomech. Eng. 105(3), 300–307 (1983)

Duck, F.A.: Physical Properties of Tissues: A Comprehensive Reference Book. Academic Press, San Diego (1990)

Edd, J., Horowitz, L., et al.: In vivo results of a new focal tissue ablation technique: irreversible electroporation. IEEE Transactions on Biomedical Engineering 53(7), 1409–1415 (2006)

Edd, J.F., Davalos, R.V.: Mathematical modeling of irreversible electroporation for treatment planning. Technology in Cancer Research and Treatment 6, 275–286 (2007)

Feng, Y., Tinsley Oden, J., et al.: A two-state cell damage model under hyperthermic conditions: theory and in vitro experiments. J. Biomech. Eng. 130(4), 041016 (2008)

Foster, K.R., Lozano-Nieto, A., et al.: Heating of Tissues by Microwaves: A Model Analysis. Bioelectromagnetics 19, 420–428 (1998)

Gabriel, B., Teissie, J.: Control by electrical parameters of short- and long-term cell death resulting from electropermeabilization of Chinese hamster ovary cells. Biochim. Biophys. Acta 1266(2), 171–178 (1995)

Gabriel, S., Lau, R.W., et al.: The dielectric properties of biological tissues: III. Parametric models for the dielectric spectrum of tissues. Phys. Med. Biol. 41(11), 2271–2293 (1996)

Gehl, J., Skovsgaard, T., et al.: Vascular reactions to in vivo electroporation: characterization and consequences for drug and gene delivery. Biochimica et Biophysica Acta 1569, 51–58 (2002)

Heisler, M.P.: Temperature charts for induction and constant temperature heating. ASME Transactions (1947)

Henriques, F.C., Moritz, A.R.: Studies in thermal injuries, V: the predictability and the significance of thermally induced rate processes leading to irreversible epidermal damage. Arch. Pathol. 43, 489–502 (1947)

Hulsheger, H., Niemann, E.G.: Lethal effects of high-voltage pulses on E. coli K12. Radiat. Environ. Biophys. 18(4), 281–288 (1980)

Ivorra, A., Rubinsky, B.: Electric field modulation in tissue electroporation with electrolytic and non-electrolytic additives. Bioelectrochemistry 70(2), 551–560 (2007)

Jiang, S.C., Ma, N., et al.: Effects of thermal properties and geometrical dimensions on skin burn injuries. Burns 28(8), 713–717 (2002)

Jones, J.L., Proskauer, C.C., et al.: Ultrastructural injury to chick myocardial cells in vitro following "electric countershock". Circ. Res. 46(3), 387–394 (1980)

Kekez, M.M., Savic, P., et al.: Contribution to the biophysics of the lethal effects of electric field on microorganisms. Biochim. Biophys. Acta 1278(1), 79–88 (1996)

Krassowska, W., Nanda, G.S., et al.: Viability of cancer cells exposed to pulsed electric fields: the role of pulse charge. Ann. Biomed. Eng. 31(1), 80–90 (2003)

Lavee, J., Onik, G., et al.: A Novel Nonthermal Energy Source for Surgical Epicardial Atrial Ablation: Irreversible Electroporation. The Heart Surgery Forum 10(2), E162–E167 (2007)

Lee, E.W., Loh, C.T., et al.: Imaging guided percutaneous irreversible electroporation: ultrasound and immunological correlation. Technology in Cancer Research and Treatement 6(4), 287–294 (2007)

Lee, R.C.: Cell Injury by Electric Forces. Annals of the New York Academy of Sciences 1066, 85–91 (2005)

Lee, R.C., Despa, F.: Distinguishing Electroporation from Thermal Injuries in Electrical Shock by MR Imaging. In: Engineering in Medicine and Biology 27th Annual Conference, Shanghai, China. IEEE, Los Alamitos (2005)

Lee, R.C., Kolodney, M.S.: Electrical injury mechanisms: Electrical breakdown of cell membranes. Plast. Reconstr. Surg. 80(5), 672–679 (1987)

Lee, R.C., Zhang, D., et al.: Biophysical Injury Mechanisms in Electrical Shock Trauma. Ann. Rev. Biomed. Eng. 2, 477–509 (2000)

Lubicki, P., Jarayam, S.: High voltage pulse application for the destruction of the Gram-negative bacterium. Bioelectrochem. Bioenerg. 43, 135–141 (1997)

Maor, E., Ivorra, A., et al.: The effect of irreversible electroporation on blood vessels. Technology in Cancer Research and Treatement 6(4), 307–312 (2007)

Martin, G.T., Pliquett, U.F., et al.: Theoretical analysis of localized heating in human skin subjected to high voltage pulses. Bioelectrochemistry 57(1), 55–64 (2002)

Miklavcic, D., Beravs, K., et al.: The importance of electric field distribution for effective in vivo electroporation of tissues. Biophysical Journal 74(5), 2152–2158 (1998)

Miklavcic, D., Sel, D., et al.: Sequential Finite Element Model of Tissue Electropermeabilisation. In: Proceedings of the 26th Annual International Conference of the IEEE EMBS, San Francisco, CA (2004)

Miller, L., Leor, J., et al.: Cancer cells ablation with irreversible electroporation. Technol. Cancer Res. Treat. 4(6), 699–705 (2005)

Mir, L.M., Orlowski, S.: Mechanisms of electrochemotherapy. Advanced drug delivery reviews 35, 107–118 (1999)

Okino, M., Tomie, H., et al.: Optimal electric conditions in electrical impulse chemotherapy. Jpn. J. Cancer Res. 83(10), 1095–1101 (1992)

Onik, G., Mikus, P., et al.: Irreversible electroporation: implications for prostate ablation. Technol. Cancer Res. Treat. 6(4), 295–300 (2007)

Pavlin, M., Miklavcic, D.: Effective Conductivity of a Suspension of Permeabilized Cells: A Theoretical Analysis. Biophysical Journal 85, 719–729 (2003)

Pennes, H.H.: Analysis of tissue and arterial blood temperatures in the resting forearm. J. Appl. Physiol. 1, 93–122 (1948)

Ramirez, L.H., Orlowski, S., et al.: Electrochemotherapy on liver tumours in rabbits. Br. J. Cancer 77, 2104–2111 (1998)

Rubinsky, B.: Irreversible Electroporation in Medicine. Technology in Cancer Research and Treatement 6(4), 255–260 (2007)

Rubinsky, B., Onik, G., et al.: Irreversible electroporation: A new ablation modality – clinical implications. Technology in Cancer Research and Treatment 6(1), 37–48 (2007)

Sale, A.J., Hamilton, W.A.: Effects of high electric fields on micro-organisms. 1. Killing of bacteria and yeasts. Biochimica et Biophysica Acta 148, 781–788 (1967)

Sapareto, S., Dewey, W.: Thermal dose determination in cancer therapy. Int. J. radiation oncology Biol. Phys. 10, 787–800 (1984)

Schneider, P.J.: Conduction Heat Transfer. Addison Wesley, Reading (1955)

Schoenbach, K.H., Peterkin, F.E., et al.: The effect of pulsed fields on biological cells: Experiments and applications. IEEE Trans. Biomed. Eng. 25, 284–292 (1997)

Sersa, G., Cemazar, M., et al.: Tumour blood flow changes induced by application of electric pulses. Eur. J. Cancer 35, 672–677 (1999)

Shafiee, H., Garcia, P.A., et al.: A Preliminary Study to Delineate Irreversible Electroporation From Thermal Damage Using the Arrhenius Equation. ASME J. of Biomechanical Engineering 131 (2009)

Somersalo, E., Cheney, M., et al.: Existence and Uniqueness for Electrode Models for Electric-Current Computed-Tomography. SIAM Journal on Applied Mathematics 52(4), 1023–1040 (1992)

Swarup, A., Stuchly, S., et al.: Dielectric properties of mouse MCA1 fibrosarcoma at different stages of development. Bioelectromagnetics 12, 1–8 (1991)

Tropea, B.I., Lee, R.C.: Thermal injury kinetics in electrical trauma. J. Biomech. Eng. 114(2), 241–250 (1992)

Vernhes, M.C., Cabanes, P.A., et al.: Chinese hamster ovary cells sensitivity to localized electrical stresses. Bioelectrochem. Bioenerg. 48(1), 17–25 (1999)

Weaver, J.C.: Electroporation of cells and tissues. IEEE Transactions on Plasma Science 28(1), 24–33 (2000)

Weaver, J.C.: Electroporation of Biological Membranes from Multicellular to Nano Scales. IEEE Transactions on Dielectrics and Electrical Insulation 10(5), 754–768 (2003)

Weaver, J.C., Chizmadzhev, Y.A.: Theory of electroporation: a review. Bioelectrochem. Bioenerg. 41, 135–160 (1996)

Weaver, J.C., Powell, K.T.: Theory of electroporation. In: Neumann, E. (ed.) Electroporation and Electrofusion in Cell Biology, vol. 7. Plenum Press, New York (1989)

White, F.M.: Heat and Mass Transfer. Addison-Wesley Publishing Company, Reading (1991)

Experimental Studies on Non-thermal Irreversible Electroporation in Tissue

Boris Rubinsky

Department of Mechanical Engineering, University of California, Berkeley, CA 94720 USA
and
Center for Bioengineering in the Service of Humanity and Society, School of Computer Science and Engineering, Hebrew University of Jerusalem, Givat Ram Campus, Jerusalem, Israel
Corresponding author E-mail: rubinsky@me.berkeley.edu, rubinsky@cs.huji.ac.il

Introduction

In the past, irreversible electroporation of tissue was studied for two applications: a) in the food industry for food processing and b) as a means to determine the upper limit of electrical parameters for reversible electroporation of tissue. Non-thermal irreversible electroporation (NTIRE) is a new modality for tissue ablation that applies the irreversible electroporation pulses in such a way as to avoid thermal damage to tissue components while irreversible affecting the cells. This particular aspect of irreversible electroporation was not studied before. This chapter reviews experimental studies on non-thermal irreversible electroporation in tissue done by our group. The studies will be discussed in the chronological order in which they were done. In all the studies discussed in this chapter, treatment planning was done prior to performing the experiments. Treatment planning is done to identify the appropriate sequence of electrical pulses, which produce the desired cell ablation without causing damage to the remaining tissue structure, [1]. It is not trivial as the range of parameters used can vary from tissue type to tissue type and from circumstance to circumstance. In the absence of other information, in these earlier studies modeling the electrical field and the temperature distribution caused by Joule heating during the application of the electrical pulses identified the treatment planning. It is possible that in the future it will be also valuable to model that change in pH in the tissue, as they could also affect the tissue components, e.g. [2]. It should be emphasized that when thermal or chemical damage do occur during irreversible electroporation, the outcome of the procedure will most likely be different from those described in this chapter.

NTIRE of the Liver

The first two papers in this area have studied NTIRE in the liver, [3], [4]. The liver is a highly vascular organ and has served often as the preferred organ for studying methods for minimally invasive surgery, e.g. [5], [6]. Main reasons for studying minimally invasive ablation techniques in the liver, in addition to its relative homogeneity, are the value of minimally invasive tissue ablation in treatment of liver cancer and the difficulties in performing conventional resection surgery in this organ.

B. Rubinsky (Ed.): Irreversible Electroporation, BIOMED, pp. 155–181.
springerlink.com

The experiments reported in [3], were performed on the liver of Spraque Dawley rats. The 10 mm diameter and one mm thick electroporation electrodes were made of sintered Ag/AgCl (E255, *In Vivo Metric*, Healdsburg, CA). While in conventional electroporation it is more common to use stainless steel electrodes, in this study Ag/AgCl was used because one aspect of our research is to evaluate the changes in electrical properties of tissue during and after electroporation for use in monitoring and controlling electroporation [6], [7]. The lobe of the liver was clamped between the two electrodes placed in contact with the liver in such a way that the liver was sandwiched between the two parallel cylindrical surfaces of the electrodes. The NTIRE protocol was developed through treatment planning and consists of one 20 ms pulse of 400 V delivered across 4 mm of liver. Roughly this corresponds to 1000 V/cm. However, it should be emphasized that in the complex geometry of real tissue the geometry and configuration of the tissue and of the electrodes should be carefully considered and analyzed. Evaluating the electrical fields only from the voltage difference between the electroporation electrodes is only roughly indicative of the actual electrical fields, the temperature and the thermal damage which occurs during the electroporation [1], [9], [10].

The pulse was delivered with a commercial pulse generator (ECM 830, *Harvard Apparatus*; Holliston, MA). The animal was observed continuously while in thermal incubation until three hours after pulsing. At this point, the animal was euthanized. To fix the liver at its current state for microscopic viewing, we flushed the vasculature with physiological saline for ten minutes at a hydrostatic pressure of 80 mmHg from an elevated IV drip. Injecting the fluid into the left ventricle and letting it exit from a cut made in the right atrium accomplished this. Dr. Narayan Raju designed this tissue fixation type procedure. We have found that flushing the tissue prior to fixation can produce valuable insight into the mechanisms of action of NTIRE in relation to its effect on the blood vessels. Immediately following saline perfusion, a formaldehyde fixative was perfused in the same way for ten minutes. The treated liver lobe was then removed and stored in the same formaldehyde. Hematoxylin-eosin staining was then performed on cross-sections through the center of the treated region to elucidate the microscopic morphology.

Figure 1 presents the highlights of our findings. Figure 1a, shows a macroscopic view of the treated liver (one half plane). The dark circular region is the area affected by NTIRE. It corresponds to the dimension of the electroporation electrodes. A rim of thermal damage was observed on the surface of the liver. It corresponds to the edges of the electrodes and it is due to the high local fields. It could be avoided by using rounded edge electrodes. Figures 1b and 1 c show the mathematical calculations of the electrical field and the temperature distribution superimposed on a macroscopic tissue cross section made through the central plane of the lesion and demonstrates that the lesion is non-thermal irreversible electroporation. The figure shows that the ensuing electric field from the pulse is fairly uniform in the area directly between the electrodes at about 1kV/cm. The electric field drops precipitously as the distance from the electrode axis exceeds the electrode radius and approaches zero near the outer edges of the liver. When analyzing the electric field contours it is important to notice the 600 V/cm to 700 V/cm contours. This has been also determined in previous studies with the rabbit liver as the parameters of demarcation between reversible and irreversible

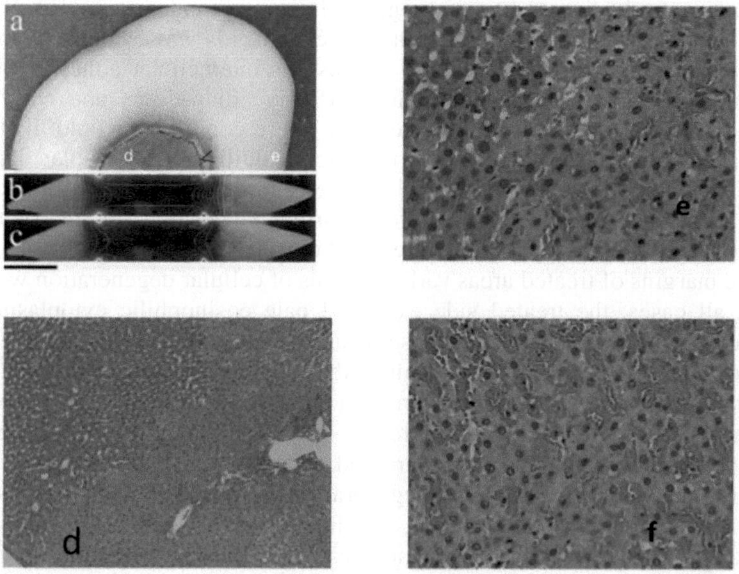

Fig. 1. NTIRE treated liver. (1a) Top view of the treated liver. Thin semicircle of thermally ablated tissue is visible near the extents of the tissue ablated by irreversible electroporation (dark area). (1b) Electric field strength resulting from the 400 V pulse. Lines of constant electric field strength are superimposed upon the side view of this liver. The lines represent 100 V/cm increments from 100 V/cm to 2000 V/cm, where the lowest values are at the far left and right edges and the highest occur within an annular region corresponding to the semicircle referred to in the top view. Two lines that cross the central region are 1000 V/cm contours. (c) Temperature rise due to Joule heating at the end of the 20 ms pulse. Lines of constant temperature are superimposed upon the same side view. The contours represent 1 °C increments from 38 °C to 50 °C where the lowest values are at the far left and right edges (ΔT is proportional to the square of the local electric field strength). The annular region where temperatures exceed 50 °C corresponds to the area that experienced thermal ablation (arrow), while the remaining effects were non-thermal in nature. Bar indicates 5 mm in all photos. Figures 1d,1e,1f: Histological effects of the irreversible electroporation pulse. (1d) and (1f) show the margin of ablation in 10x and 40x resolutions respectively. (1f) shows the core of the ablated region.

electroporation [11]. This indicates that irreversible electroporation was likely to have occurred in the region directly between the electrodes. Figs 1d and 1e, show the margin of the treated region (upper left hand corner is the untreated region). The figures show that the pale regions in Figs 1a, 1b and 1c correspond to a lack of red blood cells caused by the fixation procedure which flushed the vasculature with saline and then formaldehyde. Indeed, histology shows no damage to the hepatocytes or endothelial cells in this region. The cytosol is normal, the plasma membranes are distinct and the nuclei are healthy. In addition, the endothelial cells appear to have maintained their structural integrity since there were no leaks of red blood cells into the surrounding sinusoids. Moreover, this area resembles closely the control samples of liver in every way. Histological analysis revealed acute damage to the tissue in treated regions. In central treated areas, Fig 1f, widespread pale eosinophilia and congestion in the sinusoids was evident. In some areas this congestion was so dense as to suggest the intercellular adhesion between adjacent hepatocytes had been impaired. Massive diapedeses and early fibrin deposition also occured within the sinusoids. These conditions would likely have caused coagulative necrosis of the hepatocytes. Additionally, vacuolar degeneration of hepatocytes was present throughout the central treated region, but occurred most significantly in centrilobular areas, further suggesting an interruption of blood flow and oxygen supply which resulted in ischemic damage.

At the margins of treated areas various stages of cellular degeneration were present. In all cases, the treated side exhibited pale eosinophilic cytoplasm when compared with the untreated side. In some areas, moderate pyknosis of the nuclei and sinusoidal congestion were present in the treated side and conspicuously absent in the other side. Cell borders were also clearly visible on the non-treated side and nearly indistinguishable in the treated side. This suggests that the primary expected consequence of the procedure had occurred: cell membrane disruption. In other areas, obvious vacuolar degeneration marked the boundary between healthy and pathological parenchyma. The line of demarcation was often surprisingly narrow (between adjacent hepatocytes), but did not correspond to lobular boundaries. Throughout the treated area, insult to endothelial cells was evident. In nearly all cases, endothelial necrosis was present and large numbers of neutrophils were marginated along and had infiltrated within the vessel walls, indicating acute vascular inflammation. Erythrocytes were also seen pooling immediately beneath the endothelium. Notably, at three hours after treatment, this vascular necrosis had not destroyed large blood vessel architecture as indicated by the successful flushing of blood from these vessels (clear lumen), Fig 1d.

The goal of this first NTIRE study was to determine through in-vivo experiments in rat liver if sufficiently strong electric pulses cause non-thermal and controllable ablation as a consequence of irreversible electroporation alone. The results suggest that irreversible electroporation can be used to ablate large volumes of tissue in a controlled manner and with a sharp boundary between necrotic and unaffected tissue. In addition, the tissue fixation procedure we have employed showed that irreversible ablation can cause occlusion of small blood vessels (sinusoids) and their congestion with crenated red blood cells. Related effects of reversible electroporation on blood flow were found previously, though if reversible electroporation has occurred in those cases they differ with what we have found [12,13,14,15]. This finding has important applications. It suggests that in the irreversible electroporated region any cell that may survive electroporation would experience ischemia and death through oxygen depravation. An intriguing use of the congestion of red blood cells could be in the entrapment of any type of drug inside the irreversible electroporated tissue. Injection of any drug or medical imaging contrast agent in the vasculature or in the interstitial space prior or during irreversible electroporation will cause the entrapment of the drug in that area. This way a bolus of high concentration drugs could continuously diffuse from the tissue in which it is entrapped and act focally in the irreversible electroporated area and around it or a concentrated area of imaging contrast agents could identify the treated region.

Perhaps the most important finding of this study is that the large blood vessels architecture is intact after non-thermal irreversible electroporation, while the cells around it are ablated. This is the unique consequence of non-thermal ablation and is perhaps the most important attribute of NTIRE, which distinguishes it from other ablation techniques. Other tissue ablation techniques non-discriminately affect all the organic molecules in the treated volume. As a consequence the blood vessel architecture is also damaged. Because NTIRE affects only the cell membrane the extracellular architecture remains intact and the scaffold of the blood vessels, ducts and any other structures that are not cells remains intact. The mechanical integrity of the tissue is retained. This has major significance in treatment of cancer. One of the major difficulties in ablating cancer abutting large blood vessels or important duct structures is the fact that thermal treatment cannot be used to affect cells near blood vessels because of the thermal effect of the vessels on one hand and because the thermal ablation impairs the larger blood vessels on the other. The ability of irreversible ablation to cause cell necrosis near the vessels while preserving the large vessel architecture could be used to treat tumors near the larger blood vessels and complex structures of the tissue. This result also suggested to us the use of NTIRE in treatment of restenosis, a study that we will discuss later.

The first long term, large animal study of NTIRE was reported in [4]. The goal of that study was to test the NTIRE tissue ablation treatment planning based methodology in the pig liver and to provide results on long term histopathology of NTIRE ablated tissue. Prior to experiments, treatment planning was done through a mathematical analysis of the electrical field in the treated area as well as the temperature distribution, which is produced by Joule heating during treatment (Figs 2a,2b). This study was conducted in accordance with Good Laboratory Practice regulations as set forth by the 21 Code of Federal Regulations (CFR) Part 58. Each procedure started with anesthetization of the animal under general anesthesia per SOP #33156. It is very important to notice that upon application of the NTIRE pulse a variable degree of generalized muscle contraction occurred in each animal, from no contraction to mild to moderate contraction. The degree of contraction appeared to be related to the level of anesthesia of the animal the degree of muscle blocking agents given, if any, and the voltage of the pulse used. In general we found that when Pancuronium was administered to the pig the contractions were manageable even when fields of 3kV were applied to the electrodes. It appears that voltages lower than 1.5 kV produced little to no contraction in anesthetized animals. Therefore, pancuronium (0.1 mg/kg, at a dose of 1 mg/ml) was administered through an IV prior to the procedure, to reduce muscle contractions during the application of the electrical pulses. Pancuronium (0.05 mg/ml at 1 mg/ml) was administered throughout the procedure as needed. The liver was exposed via a midline incision (see Fig. 1C). IRE was performed using 18 gauge stainless steel electrodes (Oncobionic Inc., USA) with a sliding insulating sheath exposing 1 cm of electrode. The electrodes were inserted in the liver under ultrasound monitoring in configurations of either two or four electrodes in a roughly axial parallel configuration. Figure 1D illustrates the insertion of the four needle electrodes using an insertion template. Square DC pulses were applied to the liver through the electrodes using a DC pulse generator (IGEA Co., Italy). The pulse was delivered in a bipolar manner between two electrodes. Lesions made using the four electrode configurations had overlapping lesions created by activating the electrodes in different bipolar configurations between the four electrodes. The parameters that were varied in this study were the voltage, the distance between the electrodes, and the number of pulses. The number of pulses varied between two and eight, 100 microsecond pulses. Between four and twelve lesions were produced in the liver of each animal. The treatment planning and the introduction of the electrodes in this study are shown in Fig 2. An explanation of the results is given in the figure legend.

An interesting finding with significance to NTIRE is shown in Fig 3. The insertion of the needles was done under ultrasound monitoring and the procedure was subsequently followed by ultrasound. Since NTIRE is a minimally invasive procedure the use of medical imaging for treatment of tissues inside the body is highly recommended. In Figure 3A axial ultrasound shows four bright

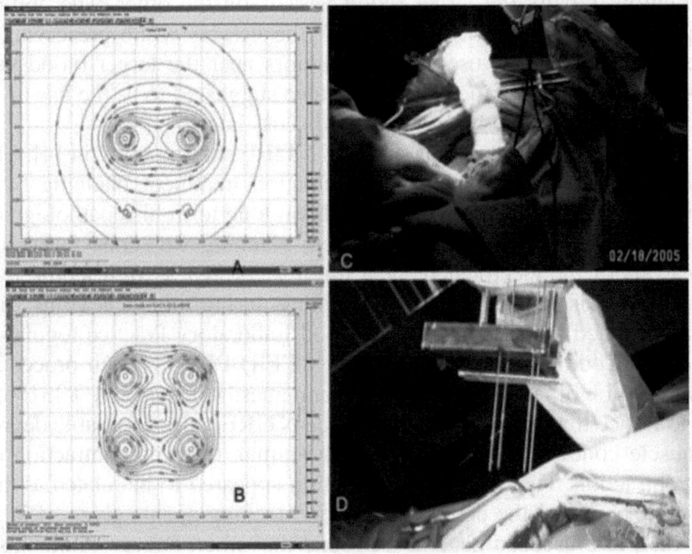

Fig. 2. Treatment planning and surgical protocol for irreversible electroporation ablation in the pig liver. IRE pulse parameters are bipolar electroporation, 2.5 kV pulses applied in a train of eight 100 microsecond pulses separated by 100 milliseconds. Four electrodes 18 gage stainless steel electrodes, 1.5 cm distance between probes. (A) Results of mathematical analysis showing constant electrical field magnitude isolines during the application on an electroporation pulse between two electrodes. The numbers on the figures indicate the magnitude of the electrical field in increments of 100 V/cm starting from 100 V/cm (outer isoline). The power delivered during each pulse is 1.2 J per cm depth of electrode. (B) Results of mathematical analysis show constant electrical field magnitude isolines due to the superposition of the application of the IRE pulses between the four pairs of IRE electrodes. The numbers on the figures indicate the magnitude of the electrical field in increments of 100 V/cm starting from 600 V/cm (outer isoline). Experiments have shown that 600 V/cm induces irreversible electroporation in liver. (C) Photograph of application of electroporation probes with ultrasound, (D) insertion of a set of four electrodes. [4]

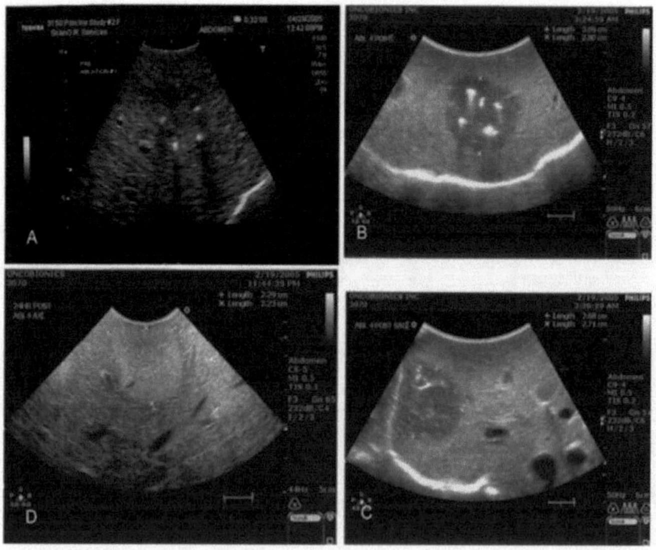

Fig. 3. Ultrasound images of an IRE process: a) during the insertion of the needles (bright spots), b) and c) Immediately after the delivery of the electroporation pulses d) 24 hours later. [4]

hyperechoic dots representing the four needle tracks. Most interesting was the finding that immediately following pulse application ultrasound showed a markedly hypoechoic lesion in the expected location of the IRE lesion [Figs. 3B (axial) and 3C (sagital)]. Subsequent studies show that this effect is more pronounced in vascular tissues and can be seen on any imaging technology, such as CT, ultrasound and MRI. At 24 hours the ultrasound image showed the hyperechoic lesion had changed character and was now uniformly hyperechoic. Figure 3D is the sagital view after 24 hours that corresponds to Figure 3C.

To fix the liver in its current state for microscopic viewing, a Foley catheter was placed into each branch of the portal vein and hepatic veins and the liver was flushed the with physiological saline for ten minutes at a hydrostatic pressure of 80 mmHg from a pressurized IV drip. Immediately following saline perfusion, a 5 % formaldehyde fixative was perfused in the same way for ten minutes. The liver lobe in which the IRE lesion was made, was then removed and stored in the same formaldehyde solution. For macroscopic analysis the tissue was bread loafed perpendicular to capsule surface and parallel to the needle tracts. All cassettes were processed routinely from 10% Phosphate Buffered Formalin to wax blocks. Five micrometer sections were made from each block and stained with Hematoxylin and Eosin for histologic examination. Some cassettes were also stained with Von Kossa stain for calcium detection. The goal of the histopathology was to verify the extent and the nature of tissue ablation with IRE; in particular, in relation to the margins of the treated zone, blood vessels and other vascular structures, and the long-term resolution of the lesions.

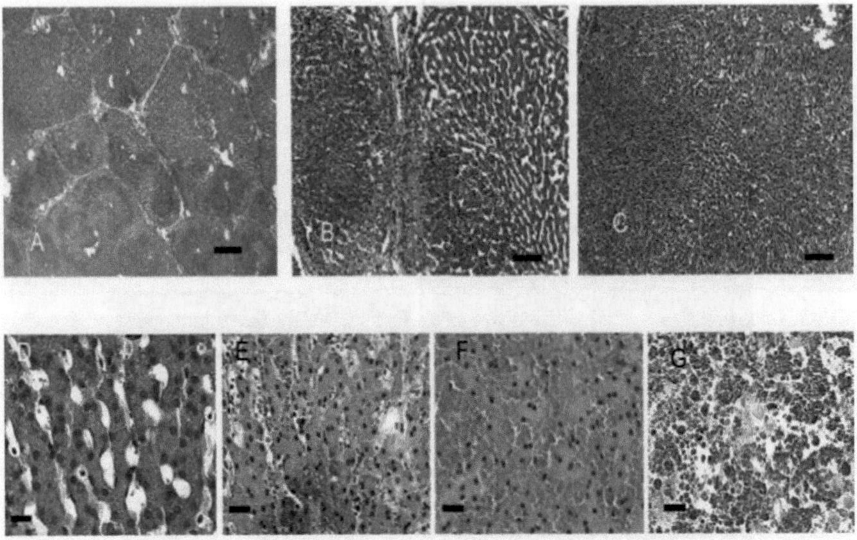

Fig. 4. Microscopic histology of IRE ablation in the pig liver, 24 hours post IRE. (A) H&E stained section. Margin of ablated area (bottom half) and unaffected area (top half). Focal dark areas are necrotic hepatocytes with calcification. Scale bar 500 micron. (B) H&E stained section. Margin of ablated area (left) and unaffected area (right). Focal dark areas are necrotic hepatocytes with calcification. Scale bar 100micron. (C) H&E stained section. Central area of ablation. Hepatocytes have hypereosinophilic cytoplams and pyknotic nuclei. Sinusoids are congested. Scale bar 100 micron. (D) H&E stained section. Normal area of hepatic parenchyma for comparison to ablated region. Scale bar 25 micron. (E) H&E stained section. Area of hepatic necrosis with hypereosinophilic cytoplasm and pyknotic nuclei. Aggregate of hepatocytes at bottom also contain calcium. Scale bar 25 micron. (F) H&E stained section. Area of hepatic necrosis with hypereosinophilic cytoplasm and pyknotic nuclei. Hepatocytes are separated by hemorrhage. Scale bar 25 micron. (G) Von Kossa stained section. Brown granular pigment is calcium in areas of hepatic necrosis of ablated region. Scale bar 25 micron. [4]

Figure 4 shows the microscopic appearance of the ablated tissue at different locations and magnifications at day one after the procedure. The gross lesion appears hemorrhagic. Histology shows the ablated area with diffuse necrotic hepatic parenchyma and no obvious viable tissue within the ablated zone. Needle tracts form 1-2 mm holes filled with blood and fibrin. The ablated zone is well demarcated and non-encapsulated from the immediately adjacent unaffected hepatic parenchyma with an abrupt transition between necrotic hepatic parenchyma in the ablation area and adjacent normal hepatic parenchyma. The central region of the treated area has coagulation necrosis of hepatocytes. The hepatocytes have hypereosinophilic cytoplasm and pyknotic to karyolitic nuclei and in some areas contain basophilic granular intracytoplasmic material (calcification) and loss of cellular detail. Sinusoids in this area had congestion admixed with fibrin and small to moderate numbers of inflammatory cells (neutrophils and eosinophils). Within the lesion, portal vessels and bile ducts had variable signs of wall necrosis but appeared structurally intact. Many larger bile ducts were intact. Necrotic vessels in the ablated area had variable loss of endothelium and are surrounded by scattered red blood cells (hemorrhage), neurophils, and eosinophils and edematous stroma. In some areas the vessel walls are hypereosinophilic and homogenous consistent with fibrinoid necrosis. Occasional affected vessels have organized intraluminal fibrin thrombi. Moderate mixed inflammatory cells and edema are present in portal stroma and adjacent to affected portal vessels with extension into perilobular stromal connective tissue. The peripheral region of the ablated zone has similar necrotic lobules as described more centrally. In some areas, necrotic hepatocytes had intracytoplasmic basophilic granular material consistent with dystrophic calcification. This is most prominent at the margin of the ablated area with some cells completely replaced by calcification. The region outside the ablated zone is unremarkable except for occasional portal zones showing occasional lymphocytic infiltrates. The margin of the ablated area has increased inflammatory cells characterized predominantly by moderate to large numbers of neutrophils and lesser numbers of eosinophils, lymphocytes and macrophages.

Figure 5 illustrates the long-term effects of NTIRE on the liver. Day 1, post IRE ablation results are consistent with those above. Particular here is a large vein at the deep margin of the ablated area. Ablation extends up to this vein. The vessel is unremarkable. The regional lymph node has mild cortical reactivity and expansion of medullary sinuses consistent with a drainage reaction. Medullary sinuses are expanded by red blood cells, eosinophils, neutrophils, and macrophages. Paracortical regions are expanded by increased lymphocytes. Occasional lymphoid follicles have reactive germinal centers. This is surrounded by fibrosis admixed

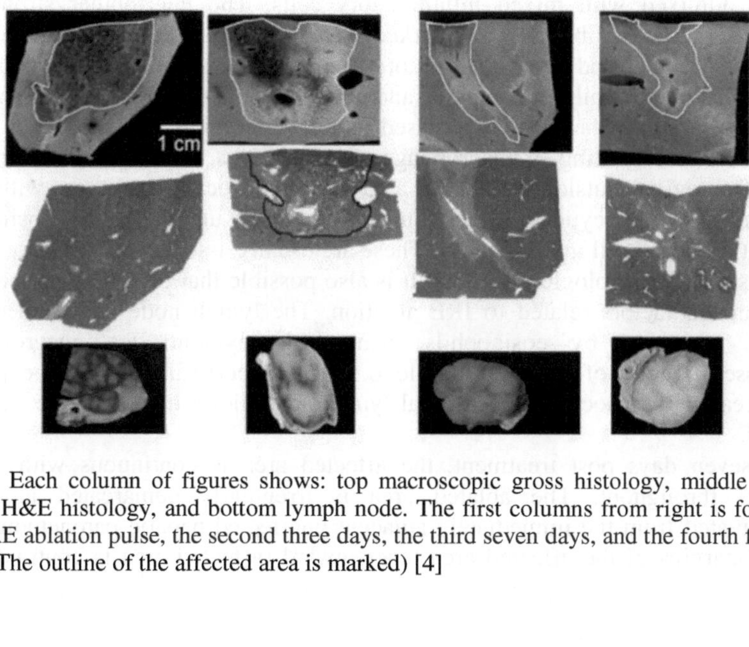

Fig. 5. Each column of figures shows: top macroscopic gross histology, middle macroscopic H&E histology, and bottom lymph node. The first columns from right is for 1 day post IRE ablation pulse, the second three days, the third seven days, and the fourth fourteen days. (The outline of the affected area is marked) [4]

with macro-phages and lymphocytes. Within this particular lymph node, there is a well demarcated chronic, eosinophilic granuloma measuring approximately 0.5mm in diameter surrounding a mineralized core. The origin of the mineralization is unknown, but is compatible with a necrotic parasite reaction. In this liver sample, eosinophils were also part of the inflammation. These are usually associated with parasitic infections and dermatologic disorders and the mineralization may explain the eosinophils in this animal series. We observed substantially enlarged lymph nodes in all our animal experiments. Therefore, it is possible that the eosinophils are mediated by unknown immunogenic factors related to the IRE treatment. Nevertheless, because of the results in this animal we cannot determine the cause with certainty.

Day 3, post treatment affected area is continuous with a central hemorrhagic, necrotic zone and variable loss of hepatic architecture. The continuous area is irregularly demarcated and non-encapsulated from adjacent unaffected hepatic parenchyma. Within the deep part of the ablated area and at one lateral margin there are two large veins. The ablation changes extend up to the vessels. Vessels are patent, surrounded by stromal edema and have variable necrosis of endothelial cells lining the intima. The integrity of the larger blood vessels is typical to NTIRE and it is an important attribute of this technique. The central hemorrhagic regions has diffusely necrotic lobules with marked congestion (hemorrhage), fibrin, and variable loss of hepatocyte cellular architecture. There are also multifocal chords of necrotic hepatocytes with diffuse replacement by basophilic material (calcium). The more peripheral affected regions consist of lobules with increased cellularity and loss of sinusoidal architecture (regenerative nodules). The regenerative nodules replace lobules that were overall necrotic at day 1, post treatment. Cells within these lobules are pleomorphic (spindle to oval to polygonal) and are consistent with regenerative hepatocytes that are extending from the lobule/stroma margin admixed with mixed inflammatory cells. The interlobular stroma has marked increase in fibrosis with marked bile duct proliferation throughout. Bile ducts are irregular and occasional mitotic figures are present. Most ducts contain intraluminal neutrophils, eosinophils, and necrotic cellular debris. The stroma contains proliferative fibroblasts, increased small veins and capillaries, and mixed inflammation consisting of macrophages, lymphocytes, eosinophils, and neutrophils. The regions outside the ablated area have multifocal portal areas with small numbers of lymphocytic infiltrates but are otherwise unremarkable. Eosinophils are part of the overall inflammation. These are usually associated with parasitic infections and dermatologic disorders. It is also possible that they may be mediated by unknown factors related to IRE ablation. The lymph node shows medullary sinuses expanded by eosinophils, neutrophils, lymphocytes, macrophages, and lesser numbers of scattered red blood cells. Paracortical regions are expanded by increased lymphocytes. Occasional lymphoid follicles have reactive germinal centers.

At seven days post treatment, the affected area is continuous with similar changes throughout. The ablated area is irregularly demarcated and non-encapsulated from the immediately adjacent unaffected hepatic parenchyma. The lateral margins of the affected area taper inward (affected area is contracted) as expected with increased fibrosis (scarring). Hepatic lobules have been replaced by regenerative hepatic nodules that are undergoing hepatic vacuolar degeneration. Nodules are variable in size surrounded by dense fibrous stroma. Nodules consist of diffuse hepatocytes with no sinu-soidal architecture. Hepatocytes are swollen, have increased clear to granular cytoplasm and pyknotic nuclei (vacuolar degeneration). The stroma contains increased bile ducts that are overall intact and unremarkable (less proliferative than at three days post treatment). It also contains increased small to medium size veins and capillaries (compared to three days post treatment). The stroma shows proliferative fibroblasts and mixed inflammation consisting of macrophages, and lymphocytes. A large vein runs through much of the ablated area. This vein has variably necrotic endothelium with some sloughing of endothelial cells. The lumen is patent. Ablated area extended to vessel wall. The regions outside the ablated area have multifocal portal areas with small numbers of lymphocytic infiltrates but are other wise unremarkable. The lymph node shows medullary sinuses that are expanded by eosinophils, neutrophils, lymphocytes, macrophages, and lesser numbers of scattered red blood cells. Paracortical regions are expanded by increased lymphocytes. Multifocal lymphoid follicles have reactive germinal centers.

By 14 days the ablated lesions are difficult to identify. The affected area is almost completely replaced by fibrous scar tissue. Scattered throughout the fibrous stroma are many profiles of bile ducts as well as many variably sized vessels consisting predominently of veins and capillaries. The stroma in the affected area contains moderate aggregates and scattered inflammatory cells consisting of lymphocytes, plasma cells, and macrophages admixed in some areas with red blood cells. Rare individual and aggregates of viable hepatocytes are present within the fibrous stroma (in most areas close to the margin of viable hepatic lobules). The stroma also contains scattered individual and aggregates of swollen vacuolated cells with pyknotic nuclei (hydropic degeneration) consistent with hepatocytes seen in regenerative nodules at seven days post treatment. It was felt by our pathologist that this was part of the regenerative process, but the possibility that these were untreated viable heptocytes could not be ruled out. The adjacent viable hepatic lobules had peri-lobular fibrosis and mild to moderate chronic peri-portal inflammatory infiltrates. Focal areas of mineralization were identified throughout the scar and were consistent with dystrophic calcification or osseous metaplasia. The periportal lymph nodes were again enlarged with similar features to those at seven days; however, germinal centers contained greater numbers of macrophages consistent with a chronic lymphoid reactivity.

In summary, histology shows that the IRE ablated area is continuously necrotic, with an abrupt transition, several cell layers thick, between treated necrotic hepatic parenchyma and the adjacent untreated normal hepatic parenchyma. Larger vascular structures are mainly intact. A unique aspect of IRE is the mechanisms of cell ablation. In IRE when cells are exposed to a pulsed electrical field with the proper parameters, irreversible nano-scale defects are created in the cell membrane and the cell loses its} homeostatic mechanisms. However, IRE does not affect connective structures, or denatures molecules, and collagen. This selective targeting of the cell membrane by IRE has important clinical implications. Since IRE affects only cell membranes injuries to tissue scaffold are eliminated. Our pathology demonstrated intact bile ducts within the lesions. If successfully translated into clinical practice this could have major implications by markedly decreasing the incidence of biliary complications associated with hepatic tumor ablations. Other organs could also benefit from this property of IRE. In the prostate this could translate into significantly less chance of urethral damage and in the kidney less chance for damage to the collecting system. IRE could also be promising for treatment of pancreatic cancer by producing no effects on the structures surrounding the pancreas. In the brain it could be used to treat tumors near large blood vessels.

Preservation of vasculature in an IRE lesion also has important clinical implications. Due to this intact vasculature lesions can heal throughout their volume simultaneously with amazing rapidity. Our experiments show almost complete lesion resolution in two weeks time. It would be expected that tumors with significant vascularity such as renal, breast, prostate, and hematomas might have very rapid resolution, while a vascular tumors such as colon metastases might resolve in longer periods of time. Rapid lesion resolution beyond making treatment efficacy easier to assess has significant clinical implications in situations where rapid decrease in the volume of the ablated zone is desirable such as the prostate, benign fibroadenomas of the breast, and myomas of the uterus. Rapid resolution of lesions without protein denaturation results in another unexpected pathologic finding related to an immunologic reaction in the lymph nodes draining the area of the ablation. All animals showed an obvious peripheral lymphadenpathy in the drainage area of the ablated tissue with microscopic evidence for associated inflammatory response. If an IRE treated tumor specific reaction in draining lymph nodes can be harnessed it could have implications for tumors that are known to spread via draining lymph nodes such as breast and prostate cancer. Such a reaction in lymph nodes draining an area of cancer could result in destruction of micrometastasis in the effected lymph nodes, which could significantly effect survival in melanomas, colon cancer metastatic to liver, breast cancer, and prostate cancer, to name a few.

The finding that IRE produces a lesion within liver with uniform necrosis throughout and to the very margins of the lesion and with a very narrow zone of transition from complete necrosis to normal liver is consistent with the selective mechanism of ablation of IRE and should provide accuracy in clinical tissue ablation. IRE lesions show complete destruction of tissue extending directly up to the vessel wall without sparing of tissue adjacent to the vessel. In addition, this was accomplished without vessel destruction or occlusion. This has major implications, in the liver, where local recurrences near vessels have been an ongoing clinical limitation of thermal ablation and to other tumor situations such as in centrally located kidney lesions where major vascular structures have also been problematic in producing a reliable ablation.

In addition to the histological findings the results of the long term study have confirmed that the extent of tissue damage by IRE is consistent with mathematical predictions and therefore, treatment planning is valuable for the design of IRE treatments. Additional findings are that there is value in increasing the number of pulses to increase the size of the treated tissue, up to 100 pulses delivered in such a way as to not produce thermal effects were reported in that study.

NTIRE of the Prostate

A potentially important application of NTIRE is in the treatment of prostate cancer and benign prostate hypertrophy (BPH). An experimental study on the use of NTIRE in the prostate was reported in [16]. The study reported in [16] is limited in the number of animals tested and the parameters used and much more work remains to be done in developing optimal NTIRE parameters for the prostate.

Six male beagle dogs had their prostates treated using IRE. Each procedure started with anesthetization of the animal under general anesthesia per SOP #33156. In addition, as with all large animal experiments, pancuronium (0.1 mg/kg,) was administered through an IV prior to the procedure, to reduce muscle contractions during the application of the electrical pulses. Pancuronium (0.05 mg/ml at 1 mg/ml) was administered throughout the procedure as needed. Using trans-rectal ultrasound 18 gauge electrodes were placed transperennially into the prostate in 3 dogs. Two to 4 probes were placed into one lobe of the prostate in 5 dogs. In 2 dogs, 4 probes were placed in order to create a hemi-ablation of the prostate. The probes were used in pairs of two, and each pair applied 80 pulses of 1500, with a pulse length of 100 microseconds. In the sixth dog pairs of probes were placed to purposely create a lesion in certain sensitive structures such as the urethra, rectum and neurovascular bundles. Pulses applied in the safety study at each anatomical structure were 80 pulses of 2000 V with a pulse length of 100 microseconds. Pulses were applied using a DC generator (IGEA Inc.) that delivered pulses in the microsecond range of duration, with a variable pulse interval and voltage range. When IRE was delivered in a bipolar manner between two probes they were separated by between 1 and 1.5 cm. In three subjects a single probe with both electrodes incorporated into it and separated by 5mm was placed trans-rectally using ultrasound guidance. Treatment parameters of the single probe were 1 or 8 pulses of 1000 V with a pulse length of 100 microseconds. Pulses delivered per lesion varied in number. Three animals sacrificed at one day had 8 pulse delivered at 1 kV and two animals sacrificed at 2 weeks had 80 pulses delivered per pair of electrodes using 1.5 kV. The duration of the pulses was 100 microsecond with an interval between pulses of 100-200 milliseconds. Upon application of the IRE pulse a variable degree of generalized muscle contraction occurred in each animal, from no contraction to mild to moderate contraction. The degree of contraction appeared to be related to the level of anesthesia of the animal, the degree of muscle blocking agents given, if any, and the voltage of the pulse used, with the higher voltages causing greater contractions.

At day one the lesion made at 1 kV with 8 pulses appeared grossly to be hemorrhagic in the treated area (Figure 6A). Histology showed the ablated area to have diffuse necrotic glandular tissue with no obvious viable tissue within the ablated zone (Figure 6C). The ablated zone was well demarcated from the immediately adjacent unaffected prostate parenchyma with an abrupt transition between necrotic glandular tissue in the ablation area and adjacent normal glandular tissue

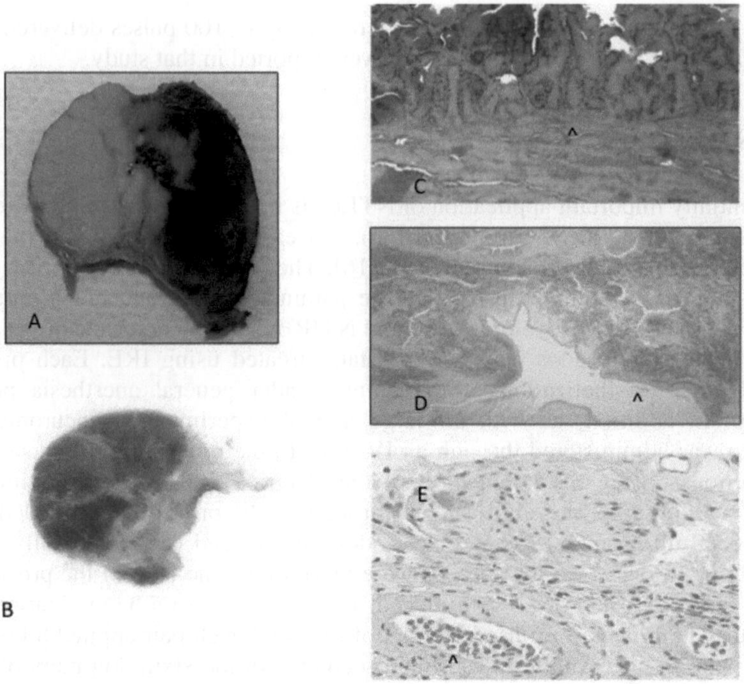

Fig. 6. 6A -Gross pathology of the IRE lesion at 24 hrs. The right side of the gland is hemorrhagic. (Pulses=8, 1.5 kV), 6B- Whole mount slide of a prostate where the right side of the gland was electroporated 2 weeks prior. There is marked shrinkage of the lobe with replacement by fibrous tissue (Pulses=80, 1.5 kV), 6C- Photomicrograph of prostate tissue that has been electroporated at 24 hrs. Photomicrograph at the margin of the IRE lesion. A very narrow zone of transition between normal and necrotic tissue is noted at the margin No glandular elements are visible (Pulses=8, 1.5 kV), 6D- The urethra is noted at the center of the micrograph as the open space at 24 hrs. Sub-mucosal hemorrhage is noted but the integrity of the urethra is still intact Pulses =8, 1.5 kV. 6E- Photomicrograph of the neurovascular bundle after electroporation at 2 weeks (Pulses=80, 1.5 kV). It can be seen that both the vessel and the nerve trunk show no evidence for necrosis. [16]

(Figure 6C). Necrotic glandular tissue was noted adjacent to the urethra however the urethral structural integrity remained intact without evidence for necrosis within the sub-mucosa (Figure 6D), even when the urethra was subjected to direct ablation during the safety portion of the study. In the neurovascular bundle areas, including the neurovascular bundle that was directly ablated during the safety portion of the study, vessels had variable loss of endothelium and were surrounded by scattered red blood cells (hemorrhage), neurophils, and edematous stroma. In some areas the vessel walls were hyper eosinophilic and homogenous consistent with fibrinoid necrosis. The lumen of affected vessels remained patent however without evidence for thrombosis (Figure 6E). In addition, no heat sink effect was evident adjacent to vessels with complete necrosis adjacent and often surrounding patent vasculature.

Nerves within the neurovascular bundles appeared to be intact and unaffected (figure 6E). Even ganglion cells showed no evidence of cells death.

The regional lymph nodes were enlarged in the drainage area. There was mild cortical reactivity and expansion of medullary sinuses consistent with a drainage reaction. Medullary sinuses are expanded by red blood cells, eosinophils, neutrophils, and macrophages. Para-cortical regions are expanded by increased lymphocytes. Occasional lymphoid follicles have reactive germinal centers.

At 2 weeks after the procedure the hemorrhagic changes had mostly resolved within the ablated tissue. The volume of tissue had already been markedly reduced compared to the untreated tissue. The contracted region of the previous lesion consisted primarily of collagenous tissue consistent with early scar formation (Figure 6B). Once again, the urethral wall adjacent to the lesion showed no evidence of necrosis and the urethral mucosa was intact. The nerves and vessels within the NVB on the side of the lesion were intact and patent. All layers of the rectum adjacent to the lesion appeared to be viable with no evidence for fistula formation.

The results of this study point to some of the advantages of NTIRE. First, the procedure obviously can produce prostate tissue ablation. The ability of NTIRE to ablate cells without destroying the extracellular matrix has major advantages in the case of the urethra, rectum and blood vessels. The sparing of the nerve may be related to the insulating nature of the myelin layer surrounding some nerves. Obviously, if this is the mechanism through which the nerves are spared any mechanical damage done to this layer or if pulses that break the insulating properties of the layer are delivered through a capacitance effect, the nerve cell may not be spared as evident from the early studies of [17].

NTIRE of the Heart and the Cardiovascular System

The ability of NTIRE to ablate cells without affecting the extracellular matrix may find important applications in ablation of cells for the cardiovascular system. One study in this area was performed by Dr. Liron Miller in the laboratory of Professor Jonathan Leor at the Sheba Hospital in Israel and was not reported in a reviewed publication.

Male Sprague-Dawley rats (150-200g) were obtained from (Harlan Lab. Jerusalem Israel) through the Office of Laboratory Animal Care at the Sheba Hospital, Israel. They received humane care from a properly trained professional in compliance with both the *Principals of Laboratory Animal Care* and the *Guide for the Care and Use of Laboratory Animals*, prepared and formulated by the Institute of Laboratory Animal Resources and published by the NIH. Four animals were treated. Each procedure started with anesthetization of the animal via intraperitoneal injection of 40 mg/kg ketamine and 10 mg/kg xylazine. Intubation was performed and the rats were ventilated with room air using a rodent ventilator. The chest was shaved under sterile conditions and surgically opened by left thoracotomy through the fourth intercostals space to expose the beating heart. Two electrodes were inserted into the heart at a depth of 4 mm. The 25 gage parallel electrodes were at a distance of 5 mm. The electrical fields were calculated as

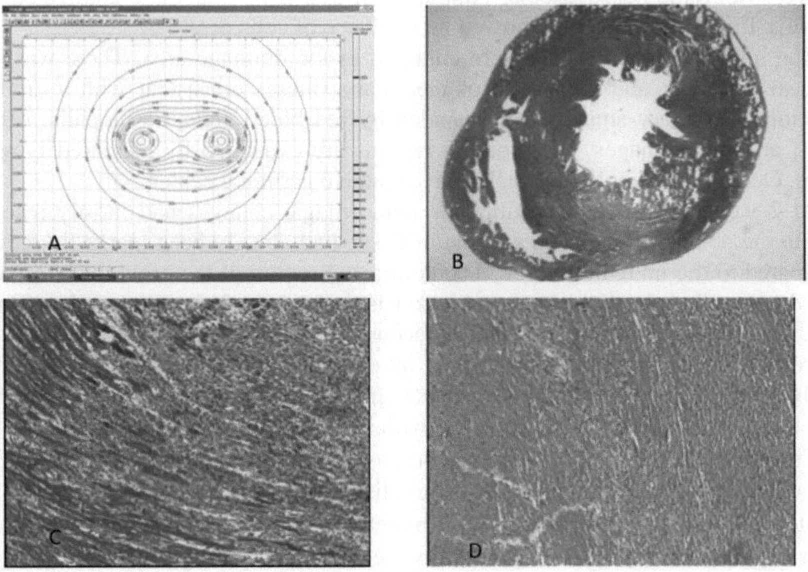

Fig. 7. 7A) Electrical fields that develop in the heart with two electrodes of gage 25, separated by 5 mm and a pulse of 750 V for 400 ms. The pulses were applied in the direction of the muscle fiber. 7B) Typical macroscopic cross sections through the heart in a plane parallel to that of electrode needles. He&E staining shows the necrotic region with a lighter color. 7C,7D Microscopic images of treated tissue in the heart. Figure 7c shows the margins of the treated areas. The muscle fibers that are alive ar shown as dark (red). The treated area has lost after 48 hours any living muscle fiber. It is evident from both that the entire treated area has been ablated and that the margins are very sharp, several cells thick.

shown in Fig. 7A and determined to produce non-thermal IRE. Two 400 microsecond pulses of 750 V, separated by 80 milliseconds were applied to the heart. The needles were placed perpendicular to the direction of the muscle fibers. The incision was sutured close and the animals recovered from anesthesia and were returned to their cages. The animals were maintained under observation for two days (48 hours) and than euthanized with an overdose of ketamine and xylazine. The hearts were removed, sectioned in the treated are and analyzed after embedding in paraffin and with imunohistochemical staining. Hematoxylin-eosin staining was performed on cross-sections through the center of the treated region to elucidate the morphology.

All the animals survived the procedure. The results are illustrated by Figs7 B,7C,7D. In Fig 7B, the cross section is through the heart in a plane parallel to the direction of the needles. The hematoxylin and eosin staining shows areas of heart tissue that are lighter. These areas correspond to the plane in which the needles were placed. Microscopic analysis, Figs 7C,7D shows that the lighter areas correspond to ablated tissue. Histological analysis revealed acute damage to the tissue in treated regions. Figure 7C shows the margins of the treated areas. The muscle fibers that are alive are shown as dark (red). The treated area has lost after 48 hours any living muscle fiber. It is evident from both Figures 7C, 7D that the entire treated area has been ablated and that the margins are very sharp, several cells thick. The fact that the animals survived the procedure and where alive 48 hours later while in the treated area the cells became completely ablated is indicative of the unique attributes of NTIRE, which retains even crucial structures such as the extracellular structure of the heart while ablating cells.

Another important application of NTIRE in the heart is for tissue ablation in treatment of arrhythmias. A study on this application was reported in [17]. Five pigs underwent beating heart surgical epicardial ablations of their right and/or left atrial appendages, utilizing a sequence of 8, 16, or 32 direct current pulses of 1500 to 2000 V, 100 μs each, at a frequency of 5 per second, applied between two 4-cm long parallel electrodes (Figure 8A) with an IRE pulse generator. In 3 of the pigs, 2 ablations were performed on the left atrial appendage, 1 cm apart from each other, for a total of 10 atrial ablations. Temperature was measured at the left and right atrial appendages tips before, during, and after the pulse sequence applications with an EasyWay 15 Thermometer Datalogger (Extech Instruments, Waltham, MA, USA) and a type-K thermocouple. In 2 pigs, an electrical isolation study was performed following IRE ablation by pacing the tips of the left and right atrial appendages distal to the ablation lesions, at a stimulus strength of 20 mA, and capturing the heart response. All ablation pulse sequences were completed within 1 to 4 seconds, depending on the number of pulses within each sequence, and caused no permanent arrhythmia or any untoward rhythm disturbance apart from rapid atrial pacing during the pulse sequence application (Figure 8B). Local heat measurements at the left and right atrial appendages tips showed no temperature changes during the pulse sequence applications in any of the pigs. In the 2 pigs in which electrical isolation study was performed, complete electrical isolation was confirmed by pacing the tips of the left and right atrial appendages, distal to the ablation lesions, at a stimulus strength of 20 mA, and failure of the heart to capture. The sternotomy incision was then closed, and the pigs were extubated and followed for 24 hours, at which point they were euthanized. Animal hearts were excised 24 hours after surgery and stained with Masson trichrome staining to elucidate the myocardial damage. Degrees of myocardial damage (nuclear staining, homogeneity of myocytes striation), hemorrhage, or amount of inflammatory cells were studied. Complete myocardial necrosis was characterized by rounded, homogenous myocytes with loss of nuclear staining. Histological evaluation of lesion transmurality was based on maximum depth of myocardial necrosis. Complete transmural lesion was defined as myocardial necrosis throughout the atrial wall. On gross inspection of all 10 atrial lesions 24 hours after ablation, a clear demarcation line between ablated and normal tissue was observed, with no tissue disruption or charring (Figure 8C, 8D). Masson trichrome staining results of all 10 atrial ablations showed continuous transmural destruction of the atrial tissue—from epicard to endocard—at the site of the electrode application (Figure 8D). The 10 atrial lesions had a mean depth of 0.9 cm (range, 0.4-1.4 cm), mean width of 0.66 cm (range, 0.5-1.0 cm), and measured 3 to 3.5 cm in length. Higher

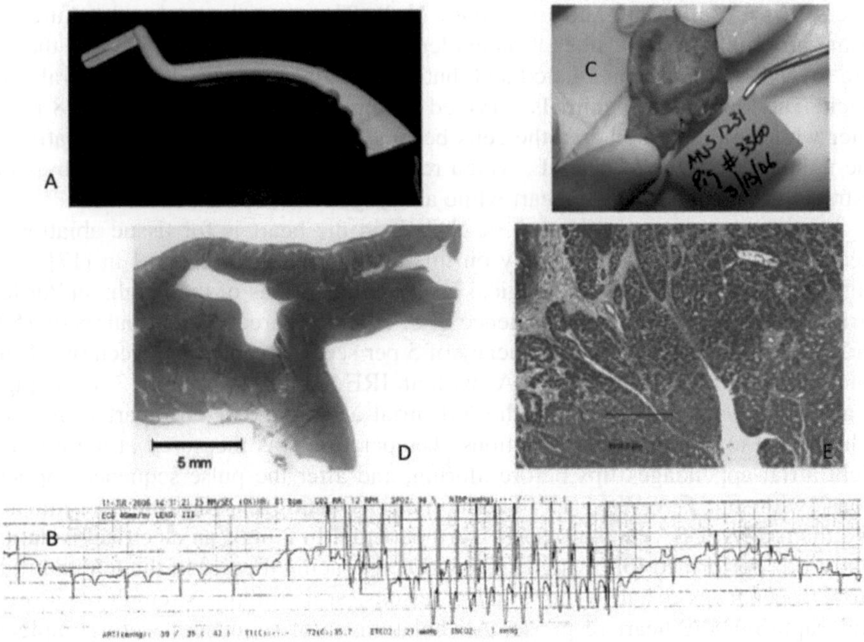

Fig. 8. 8A) A specially designed hand-held clamp containing 2 parallel electrodes embedded in its end jaws, which can be rotated at their handle articulation up to 270° to allow shaping of the device as the surgical situation requires and provide a clear visual field of the ablation site. 8B) Electrocardiographic tracing showing rapid atrial pacing during a 16-pulse irreversible electroporation ablation sequence, with a frequency of 5 pulses per second, and immediate resumption of sinus rhythm following ablation. 8C) Gross specimen of the left atrial appendage spread open, showing 2 clear linear ablation lesions on both sides of the appendage. 8D). Stereomicroscopy pictures of Masson trichrome stainings (original magnification ×2). Left atrial appendage lesion showing trans-mural lesion (in purple) from epicard to endocard. 8E) Higher magnifications of the ablated areas showing a sharp demarcation line between the injured necrotic myocardial tissue and the surrounding normal myocardium.

magnifications of the ablated areas have repeatedly demonstrated a sharp demarcation line between the injured necrotic myocardial tissue and the surrounding normal myocardium (Figure 8E). The study was the first to demonstrate the ability of IRE to serve as a surgical modality to create epicardial atrial ablation. Each of the 3- to 3.5-cm long transmural ablation lines were created in 1 to 4 seconds and caused no peripheral thermal damage. These features demonstrate the potential superiority of the IRE nonthermal modality over the currently used hyperthermic or hypothermic energy sources for future clinical application in the surgical treatment of atrial fibrillation.

An important aspect of NTIRE is that while ablating cells it keeps the extracellular matrix intact. In the case of blood vessels it means that the extracellular matrix remain mechanically intact with an intact molecular structure and mechanical properties. This has suggested to us that NTIRE is an ideal modality for treating blood vessels for various applications, such as restenosis. A few studies on NTIRE for blood vessels were reported in [19], [20], [21]. A more comprehensive summary of the research can be found in the PhD thesis of Elad Maor [22]. All the animal studies were done performing first a mathematical treatment protocol to verify that the procedure is indeed non-thermal. The rat studies were performed as described in [19], [20], [21]. Animals were anaesthetized with an intramuscular injection of Ketamin and Xylazine (90mg/Kg and 10 mg/Kg, respectively). The left common carotid artery of each animal was exposed and a custom made electrode clamp with two parallel disk electrodes (diameter = 5 mm) was applied on the left common carotid artery (Figs 9a, 9b). A variety of studies on NTIRE are described in [19], [20], [21], [22]. One typical study evaluated the effectiveness of various NTIRE protocols. In that study, eight IRE protocols with different ways of delivering sequences of 100 μs in length were compared: 1-4) 10 pulses at a frequency of 10 Hz with electric fields of 3500, 1750, 875 and 437.5 V/cm and 5-8) 45 and 90 pulses at a frequency of 1 Hz with electric fields of 1750 and 875 V/cm. All groups had their left common carotid artery treated with NTIRE and their right common carotid artery used as a control. NTIRE was performed by applying short electric pulses between the electrodes with a high voltage pulse generator intended for electroporation procedures (ECM 830, Harvard Apparatus, Holliston, MA). The procedure was repeated in three successive locations along the common carotid artery, thus treating approximately 1.5 cm of the left common carotid artery. At the end of the procedure the skin incision was suture closed and animals were kept alive for various periods of time. For histological examination, animals were euthanized with an overdose of Phenobarbital followed by bilateral chest dissection. Gross inspection of carotid arteries was used to identify arterial wall integrity or intraluminal massive thrombus formation. The arterial tree was perfused with 10% buffered formalin, and the left and right carotid arteries were harvested near the bifurcation of the internal and external carotid arteries. The treated area was cut to two or three consecutive slices. One section from each slice was used for histological analysis. Each slice was fixed with 10% buffered formalin, embedded in paraffin, and sectioned with a microtome (5-μm-thick). Sections were stained with hematoxylin and eosin. Each section was photographed at ×200 magnification, and the following parameters were quantitatively evaluated: number of (vascular smooth muscle cells) VSMC nuclei in each of the three layers of the Tunica Media, total area of the Tunica Media, and the average thickness of the Tunica Media based on 5 different measurements in each section. Most efficient protocols were 10 pulses of 3500 V/cm at a frequency of 10 Hz and 90 pulses of 1750 V/cm at a frequency of 1 Hz, with ablation efficiency of 89±16 % and 94±9 % respectively. Extra-cellular structures were not damaged and the endothelial layer recovered completely. Typical histological results at one week after the application of successful NTIRE protocols that are shown in Fig 9d. In comparison Fig 9c shows a normal artery. The main aspect is that the VSCM cell were completely ablated and after one week, while there were no VCM cells, the endothelial cells started to regrow, in a normal fashion. This indicates the value of NTIRE and of the non-thermal aspect of the procedure. The extracellular matrix is retained and it

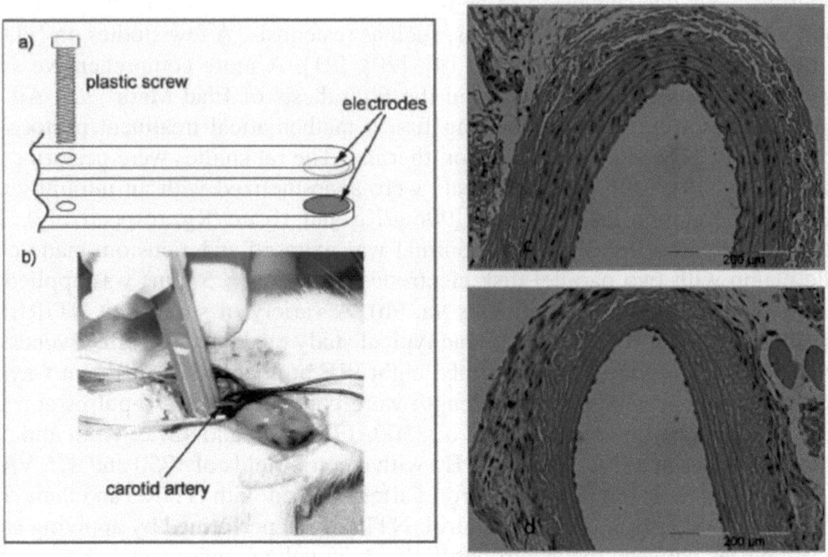

Fig. 9. 9a) a schematic of the electrodes for extra vascular treatment of the blood vessel. 9b) a typical procedure in which the blood vessels was clamped between the electrodes. 9c) a cross section through a control artery, 9d) a cross section of a treated artery a week after the treatment. Note the repopulation of the endothelial layer compared with the complete absence of the vascular smooth muscle cells. This is possible because of a unique aspect of NTIRE which ablates cells while retaining the extracellular matrix intact.

Successful NITRE ablation of VSMC induced a reduction in media thickness: 25±17% reduction in Group 1 (45±10 vs. 59±8 μm) and 27±7% reduction in Group 4 (37±4 vs. 51±6 μm) .No Change in media thickness was induced in the two non-successful NTIRE groups (61±9 vs. 58±7 μm in Group 5, 61±9 vs. 60±6 μm in Group 8). Endothelial cells of treated arteries were similar in number and morphology to those of non treated control arteries. Elastic fibers and wall integrity were preserved and similar to control arteries. There were no signs of local inflammation in treated arteries. There were no cases of bleeding, thrombosis and animal mortality.

In addition to the experiments described earlier we have performed a similar set of experiments with intravascular NTIRE electrodes in the rabbit (Figs 10). We tested several types of electrodes such as those shown in Figs 10 A and 10 B. Figure 10 C shows an angiograme of the left and right carotide artery four weeks after an angiography. The left was not treated with NTIRE after angiography while the right was. It is evident that the left artery has become completely occluded by the wild growth of smooth muscle cells while the right one is intact. This shows that NTIRE not only prevents neointimal formation but also negative vascular remodeling, without the need for a stent!

The evidence that NTIRE can ablate tissue around blood vessels suggests a wide range of applications related to the ablation of tissue around blood vessels without affecting the extracellular matrix. For instance, tumors around blood vessels in the kidney, brain or pancreas could be treated in the same way, as well as direct treatment of soft plaque.

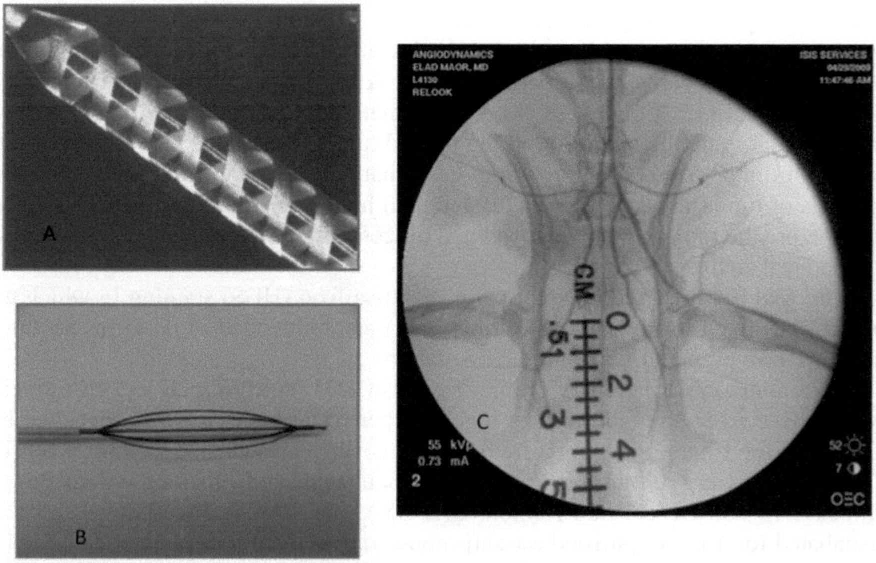

Fig. 10. 10A), 10B), typical intravascular NTIRE electroporation electrodes, 10 C) and angiogram of the rat carotid arteries one month after they were damaged by angiography. The right was treated with NTIRE and the left not. Note that the left has become closed and the right is intact.

NTIRE in the Treatment of Tumors

The first successful use of NTIRE irreversible electroporation for the minimally invasive treatment of aggressive cutaneous tumors implanted in mice was performed in the laboratory and under the direction of Lluis Mir and reported in [23]. Mathematical models of the electrical and thermal fields that develop during the application of the pulses were used to design an efficient treatment protocol with minimal heating of the tissue. Tumor regression was confirmed by histological studies, which also revealed that it occurred as a direct result of irreversible cell membrane permeabilization. Parametric studies show that the successful outcome of the procedure is related to the applied electric field strength, the total pulse duration as well as the temporal mode of delivery of the pulses. The best results were obtained using plate electrodes to deliver across the tumor 80 pulses of 100 μs at 0.3 Hz with an electrical field magnitude of 2500 V/cm. These conditions induced complete regression in 12 out of 13 treated tumors, (92%), through irreversible electroporation and in the absence of tissue heating. The study was performed on C57Bl/6 female mice, 6–8 weeks old, inoculated subcutaneously in the left flank with $1x10^6$ cells from a LPB cell line, a methylcholanthrene-induced C57Bl/6 mouse sarcoma cell line, producing in 9 days tumors of 4 to 5mm in diameter. For electroporation, the mice were anaesthetized using a mixture of xylazine 12.5 mg.kg^{-1} (Bayer Pharma, Puteaux, France) and ketamine 125 mg.kg^{-1} (Parke Davis, Courbevoie, France), an incision was performed on the skin near the tumor and the skin flap containing the tumor was lifted and stainless-steel plate electrodes were placed in direct contact with both sides of the cutaneous tumor, with the tumor sandwiched between the parallel plates. After delivery of the electroporation pulses, the skin incisions were closed with metallic clips, the mice were returned to their cages and the evolution of the treated tumors was followed with measurements of tumor size every second day. Alternatively, mice were kept for different periods of time (between 1 and 72h) and then humanely sacrificed by CO_2 inhalation before the tumors were removed and processed for histological or immunohistochemical analysis.

The analysis involved haematoxylin–eosin-saffron (HES) staining in which tumors were fixed in Finefix (Milestone, Italy) and embedded in paraffin. Sections of 4μm were prepared for routine HES staining.

For immunohistochemistry of microvessels CD31 paraffin sections (4μm-thick) were dewaxed and rehydrated. Endogenous peroxidase activity was quenched by 3% H_2O_2 for 10min. Sections were placed in cover-plates (Shandon, Life Sciences Technology, Cergy-Pontoise, France) and incubated with blocking serum Power Block 1:10 (BioGenex, San Ramon, CA, USA) for 10min. The slides were then incubated for 1h with purified rat-anti-mouse monoclonal anti-platelet endothelial cell adhesion molecule (PECAM-1 also called CD31), dilution 1:300, (PharMingen, Heidelberg, Germany) followed by rabbit anti-rat immunoglobulins (Dako Denmark) dilution 1:200, for 30min, followed by PowerVision poly-HRP anti-Rabbit IgG (Immuno Vision Technologies, Brisbane, CA) for 20min. Finally, slides were exposed to diaminobenzidine chromogenic substrate (DAB PowerVision Histostaining Kit; ImmunoVision Technologies) for 10min, washed with distilled water, counterstained with Mayer's hematoxylin, and mounted in permanent medium (Pertex). All slides were immunolabelled the same day to ensure standardized intensities of immunochemical signals and counterstaining.

Double Strand DNA breaks, which are often associated with cell apoptosis, were detected using the "In Situ Cell Death" Detection kit (Roche; Mannheim, Germany) (TUNEL*(Terminal deoxynucleotidyl transferase (TdT)-mediated dUTP Nick End-Labeling)* method) performed according to the manufacturer's instructions. De-paraffinized sections were incubated with Citrate buffer, pH 6 and placed in a water-bath, 98°C for 40min and all sections were treated with TUNEL reagents (TUNEL mixture: 1 hour at 37°C under a coverslip) except for one where the enzyme was omitted (negative control). After washings with Rince Buffer Biogenex, slides were incubated with the secondary anti-fluorescein-AP conjugate, and the signal was revealed with Fast Red substrate solution for 20min. Slides were lightly counterstained with hematoxylin prior to aqueous mounting by Aqua-Perm (Shandon Aqua-Perm™ Thermo Electron IVDD Compliant, Waltham, MA).

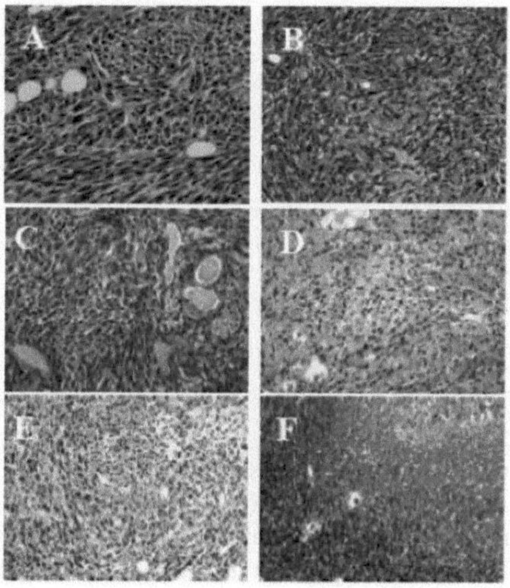

Fig. 11. Analysis of tumor evolution by HES histological staining after IRE. A: control; B, C, D, E and F: respectively 1, 2, 6, 24 and 48h after irreversible electroporation.

Figures 11 shows typical results from HES staining experiments. In the untreated control tumors, cells display a large nucleus surrounded by a well marked cytoplasm and a well defined cell membrane (Fig 11A). The slides stained with the classical HES staining revealed that 5 min after the EP, no change in tumour cell morphology was detectable. However, the congestion of blood in the tumour vessels was evident. This effect is similar to the irreversible electroporation consequences in the liver described in [3]. Few changes in tissue architecture and cell morphology were detected 1h and 2h later (Fig 11B and 11C). However, 6h after the EP, dramatic changes were observed (Fig 11D). Cytoplasmic limits between the cells were barely distinguishable in several parts of the tumour. The nuclei were still visible, but in a sort of syncitium in these parts of the tumour. At 24h, almost the entire tumor presented this image (Fig 11E). The nuclei, extremely pycnotic, appeared to be all in the same "cytoplasm" as no limit was detectable between the original cells. At 48h (Fig 11F) and 72h, the nuclei were even smaller and tissue necrosis still more evident.

Figures 12 show the results from both, Tunel staining, left panel and CD31 staining right pannel. TUNEL staining revealed that in the control slides, very few cells were stained positive. The staining was strictly localized at the level of the cell nucleus (Fig 12A). Background was very clear (Fig 4A). However, as early as 5min after the treatment, changes began, albeit slightly noticeable: a larger number of nuclei were positive, and in many of these nuclei, staining was not contained inside the nucleus, but was clearly spreading in the cytoplasm (Fig 12B). Background was still very clear (Fig 12B). The overall staining of the slices increased 1h after EP delivery. There was a clear increase in the number of red-dye leaking cells and there was

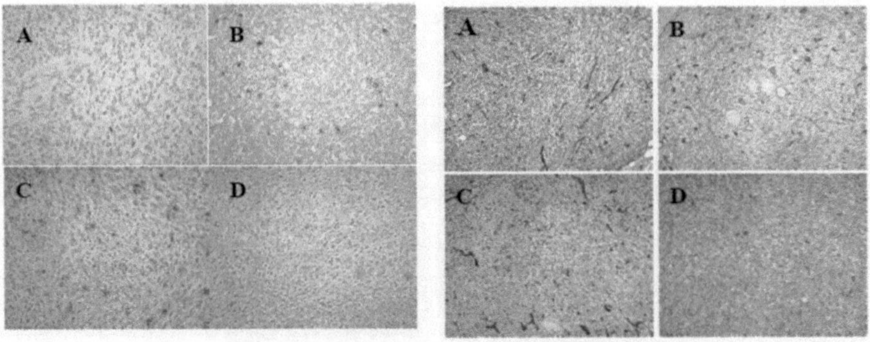

Fig. 12. Left hand side panel. TUNEL analysis at different times after the pulses delivery. A: control; B, C and D: respectively 5 min and 1 and 24h after IRE. Right hand side pannel: Immunohistochemical analysis of the tumour vasculature evolution by means of CD 31 staining in LPB tumors after treatment. A: control; B, C and D: respectively 2, 6 and 24h after IRE.

also the presence of a red background (Fig 12C). Thus, two types of staining can be clearly identified in these slices and at later times. At 6h, a large number of cells still display the diffuse staining (not shown). However, at longer times (24h, Fig 12D), there was only a heavily red-stained background, which completely diffused and continued to spread throughout the entire tumor section. At 24h, no more red-dye leaking cells were found, in agreement with the HES images that showed the complete disintegration of the cells.

Evolution of the tumor vasculature was also analyzed using antibodies against the CD31 endothelial cells specific marker, right panel. The changes in vasculature, which were observed 5 min after EP delivery were slight with respect to the controls (Fig. 12A), displayed highly branched and tortuous vasculature typical to this well vascularized fibrosarcoma tumor. However, 2h later, well established vascular congestion with dilated vessels was detectable, long vessels were not visible and a slight diffuse staining began to spread in some parts of the slices. At 6h, blood vessel walls, when still present, were either hyper- or hypo-pigmented. The diffusion of the CD31 marker in the extra-cellular interstitial spaces was much more intense and micro-vascular occlusions were also detectable. These signs indicate advanced tumour vasculature lesions. At 24h, very advanced vascular lesions were seen, with faint CD31 marker staining of the endothelial cells indicating damaged blood vessel walls, and intense diffusion of the staining in the whole tissue. What remained were blood vessel skeletons over a necrotic background.

The results achieved with the different sets of electrical parameters indicate that the main parameter affecting the results is the electric field strength. Trains of a large number of short pulses resulted in the best antitumor effects (up to 92% of tumor ablation).

This study has produced several observations concerning the mechanisms of cell ablation in the IRE method as well as evidence of the efficacy of IRE to completely destroy aggressive tumors. HES staining revealed that the IRE pulses induce vascular congestion, which should also cause tissue hypoxia and may further contribute to tumor cell death. Twenty-four hours after the application of the pulses, all treated tissue was necrotic. Interestingly, the treatment does not produce massive apoptosis. The number of nuclei stained by the TUNEL reaction slightly increases 1h after the treatment. The detection of diffused TUNEL staining first in the cytoplasm around the cell nucleus, and later around the cells, is consistent with the expected effects of IRE pulses. The observation of a diffused TUNEL staining of DNA corroborates that cells are no longer limited by their natural membrane barrier and the DNA diffuses outside the cell, which is indeed the hallmark of irreversible electroporation.

The analyses of the CD31 staining show rapid and severe lesions of the vasculature. At this stage it should be emphasized that the vasculature of these tumors is substantially different from the vasculature of normal major blood vessels. The blood vessels here have no inherent mechanical integrity and extracellular scaffolding. The results in [22] reveal that the disaggregation of the membranes starts a couple of hours after the pulse delivery and becomes very intense 6 h later, and is complete at 24 h. The localization of the membrane antigen CD 31 at the cell membrane is progressive and becomes complete. The antigen diffuses throughout the treated tumor volume, indicating the complete disruption of the membranes as well as cell necrosis. It is worthy to note that TUNEL analysis of the cell death and CD31 analysis of the tumor vasculature evolution revealed not only the necrotic pathway of the IRE-caused cell death, but also confirms the mechanisms of the method, that is the actual irreversible electroporation of the cell membrane and its consecutive disintegration.

Similar results to those discussed above were achieved with tumors transplanted in nude mice to assess whether the host immune system has a low or moderate participation in the antitumor effects observed. A supplementary study confirmed that the immune response is not instrumental for IRE ablation, which broadens the potential application space for IRE treatment to immunodepressed patients [24].

Summary of Experimental Studies

Several general conclusions emerge from the experimental studies performed so far. Non-thermal irreversible electroporation appears to be an effective tissue ablation modality. However, the procedure requires careful design of the many electroporation parameters that affect the outcome of the procedure. These involve a choice of number of pulses, pulse length, pulse amplitude and interval between pulses that will produce non-thermal irreversible electroporation. Since the number of parameters that are involved in the procedure is large, much research remains to be done to optimize the NTIRE protocol. A unique aspect of NTIRE is its ability to ablate cells while the extracellular matrix remains intact. We anticipate that on the strength of this property NTIRE will find applications in areas in which no other minimally invasive surgical procedure was successful so far. Contractions during the electroporation pulses affect electroporation in general and irreversible electroporation in particular. This area needs to be further studied and at this stage considered whenever a clinical procedure is tested. The large animal experiments show evidence of a systemic immune response. It may be valuable to consider this response and ways to take advantage of it in treatment of cancer.

References

[1] Davalos, R., Mir, L., Rubinsky, B.: Tissue ablation with irreversible electroporation. Annals Biomed. Eng. 33(2), 223–231 (2005)

[2] Granot, Y., Rubinsky, B.: Mass transfer model for drug delivery in tissue cells with reversible electroporation. Int. J. of Heat and Mass Transfer 51, 5610–5616 (2008)

[3] Edd, J.F., Horowitz, L., Davalos, R.F., Mir, L.M., Rubinsky, B.: In-vivo results of a new focal tissue ablation technique: irreversible electroporation. IEEE Trans. on Biomedical Engineering 53(4), 1409–1415 (2006)

[4] Rubinsky, B., Onik, G., Mikus, P.: Irreversible Electroporation: A New Ablation Modality - Clinical Implications. Technology in Cancer Research and Treatment 6(1), 37–48 (2007)

[5] Gilbert, J.C., Onik, G.H., Haddick, W.K., Rubinsky, B.: The Use of Ultrasonic Imaging for Monitoring Cryosurgery. IEEE Trans. of Biomed. Eng. BME-31(8), 563 (1984)

[6] Gilbert, J.C., Onik, G.M., Haddick, W., Rubinsky, B.: Real Time Ultrasonic Monitoring of Hepatic Cryosurgery. Cryobiology 22, 319–330 (1985)

[7] Ivorra, A., Rubinsky, B.: In vivo electrical impedance measurements during and after electroporation of rat liver. Bioelectrochemistry 70(2), 287–295 (2006)

[8] Granot, Y., Ivorra, A., Maor, E., Rubinsky, B.: In vivo imaging of irreversible electroporation by means of electrical impedance tomography. Phys. Med. Biol. 21, 54(16), 4927–4943 (2009)

[9] Davalos, R., Rubinsky, B.: Temperature considerations during irreversible electroporation. Int. J. of Heat and Mass Transfer 51, 5617–5622 (2008)

[10] Daniels, C., Rubinsky, B.: Electrical Field and Temperature Model of Nonthermal Irreversible Electroporation in Heterogeneous Tissues. J. of Biomech. Eng. - Trans. of ASME 131(7), Article Number: 071006 (2009)

[11] Miklavcic, D., Semrov, D., Mekid, H., Mir, L.M.: A validated model of in vivo electric field distribution in tissues for electrochemotherapy and for DNA electrotransfer for gene therapy. Biochimica et Biophysica Acta 1523, 73–83 (2000)

[12] Martin, J.B., Young, J.L., Benoit, J.N., Dean, D.A.: Gene Transfer to Intact Mesenteric Arteries by Electroporation. Journal of Vascular Research 37, 372–380 (2000)

[13] Ivanusa, T., Beravs, K., Cemazar, M., Jevtic, V., Demsar, F., Sersa, G.: MRI macromolecular contrast agents as indicators of changed tumor blood flow. Radiol. Oncol. 35, 139–147 (2001)

[14] Sersa, G., Krzic, M., Sentjurc, M., Ivanusa, T., Beravs, K., Kotnik, V., Coer, A., Swartz, H.M., Cemazar, M.: Reduced blood flow and oxygenation in SA-I tumours after electrochemotherapy with cisplatin. British Journal of Cancer 87, 1047–1054 (2002)

[15] Sersa, G., Cemazar, M., Miklavcic, D.: Tumor blood flow modifying effects of electrochemotherapy: a potential vascular targeted mechanism. Radiol. Oncol. 37, 43–48 (2003)

[16] Onik, G., Mikus, P., Rubinsky, B.: Irreversible electroporation: Implications for prostate ablation. Technology in Cancer Research and Treatment 6(4), 295–300 (2007)

[17] Stämpfli, R., Willi, M.: Membrane potential of a ranvier node measured after electrical destruction of its membrane. Experimentia 13, 297–298 (1957)

[18] Lavee, J., Onik, G., Mikus, P., Rubinsky, B.: Novel Non-Thermal Energy Source for Surgical Epicardial Atrial Ablation: Irreversible Electroporation. The Heart Surgery Forum 10(2), E162–E167 (2007)

[19] Maor, E., Ivorra, A., Leor, J., Rubinsky, B.: The effect of irreversible electroporation on blood vessels. Technology in Cancer Research and Treatment 6(4), 307–312 (2007)

[20] Elad, M., Antoni, I., Leor, J., Boris, R.: Irreversible Electroporation Attenuates Neointimal Formation after Angioplasty. IEEE Trans. Biomed. Eng. 55(9), 2268–2274 (2008)

[21] Maor, E., Ivorra, A., Rubinsky, B.: Non Thermal Irreversible Electroporation: Novel Technology for Vascular Smooth Muscle Cells Ablation. PLoS ONE 4(3), e4757 (2009), doi:10.1371/journal.pone.0004757

[22] Maor, E.: Fundamental study on the effects of irreversible electroporation pulses on blood vessels with application to medical treatment. PhD Thesis, University of California at Berkeley (2009)

[23] Al-Sakere, B., André, F., Bernat, C., Connault, E., Opolon, P., Davalos, R., Rubinsky, B., Mir, L.: Tumor Ablation with Irreversible Electroporation. PLoS ONE 2(11), e1135 (2007)

[24] Al-Sakere, B., Bernat, C., André, F., Connault, E., Opolon, P., Davalos, R.V., Mir, L.M.: A study of the immunological response to tumor ablation with irreversible electroporation. Technology in Cancer Res. Treat. 6, 301–306 (2007)

Finite Element Modeling of in Vivo Electroporation

Nataša Pavšelj and Damijan Miklavčič*

Faculty of Electrical Engineering, University of Ljubljana, Ljubljana, SI-1000, Slovenia
* Corresponding author E-mail: damijan.miklavcic@fe.uni-lj.si

Introduction

As a simple definition, a mathematical model is a representation of chosen essential aspects of a real system (may it be a living, engineering or social system), described by a set of variables and a set of equations that establish relationships between the variables. The aim of such a model is to gain knowledge about the represented system, explaining some of its phenomena, or help designing it. Mathematical modeling is a vast scientific and engineering field and is therefore not possible to explain in detail in one single book, let alone in one paper. It is used in countless scientific, engineering and even social studies and represents an important field in the study of the effects of the electromagnetic fields and accompanying coupled phenomena on cells, tissues and organs (Fear and Stuchly, 1998; Debruin and Krassowska, 1999; Debruin and Krassowska, 1999a).

As biological systems are often geometrically highly intricate, using analytical methods is, in most cases, not an option. Therefore, numerical methods are applied, the most widely used method being finite element modeling, a relatively simple but powerful tool for the analysis and the explanation of the processes taking place inside the biological systems.

Numerical models of cell and tissue electroporation proved useful for description of the underlying processes, as well as for evaluation of various influential parameters. Validated by experiments, they allow prediction of the outcome of pulse application and help in choosing the most efficient protocols and pulse parameters for specific applications. This approach also facilitates development and optimization of electrodes and their placement with respect to target tissue and can be of great help in treatment planning (Semrov and Miklavcic, 1998; Brandisky and Daskalov, 1999; Miklavcic *et al.*, 2000; Dev *et al.*, 2003; Miklavcic *et al.*, 2006a; Sel *et al.*, 2007; Corovic *et al.*, 2008; Zupanic *et al.*, 2008). A good model verified with experimental results is a powerful tool and offers useful insight into the understanding of processes modeled. However, we have to be aware of the fact that the processes involved are much more complex and the model is merely a simplified representation that can not replace experimental work. Nevertheless, it serves as a source of extra information and helps us to plan future experiments. Experimenting with such models is easier and sometimes the only possible or ethically acceptable alternative to experimenting on real biological systems.

In continuation, we explain basic aspects of finite element numerical modeling important in constructing numerical models of tissue electroporation, describe

B. Rubinsky (Ed.): Irreversible Electroporation, BIOMED, pp. 183–202.
springerlink.com

some characteristics of biological tissues and give some reference to the validation of results at the end.

Which Modeling Method to Use?

Starting point of the modeling process is deciding on the mathematical approach to adequately describe the modeled system, by first acquiring enough observable and measurable information about it. Typically, more than one modeling approach is possible and choosing the most suitable one depends on the modeler's or end user's objective needs and personal preferences, as well as physical and geometrical characteristics of the modeled system. In some cases, however, using more than one modeling approach can be beneficial in terms of model verification and validation. More than one phenomenon can determine a system, which is especially the case with biological systems. Generalizations and simplifications are possible and, in most cases, cannot be avoided. The model can refer only to some aspects of the real system, while disregarding the ones that either have a very limited influence on its accuracy or are out of scope of this particular model. Prior to building a model, we need to define its scope, apply necessary simplifications while being aware of the circumstances or the range of input variables the model is valid for.

Deciding on the right level of complexity for our model is not always an easy task as it involves a trade-off between simplicity and accuracy of the model. As a general guideline, if we are choosing from different models, giving comparable results, the simplest one is the most desirable. Namely, we need to be aware that adding complexity can make the model difficult to understand and experiment with and can pose computational problems.

Methods for solving partial differential equations: Analytical methods are rather complicated and are only feasible for use on problems where the geometry, material properties and boundary conditions can easily be described in a defined coordinate system (Cartesian, cylindrical or spherical). Simple analytical models can have certain advantages over numerical models. First, the input data needed is typically less extensive than that of numerical models. Also, analytical solutions have no numerical and discretization errors. The obvious limitation of analytical models is that only simple and uniform geometries, boundary and initial conditions can easily be modeled. In the last decades, however, analytical models have mostly been replaced by numerical models based on boundary element, finite difference, finite volume or finite element methods, due to the miniaturization and accessibility of both computer hardware and software. Of these methods, the latter is the preferred one to be used in modeling of biological systems, due to its easy implementation and its ability to handle more intricate geometries. The principle behind this method is the discretization of the geometry into smaller elements where the quantity to be determined is approximated with a function or is assumed to be constant throughout the element. Discrete elements can be of different shapes and sizes, which allows modeling of intricate geometries. In such models, the excitations can be changed easily, being that it only involves changing the boundary conditions on the same model. The model geometry, however, takes time and precision to be built and generalizations and simplifications need to be used when possible.

Finite Element Method

The finite elements method (FEM) turned out to be a very useful method for solving partial differential equations when studying electric field distributions inside biological systems. The essence of the method is the discretization of the geometry into smaller elements – finite elements – where the quantity of interest is approximated with a simple function or is assumed to be constant throughout the element (Reddy, 2004). Material properties inside each finite element are homogeneous. Mathematically, the finite element method is used for finding an approximate solution of partial differential equations (PDE) as well as of integral equations such as the heat transport equation. The solution approach is based either on eliminating the differential equation completely (steady state problems), or rendering the PDE into an equivalent ordinary differential equation, which is then solved using standard techniques such as finite differences, etc.

In solving partial differential equations, the primary challenge is to create an equation which approximates the phenomenon to be studied, but which is numerically stable, meaning that errors in the input data and intermediate computations do not accumulate and cause the resulting output to be meaningless.

In order to design a numerical model of a physical structure, the modeler must decide the appropriate resolution for modeling each component part, a task requiring considerable expertise and experience. After the right geometrical representation has been found, one of the next steps is the discretization of the geometry into finite elements – meshing. Too fine a mesh will cause unnecessary computational overheads when running the model, whereas too coarse a mesh will produce intolerable approximation errors. A simple rule is to choose a better resolution in the regions where we expect large gradients of the computed quantities or in the regions of our interest. Thin layers and small structures inevitably lead to a greater number of finite elements, as the mesh inside and around them needs to be denser. Further, borders between regions with very different material properties represent an additional problem, which again calls for higher mesh resolution. On the other hand, we are limited by computer capabilities and reducing the error by increasing the mesh resolution requires more computer time and the model can even end up being too complex to be solved. A simple way to optimize mesh density is changing its resolution and comparing the results given by models of different mesh densities. The optimal mesh is chosen as the least dense of those that yield similar results. Furthermore, using appropriate symmetries of the geometry can simplify our model a great deal.

Geometrical Considerations

When designing a numerical model, we must decide for the appropriate details to be included. Geometrically more detailed models will inevitably consume more of both, modeler's and computer time, but do not necessarily produce better quality of the results, as the inclusion of geometrical details depends on the purpose of the model. Further, taking advantage of geometrical symmetries of the system we are modeling can allow us to analyze a structure or a system by modeling only a portion of it while applying appropriate boundary conditions (see Figure 1a). This

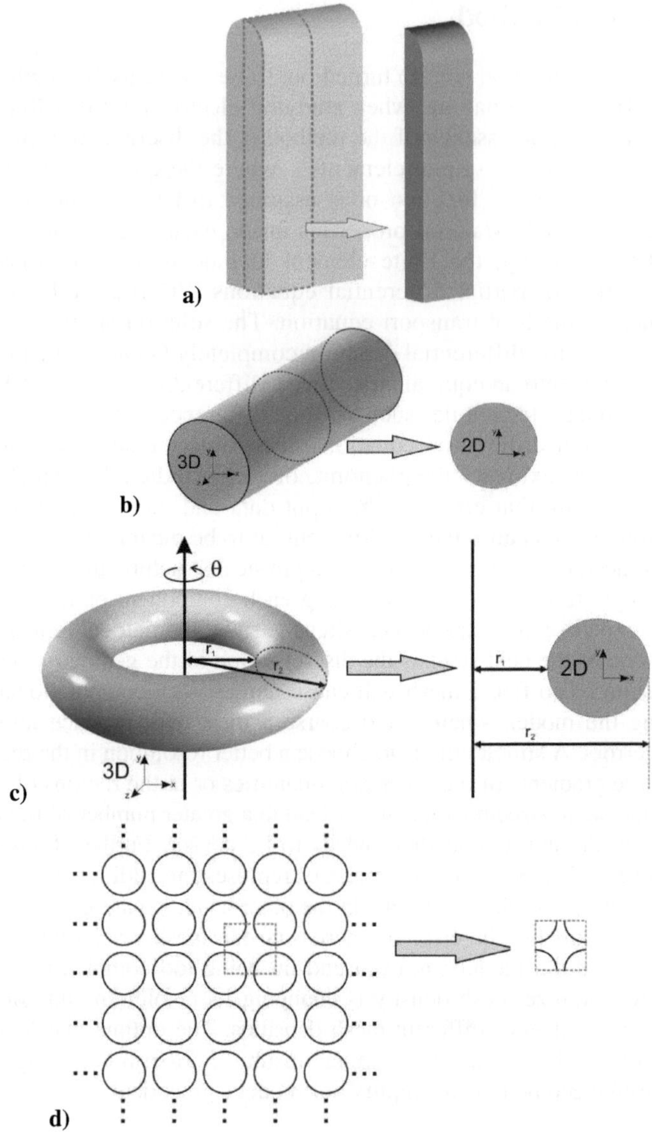

Fig. 1. Examples of using model symmetries to reduce the number of finite elements by reducing geometrical complexity. Taking advantage of such symmetries is only possible when the sources (in our case of electric current) exert the same symmetry. Boundary conditions on the section and symmetry planes and axes have to be set accordingly. a) Using only a quarter of the whole geometry to represent the whole structure. b) Two-dimensional representation of a three-dimensional geometry. In this case, the cylinder was represented by its cross-section. c) Two-dimensional representation of a three-dimensional axisymmetrical structure. d) A finite representation of an infinite two-dimensional lattice. The same approach can be used when modeling infinite structures in 3D.

approach can be used when both, the geometry as well as the sources (in our case of electric current) exert the same symmetry. It reduces the size of the model (the total number of nodes and elements), and consequently the analysis run time as well as the demands on computer resources. Alternatively, modeling only a portion of the whole geometry allows us to include more details in the model, when needed, thereby obtaining better results that would not have been possible with the full geometry. Similarly, sometimes we are able to represent a 3-dimensional structure by a single 2-dimensional plane. For example, in Cartesian coordinates, a structure stretched along a straight line (z-direction) can be represented with a structure in x-y plane, while the model is assumed to be uniform in the perpendicular z-direction (see Figure 1b). Similarly, if the structure is axisymmetrical, the plane of symmetry is the cross-section anywhere around the axis of symmetry. In this case, we are using a single 2-dimensional slice (r-z cylindrical coordinates) to represent the whole 360° of the structure (see Figure 1c). Further, in some cases, the geometrical structure we would like to model has a repetitive infinite or quasi-infinite pattern. Only a small portion of the whole array, a unit cell need be modeled by applying appropriate periodic boundary conditions (see Figure 1d).

In some cases the modeled system has no borders electrically insulating it from its surroundings; the electrical quantities are simply diminishing with increasing distance from the source. One such example is needle electrodes inserted in a tissue (Figure 2). In such cases, the outer boundaries of the model need to be far enough from the source(s), in order not to restrain the natural flow of the electric

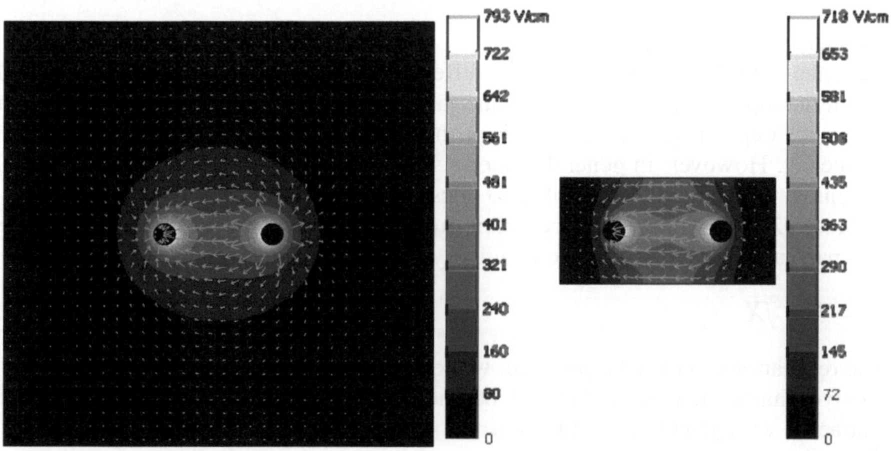

(a) (b)

Fig. 2. The electric field distribution - E (white to black contours) and the electric current density - J (white arrows) in a homogeneous material, electric pulse is delivered through needle electrodes (the two black circles). The E and J distributions are shown in a section plane cut through the material, perpendicular to electrodes. a) The insulated borders of the model are far enough from the electrodes, which allows for the natural flow of the electric current, the electric field and the electric current near model borders are very close to zero. b) The borders of the model are too close to the electrodes, the electric field close to the border is not zero, electric current is artificially constrained.

current (See Figure 2a). Namely, when modeling such a system, the borders of the model are artificially electrically insulated from the surroundings. This effectively means that the boundary condition is set in such a way that no electric current flows in or out of the enclosing box – only tangential component of the electric current exists on the outer tissue borders while the normal component equals zero. If these borders are too close to the source(s) of the electric current, such as in Figure 2b, electric field and current distribution is deformed and does not reflect true situation.

Physical Considerations

After the geometry of the model has been constructed, the next step in the modeling process is setting the physics of the model, such as underlying equations, material properties, boundary and initial conditions. Commercially available finite element software packages usually have a great number of pre-defined application modes to choose from, already containing the equations relevant for the chosen physical problem. This makes the use of such software much easier for users with less mathematical knowledge. However, this may present a trap for beginners, as understanding the physics defining our problem is crucial for the whole modeling process, from geometry building, setting the underlying physics as well as interpretation and validation of the results. In continuation we shortly describe some important aspects of designing a numerical model to be used for studying the effects of electromagnetic fields on biological material.

Frequency dependent component

First, the material's response will be different when exposed to either direct (DC) or alternating current (AC). If our material is purely resistive, the system exerts no frequency dependency; the current is proportional to the voltage irrespective of the frequency. However, in general, materials have their capacitive or inductive component so the voltage to current ratio does depend on frequency and is termed impedance (*Z*). Impedance is a complex quantity, consisting of a resistance *R* (the real part) and a reactance *X* (the imaginary, frequency dependent part):

$$\vec{Z} = R + jX$$

The resistance can only be positive, while the reactance can be either positive (inductive character, current lagging behind voltage - X_L) or negative (capacitive character, voltage lagging behind current - X_C).

$$X_L = j\omega L$$

$$X_C = -j\frac{1}{\omega C}$$

Voltage and current must be regarded as vectors in the complex plane (phasors) that are out of phase, so the voltage to current ratio – the impedance – can also be given by its magnitude and phase angle:

$\bar{Z} = |Z| \cdot e^{j\Theta}$, where

$|Z| = \sqrt{R^2 + X^2}$

$\Theta = arctg\, X/R$

In summary, resistance is only a special case of impedance, when the material we are considering exerts no or negligible capacitive or inductive character ($jX=0$). Further, if our system is exposed to direct current, the frequency dependent part – the reactance X – plays no role when the system is in steady-state, after all the transients have faded out. It does, however, dictate the course of the transient of the system. Which poses the next question in the modeling process: Are we interested only in the steady-state of our system, or, are we studying transient phenomena – changes over time from $t=0$ until the system has reached its steady-state?

Transient vs. steady-state

Transient behavior occurs when the magnitude and direction of electrical quantities change with time. On the contrary, if they are constant with time throughout the entire volume, the system is already in its steady-state. To avoid any ambiguity, the steady-state does not mean absence of movement or flow in the system! If we take electric currents in a material as an example, it simply means that the "amount" of electric current in the system does not change within observed time; the magnitude of the current exiting the system equals the current magnitude flowing into the system at any time when in steady-state. In other words, time becomes an irrelevant variable for the analysis, since recently observed behavior of the system will continue into the future.

In many systems, steady state is not achieved until enough time has elapsed after the system is started or stimulated (externally or internally). The situation after the occurrence of the described changes of the system and before all internal quantities (states) of the system reach the steady-state is defined as a transient state. As an example, in an electrical system of purely resistive character, no transients occur at $t=0$, when the electrical stimulation is turned on. However, if the imaginary part (the reactance) is present in the impedance of the electrical system, its behavior exerts inertia, meaning that the change of electrical quantities in the system is not instantaneous. The capacitive or the inductive component opposes the sudden change at $t=0$ (applied voltage or current) and enforcing the transient state onto the system that will, however, eventually fade out.

Linearity

With respect to intrinsic characteristics of a system or equations describing it, an important consideration is whether the system is linear or nonlinear. Mathematically speaking, a linear function is one which satisfies both of the following properties: i) additivity $f(x+y)=f(x)+f(y)$; and ii) homogeneity: $f(ax)=af(x)$, where x and y are independent variables and a is any rational number. In other words, a nonlinear system does not satisfy the superposition principle stating that the response of a system caused by two or more input stimuli is the sum of the responses which would have been caused by each stimulus individually. In terms of equations, a nonlinear system is any system where the variable(s) to be solved for cannot be written as a linear sum of independent components. Unfortunately, most physical systems are inherently nonlinear in their nature and, unfortunately, biological tissues are not an exception. Often, a physical property of a material is changed during a process (material's temperature coefficients, conductivity changes during tissue electroporation). Some typical types of nonlinearities often observed in nature are:

- Saturation: The condition in which, after a sufficient increase in an independent variable, its further increase produces no (or negligible) additional increase in the dependent variable (Figure 3a).
- Hysteresis: A phenomenon where the dependent and the independent variable bear a relationship which depends on prior history. More specifically, the response of the dependent variable (y) takes on different values for an increasing independent variable (x) than for a decreasing x (Figure 3b).
- Irreversibility: The irreversible processes cannot be restored to their original state once having undergone an irreversible change. Specifically, if we look at two possible states of the system – state A and state B, the transition between them is possible only in one direction (Figure 3c).

For some applications, a linear approximation of a nonlinear function can be found at (or around) a given point, for specific input values. However, if the model has to cover the whole (or larger) range of input values, the nonlinearities have to be considered in the model.

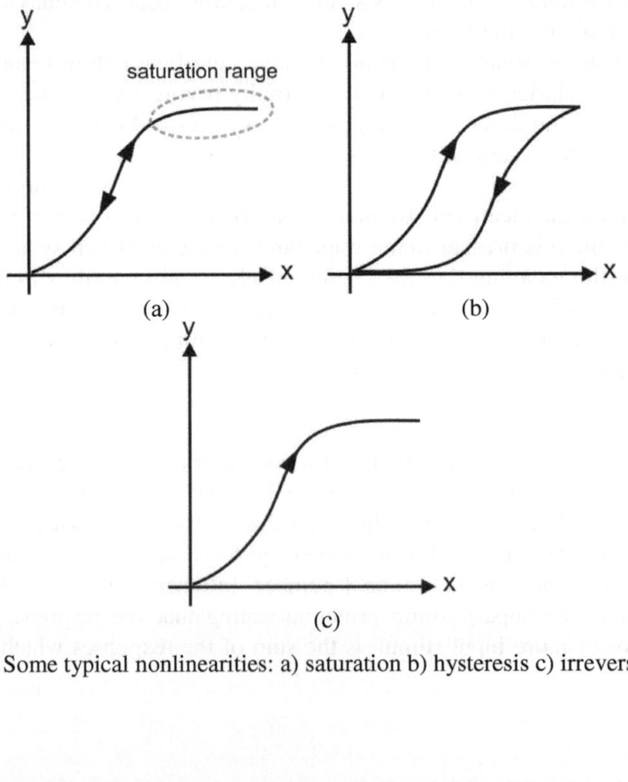

Fig. 3. Some typical nonlinearities: a) saturation b) hysteresis c) irreversibility

Multiphysics

The effects of various physical phenomena can be investigated by separately analyzing each individual phenomenon, without any regard to interaction between them. However, often we are dealing with two or more interacting, simultaneous phenomena, such as the coupling between the electric (*E*) and the magnetic field (*B*). An important coupling of physical phenomena in applications using electric pulses on biological tissues is heat transfer in tissue due to resistive heating (Tungjitkusolmun *et al.*, 2000). This coupling may give rise to tissue conductivity changes (due to temperature increase), which in turn changes the magnitude of electric current. When constructing a model, we have to estimate the influence of such interactions and, if needed to obtain accurate results, include mutual dependencies. To do so, we need data on how the material properties significant for one field (such as the electric field) vary with the value of another field (such as temperature) and vice versa.

Biological Tissues

Different biological materials perform different physiological functions so at their cellular and higher organizational level exert a number of interesting characteristics that have to be considered when representing them with a model. These differences are also clearly reflected in highly different bulk properties of biological materials (Gabriel S. *et al.*, 1996; Gabriel C. *et al.*1996; Miklavcic *et al.* 2006). They dictate the current densities and pathways that result from an applied stimulus and are thus very important in the analysis of a wide range of biomedical applications used for diagnosis and treatment. Biological tissues are inhomogeneous, nonlinear, and can, with respect to irreversible electroporation, exert saturation, hysteresis and irreversibility. We will further illuminate the theory presented so far for the case of biological tissues and give some examples.

Micro vs. bulk structure

Numerous examples could be given to illustrate either the importance or futility of including physiological details in the geometry of the model. Already at cell level, in vitro observations on cell suspensions can be represented numerically on different levels. For example, when studying the magnitude and the distribution of the electric field in a cell suspension, a material with homogeneous properties is an adequate model (Pavlin and Miklavcic, 2003). However, if the aim of our research is studying the phenomena on cell level or we are looking into interactions between cells, the influence of their size, shape, density and orientation, individual cells rather than bulk material have to be modeled (Susil *et al.*, 1998; Pavlin *et al.*, 2002; Valic *et al.*, 2003, Pucihar *et al.*, 2006, Pavlin and Miklavcic, 2008, Towhidi *et al.*, 2008). Even further, if we are using nanopulses, cell organelles must be added to the geometry (Kotnik and Miklavcic, 2006). Still, by using smart approaches, such as replacing a thin, non-zero conductivity cell membrane by a boundary condition between the cytoplasm and the exterior, getting rid of complexities while maintaining accuracy of the model is possible (Pucihar *et al.*, 2006).

Similar observations hold true on tissue level. Different tissues exert different levels of complexity and nonhomogeneity, however, including their particularities depends strongly on the purpose of the model. Skin, for example, is a very intricate tissue due to its highly inhomogeneous structure, leading to inhomogeneous electric properties. It consists of different layers, in terms of dimensions (thickness) and electrical properties: the outer thin layer of dead flat skin cells, the stratum corneum, the viable epidermis, dermis, and the subcutaneous tissue (Chizmadzhev *et al.*, 1998; Yamamoto and Yamamoto, 1976; Yamamoto and Yamamoto, 1976a). If the aim of the model is to study electroporation of skin as a target tissue, this layered structure needs to be included in our geometrical representation (Pavselj *et al.*, 2007). Moreover, even this bulk layered structure might sometimes prove inadequate. Smaller structures, such as hair follicles, sweat glands and blood vessels, or local transport regions as a result of skin electroporation (Pavselj *et al.*, 2008) may have to be added in order to study the processes on microscale, where they occur, and only then compare them to bulk observations. On the other hand, such details, unavoidably adding to overall complexity of the model, can be omitted in cases where skin electroporation is not studied directly, such as any application where electric pulses are delivered with external electrodes to tissues beneath skin (Pavselj *et al.*, 2005). However, other structures, such as major blood vessels may be important and need to be included in the model when studying electrochemotherapy of subcutaneous tumors. (Sersa *et al.*, 2008),

Alternatively, in cases where only one part of the whole system is either geometrically complex or needs better geometrical representation, the principle of embedded geometries can be used – the (usually more intricate) smaller part of the geometry is modeled separately and embedded in the model by properly setting the boundary conditions at the borders (Valic *et al.*, submitted).

<u>Conductive and capacitive component of tissue</u>

Electrical response of biological tissues when stimulated with DC electroporative pulses can be seen as quasi-stationary. Namely, for any material whose electric properties are in the range of those of biological tissues or organs and its dimensions do not exceed 1 m and the frequency of the electric field is below 1 kHz, the electrical behavior in any given moment as a response to electric current can be numerically described with a set of equations describing stationary fields. Although the impedance of biological tissue exerts capacitive component, electric field can be simplified as time independent thus the capacitive effects and the finite propagation of the electric current in the biological tissue are disregarded.

However, the transient of cell membrane charging may also be interesting to study. Namely, cell membrane charging time is in the order of microseconds, and typical pulses used for electropermeabilization of cell membrane are 100 microseconds long, with the amplitude of around 500 V/cm (Kotnik *et al.*, 1997; Kotnik *et al.*, 1998). It has been found that if pulses of much higher amplitude (e.g. 50 kV/cm) and much shorter duration are used – in the order of tens of nanoseconds – the charging effect also becomes pronounced on the membranes of intracellular organelles (Schoenbach *et al.*, 2001; Tekle *et al.*, 2005; Kotnik and Miklavcic, 2006). For a qualitative analysis of these processes, the time courses of organelle and cell plasma membrane charging become important. Thus the capacitive component describing the electrical properties of the cell, its organelle(s) and their membranes can no longer be neglected.

Tissue anisotropy

When the properties of a material are the same in all directions, the material is said to be isotropic. However, some biological materials (most typical being skeletal muscle), are distinctly anisotropic. Therefore, when referring to published electrical property data, we need to know the orientation of the electrodes relative to the major axis of the tissue during impedance measurements (longitudinal, transversal, or a combination of both).

Tissue anisotropy is often related to the physiological demands made on the tissue. For example, skeletal muscles are composed of fibers that are very large, highly elongated individual cells and are aligned in the direction of muscle contraction. Electrical conduction along the length of the fiber is thus significantly easier than conduction between the fibers (the difference is about 7-fold) (Reilly, 1998). The longitudinal conductivity is significantly higher than the transverse conductivity, especially in the low frequency range. Tissue anisotropy is frequency-dependent (Hart *et al.*, 1999). Namely, if the frequency of the current is high enough, the anisotropic properties disappear (specifically for skeletal muscle, that happens in the MHz frequency range). At higher frequencies, charge movement takes place over shorter distances so large-scale structures become less important and capacitive coupling across membranes becomes more important.

A practical problem occurs when measuring the electrical properties of anisotropic materials: how to accurately align the applied electric field and tissue fibers. Namely, it has been shown that perfect alignment is crucial for obtaining accurate longitudinal and transverse values. A study on skeletal muscle tissue showed that a 5 degree misalignment from true perpendicular or parallel orientations would result in an 18% overestimate in the perpendicular direction and a 0.4% underestimate in the parallel direction when measuring conductivity (Epstein and Foster, 1983).

Nonlinearities of biological tissues

As stated previously, nonlinear natural processes are (unfortunately, from a modeler's standpoint) quite frequent in nature. If we speak strictly about tissue properties exposed to electric current, at least two important nonlinearities need to be considered.

One is the increase in tissue conductivity (σ) due to increased electric field (E) causing cell membrane electroporation (Pliquett and Weaver, 1996; Pavselj *et al.*, 2005; Sel *et al.*, 2005). This change of material properties has two more nonlinear characteristics. First, it is considered threshold phenomenon, meaning that the electric field has to reach a certain value, termed reversible electropermeabilization threshold E_{rev} in order to cause conductivity changes. Only then can we observe conductivity changes $\sigma(E)$ between the reversible and the irreversible value (too high an electric field, causing permanent cell damage) of the electric field. Its second characteristic is that, for the duration of the pulse, this conductivity change is an irreversible phase transition process. More specifically, once the conductivity was increased in a given area of the tissue, it could not be changed back to its lower value during pulse delivery, even if the electric field strength drops below the threshold due to changed conductivities. Here, we would like to point out that one has to be careful to distinguish between the reversibility of cell electroporation (provided the electric field was below the irreversible threshold) *after* the cessation of pulsing; and the irreversible nature of the conductivity changes *during* pulse delivery – the change is only possible in one direction, tissue conductivity can only increase (Pavlin *et al.*, 2005).

The second nonlinearity comes from the electrical-thermal coupling (Pliquett, 2003). Once a part of a tissue is permeabilized, it becomes more conductive and the current density increases several times, causing resistive heating. In turn, tissue conductivity increases even more, as most biological materials exert a positive temperature coefficient of electrical conductivity in the range of 1-3% $°C^{-1}$ (Duck, 1990).

Threshold with respect to pulse parameters

Cell membrane is permeabilized when the threshold transmembrane potential is reached, thus when the external electric field is above the threshold value. This increased cell membrane permeability is reversible, provided the electric field is not too high. However, if cells are exposed to electric field above the irreversible threshold, they suffer permanent damage. For electroporation-based applications such as gene delivery (Golzio *et al.*, 2004; Andre *et al.*, 2008) or transdermal drug delivery (Prausnitz, 1999; Denet and Preat, 2003; Denet *et al.*, 2004), this is not a desirable effect, as we need the cells to be viable after the treatment. On the other hand, for applications based on irreversible electroporation, we can irreversibly destroy target cells within a narrow range while leaving neighboring cells unaffected, representing a promising new treatment for cancer, heart disease and other disease states that require removal of tissue (tissue ablation) (Davalos *et al.*, 2005; Lavee *et al.*, 2007; Onik *et al.*, 2007; Rubinsky *et al.*, 2007).

In both cases, it is important to determine the needed amplitude of electric pulses at a given electrode-tissue set-up, to achieve an electric field distribution in the tissue, adequate for a given application. Electric field reversible and irreversible thresholds are both inherent characteristics of the tissue (also different for different tissues), no matter what kind of electrodes we use or if nonhomogeneous or composed tissues are involved. Of course, the electropermeabilization process as well as cell viability depend on electrical parameters, i.e. pulse amplitude, pulse duration and the number of pulses (see Figure 4). (Macek-Lebar *et al.*, 2002; Puc *et al.*, 2003).

However, accomplishing an adequate electric field distribution in the tissue is much more complex than merely calculating the voltage we need at given electrode separation (*U/d*). Mathematically, this ratio equals electric field only when delivering pulses to a homogeneous tissue through parallel plate electrodes whose surface is large (infinite) compared to separation between the electrodes

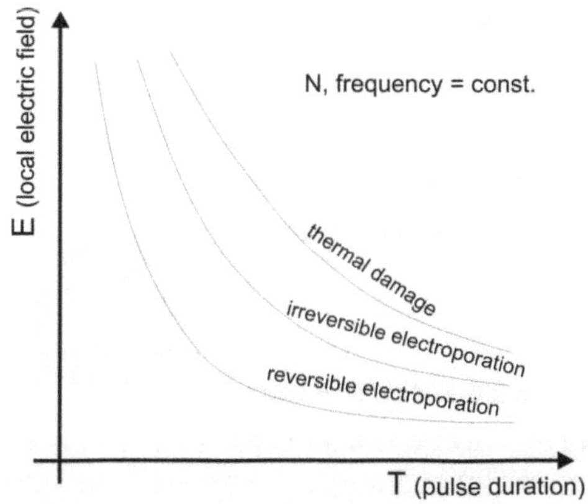

(a)

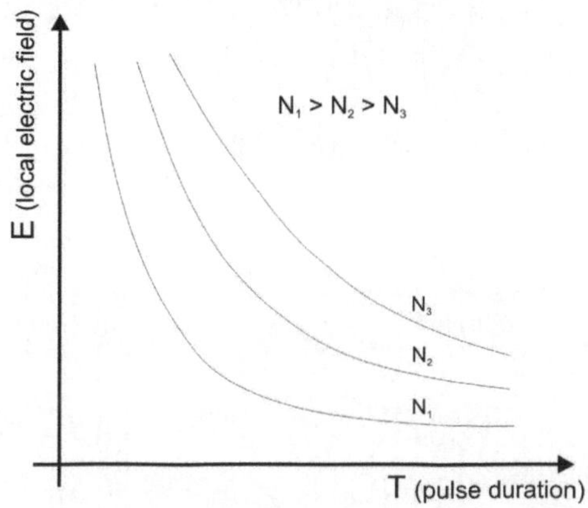

(b)

Fig. 4. Electroporation process is (for a given tissue-electrode geometry) controlled by pulse parameters. a) At constant number of pulses (N) and their frequency, lengthening pulse duration requires lower local electric field (pulse amplitude) for the same effect. If both are increased, the effects on the tissue become irreversible, or, at even higher values, tissue thermal damage can be observed, due to excessive resistive heating. b) Similarly, for any of the curves, if number of applied pulses is larger, the same effect can be achieved with a lower pulse amplitude and/or duration.

(see Figure 5a). It can still be used as an approximation of the electric field in the area away from the borders of the parallel plate electrodes of finite surface (Figure 5b). The *U/d* ratio is also often used to estimate the electric field between two parallel rows of needle electrodes; however, the approximation is very rough (Figure 5c). In the case of any other electrode geometry using plate, needle, microneedle or surface electrodes or if more than one tissue is involved, a numerical analysis has to be performed beforehand, as a part of treatment planning, in order to choose the right electrode configuration and the pulse amplitude.

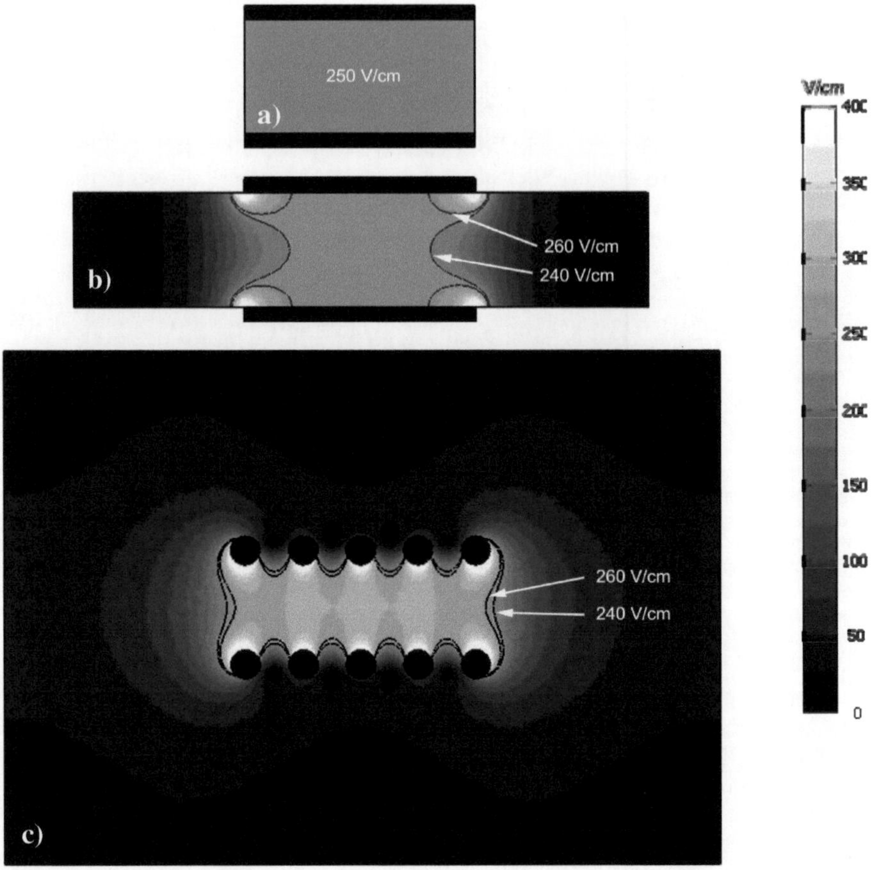

Fig. 5. Electric field distribution in a homogeneous tissue, values are given in V/cm in a section plane perpendicular to electrodes. a) The electric field equals the ratio U/d (voltage/distance between electrodes) only in the theoretical case where electric pulses are delivered through plate electrodes of infinite surface. Here, only a portion of this infinite structure is modeled with boundary conditions set to represent an infinite volume. The distance between the electrodes (d) is 4 mm, the applied voltage (U) is 100 V, and thus the electric field equals U/d=250 V/cm throughout the tissue between the electrodes. b) A real situation where electrodes are of finite surface. The electric field in the grey area between the two black iso-contours is between 240-260 V/cm, so the voltage to distance ratio in this area is a good approximation. The U/d approximation is valid in a greater portion of the tissue between the electrodes if the electrode surface is increased or the distance between them is smaller. c) Two rows of needle electrodes are used instead of plate electrodes. The length of the electrode array and the distance between the rows is the same as in b). The grey area between the two black iso-contours denoting electric field between 240-260 V/cm is very small and limited to a few narrow stripes. Throughout most of the area inside the electrode array is higher (around 300 V/cm or higher) and thus cannot be satisfactorily approximated by voltage to distance ratio.

Nonhomogeneous and composed tissues

When modeling electroporation of biological tissues, much consideration has to be given to the interpretation of the results in relation to possible simplifications in the model or inherent characteristics of different biological tissues. Namely, some simplifications might not have much effect in isotropic, homogeneous tissues, such as liver, but may yield useless results in nonhomogeneous, composed biological structures, such as layered skin or subcutaneous tumor, where electric field distribution is much more complex (Pavselj *et al.*, 2005; Pavselj and Miklavcic, 2007; Ivorra et al., 2008). To illustrate, when modeling electroporation in a homogeneous tissue, such as liver, the results are still useful and comparable to experimental data even if conductivity increase due to tissue electropermeabilization is neglected. In fact, early models did not take this nonlinear tissue behavior into account (Miklavcic *et al.*, 2000). However, when more complicated electrode-tissue set-ups were being studied with numerical models, experimentally observed phenomena could not be satisfactorily modeled in this way. Namely, upon applying electric pulses on a composed or layered tissue with a nonhomogeneous distribution of electrical conductivities, the voltage is divided among them proportionally to their electrical resistances (Pavselj and Miklavcic, 2008a). This leads to a more complex electric field distribution, meaning that some parts of the tissue, due to their low electrical conductivity (disproportionally lower than the rest of the tissue), are exposed to much stronger electric field. The electric field is the highest in the layer with the highest resistivity (lowest conductivity). In the case of the subcutaneous tumor this is the skin, which has the lowest electrical conductivity, and in the case of the skin fold, the highest electric field is in the non-conductive outermost skin layer, the stratum corneum. But more importantly, the electric field in the target tissues (tumor and viable skin layers) stays too low for successful electroporation. This fact raised the question of how is the experimentally confirmed successful permeabilization of the target tissues theoretically possible when external plate electrodes are used, which led to the inclusion of tissue conductivity changes in the numerical models.

Model Verification and Validation

The last, but nevertheless very important part of the modeling process is the verification and the validation of the constructed model, involving different aspects of evaluation. Mostly, these aspects should be taken into consideration from the very beginning of the process and can roughly be divided into three categories:

- Reaching the aim of the model (verification):
  Already in the planning phase of the modeling process, we have to set the scope of our model, the range of input data it should be valid for, as well as geometrical details to be included. However, as we build the model, some simplifications and trade-offs may have to be made. Comparing the actual result with the requirements set during the planning phase will tell us if our resulting model is still within the planned scope of the model.

- Explaining underlying physics (descriptive realism):
  In cases where almost nothing is known about the phenomena describing the modeled system, we are dealing with the so-called black box problem that can only be treated in terms of its input and output characteristics. Our only option may be finding a curve which has the best fit to a series of data points, while respecting possible constraints, without actual physical reference to the described process(es). However, different techniques of system identification (Ljung, 1999) can be applied in order to identify the physics defining our "black box", which can then be modeled. Namely, the purpose of models is gaining insight and explaining underlying phenomena, as well as using them for predicting the output at certain input data set. We should therefore direct our efforts to turn the black box into a set of equations if possible. Once again, as some trade-offs will most likely be necessary, we should assess if the modeled physical phenomena successfully explain the most important experimental observations.
- Agreement with empirical data (validation):
  One way to justify the physics used in the model (sometimes the only way) is comparing the output data obtained from the model to experimental data. Usually, or ideally, the experimental data can be divided into two groups: the training data and the validation data. The former is used to identify the process and to construct a model with its relevant parameters and constraints, while the latter is used to assess if the model is valid for any range of input parameters within the defined constraints.

Summary

Numerical modeling of cell and tissue electroporation and its accompanying phenomena has proven to be an efficient tool for the analysis and explanation of experimental observations. Moreover, it is indispensable in optimizing electrode geometries and electroporation protocols. Namely, every electroporation-based application has its own requirements and constraints. For gene electrotransfer, the cells in the treated tissue need to be viable after treatment in order to achieve high level of gene expression. This is not as important in electrochemotherapy, where killing cancerous cells is our aim. For this application, covering the entire volume of the treated tumor with a sufficiently high electric field, even if it exceeds the irreversible electroporation threshold, is of utmost importance. However, if the treated part is in the vicinity of any delicate tissue, organ or a part of an organ whose permanent damage has to be avoided at all cost, the protocol has to be carefully planned beforehand. The same goes for tissue ablation protocols using irreversible electroporation. Furthermore, any unwanted side effects, such as muscle contraction, pain or excessive tissue heating, should be minimized (Pliquett, 2003; Zupanic *et al.*, 2007; Davalos and Rubinsky, 2008). And last but not least, our efforts should also be directed towards treatment cost minimization in both, clinical as well as experimental environment.

For computer-aided optimization and treatment planning, the right representation of the real system has to be found, unavoidably including simplifications and trade-offs between simplicity and accuracy of the model (For more on irreversible electroporation treatment planning, see the chapter by Zupanic and Miklavcic.). It is impossible to construct a single model covering all aspects of cell and tissue electroporation, from subcellular level (cell organelle permeabilization using nanosecond pulses), through cell level (cell plasma membrane charging), to tissue level that again may or may not include details and tissue substructures. Further, when alternating current is used or when studying transients of the modeled system, the capacitive component of tissue electrical properties has to be taken into account. Nonlinearities, such as tissue conductivity increase during electropermeabilization must often be included in the model, as well as tissue resistive heating when longer pulses are used.

Although requirements and constraints of each application can be very different, careful analysis, parameter optimization and treatment planning is a common denominator to any electroporation-based application, allowing for efficient, safe and cost-effective procedure. With the increased accessibility and simplicity of computer software packages based on finite element method, carefully constructed numerical models can substantially contribute to increased efficiency and safety of experimental and clinical electroporation.

References

André, F., Gehl, J., Sersa, G., Préat, V., Hojman, P., Eriksen, J., Golzio, M., Cemazar, M., Pavselj, N., Rols, M.P., Miklavcic, D., Neumann, E., Teissié, J., Mir, L.M.: Efficiency of high- and low-voltage pulse combinations for gene electrotransfer in muscle, liver, tumor, and skin. Human Gene Ther. 19, 1261–1271 (2008)

Brandisky, K., Daskalov, I.: Electrical field and current distributions in electrochemotherapy. Bioelectrochemistry and Bioenergetics 48, 201–208 (1999)

Chizmadzhev, Y.A., Indenbom, A.V., Kuzmin, P.I., Galichenko, S.V., Weaver, J.C., Potts, R.O.: Electrical properties of skin at moderate voltages: contribution of appendageal macropores. Biophys. J. 74, 843–856 (1998)

Corovic, S., Zupanic, A., Miklavcic, D.: Numerical modeling and optimization of electric field distribution in subcutaneous tumor treated with electrochemotherapy using needle electrodes. IEEE Trans. Plasma Sci. 36, 1665–1672 (2008)

Davalos, R.V., Mir, L.M., Rubinsky, B.: Tissue ablation with irreversible electroporation. Annals of Biomedical Engineering 33(2), 223–231 (2005)

Davalos, R.V., Rubinsky, B.: Temperature considerations during irreversible electroporation. International Journal of Heat and Mass Transfer 51(23-24), 5617–5622 (2008)

Debruin, K.A., Krassowska, W.: Modeling electroporation in a single cell. I. effects of field strength and rest potential. Biophysical Journal 77, 1213–1224 (1999)

Debruin, K.A., Krassowska, W.: Modeling electroporation in a single cell. II. effects of ionic concentrations. Biophysical Journal 77, 1225–1233 (1999a)

Denet, A.-R., Préat, V.: Transdermal delivery of timolol by electroporation through human skin. Journal of Controlled Release 88, 253–262 (2003)

Denet, A.-R., Vanbever, R., Préat, V.: Skin electroporation for transdermal and topical delivery. Advanced Drug Delivery Reviews 56(5), 659–674 (2004)

Dev, S.B., Dhar, D., Krassowska, W.: Electric Field of a Six-Needle Array Electrode Used in Drug and DNA Delivery In Vivo: Analytical Versus Numerical Solution. IEEE Transactions on Biomedical Engineering 50(11), 1296–1300 (2003)

Duck, F.A.: Physical properties of tissue: A comprehensive reference book. Academic press, London (1990)

Epstein, B.R., Foster, K.R.: Anisotropy in the dielectric properties of skeletal muscle. Med. Biol. Eng. Comput. 21, 51–55 (1983)

Fear, E.C., Stuchly, M.A.: Modeling assemblies of biological cells exposed to electric fields. IEEE Transactions on Biomedical Engineering 45(10), 1259–1271 (1998)

Gabriel, C., Gabriel, S., Corthout, E.: The dielectric properties of biological tissues: I. literature survey. Physics in Medicine and Biology 41, 2231–2249 (1996)

Gabriel, S., Lau, R.W., Gabriel, C.: The dielectric properties of biological tissues: II. Measurements in the frequency range 10 Hz to 20 GHz. Physics in Medicine and Biology 41, 2251–2269 (1996a)

Golzio, M., Rols, M.P., Teissie, J.: In vitro and in vivo electric field-mediated permeabilization, gene transfer, and expression. Methods 33(2), 126–135 (2004)

Hart, F.X., Berner, N.J., McMillen, R.L.: Modelling the anisotropic electrical properties of skeletal muscle. Phys. Med. Biol. 44, 413–421 (1999)

Ivorra, A., Al-Sakere, B., Rubinsky, B., Mir, L.M.: Use of conductive gels for electric field homogenization increases the antitumor efficacy of electroporation therapies. Phys. Med. Biol. 53, 6605–6618 (2008)

Kotnik, T., Bobanovic, F., Miklavcic, D.: Sensitivity of transmembrane voltage induced by applied electric fields – a theoretical analysis. Bioelectrochem. Bioenerg. 43, 285–291 (1997)

Kotnik, T., Miklavcic, D.: Theoretical evaluation of voltage inducement on internal membranes of biological cells exposed to electric fields. Biophys. J. 90, 480–491 (2006)

Kotnik, T., Miklavcic, D., Slivnik, T.: Time course of transmembrane voltage induced by time-varying electric fields – a method for theoretical analysis and its application. Bioelectrochem. Bioenerg. 45, 3–16 (1998)

Lavee, J., Onik, G., Mikus, P., Rubinsky, B.: A Novel Nonthermal Energy Source for Surgical Epicardial Atrial Ablation: Irreversible Electroporation. The Heart Surgery Forum 10(2), 96–101 (2007)

Ljung, L.: System Identification - Theory For the User, 2nd edn. PTR Prentice Hall, Upper Saddle River (1999)

Macek-Lebar, A., Sersa, G., Kranjc, S., Groselj, A., Miklavcic, D.: Optimisation of pulse parameters in vitro for in vivo electrochemotherapy. Anticancer Res. 22, 1731–1736 (2002)

Miklavcic, D., Pavselj, N., Hart, F.X.: Electric Properties of Tissues. Wiley Encyclopedia of Biomedical Engineering. John Wiley & Sons, New York (2006)

Miklavcic, D., Corovic, S., Pucihar, G., Pavselj, N.: Importance of tumour coverage by sufficiently high local electric field for effective electrochemotherapy. Eur. J. Cancer Suppl. 4, 45–51 (2006a)

Miklavcic, D., Semrov, D., Mekid, H., Mir, L.M.: A validated model of in vivo electric field distribution in tissues for electrochemotherapy and for DNA electrotransfer for gene therapy. Biochimica et Biophysica Acta 1523, 73–83 (2000)

Onik, G., Mikus, P., Rubinsky, B.: Irreversible electroporation: Implications for prostate ablation. Technology in Cancer Research & Treatment 6(4), 295–300 (2007)

Pavlin, M., Kanduser, M., Rebersek, M., Pucihar, G., Hart, F.X., Magjarevic, R., Miklavcic, D.: Effect of cell electroporation on the conductivity of a cell suspension. Biophys. J. 88, 4378–4390 (2005)

Pavlin, M., Miklavcic, D.: Effective conductivity of a suspension of permeabilized cells: a theoretical analysis. Biophys. J. 85, 719–729 (2003)

Pavlin, M., Miklavcic, D.: Theoretical and experimental analysis of conductivity, ion diffusion and molecular transport during cell electroporation — Relation between short-lived and long-lived pores. Bioelectrochemistry 74, 38–46 (2008)

Pavselj, N., Bregar, Z., Cukjati, D., Batiuskaite, D., Mir, L.M., Miklavcic, D.: The course of tissue permeabilization studied on a mathematical model of a subcutaneous tumor in small animals. IEEE Trans. Biomed. Eng. 52, 1373–1381 (2005)

Pavselj, N., Miklavcic, D.: Numerical models of skin electropermeabilization taking into account conductivity changes and the presence of local transport regions. IEEE Trans. Plasma Sci. 36, 1650–1658 (2008)

Pavselj, N., Miklavcic, D.: Numerical modeling in electroporation-based biomedical applications. Radiol. Oncol. 42, 159–168 (2008a)

Pavselj, N., Préat, V., Miklavcic, D.: A numerical model of skin electropermeabilization based on in vivo experiments. Annals Biomed. Eng. 35, 2138–2144 (2007)

Pliquett, U.: Joule heating during solid tissue electroporation. Medical & Biological Engineering & Computing 41(2), 215–219 (2003)

Pliquett, U., Weaver, J.C.: Electroporation of human skin: Simultaneous measurement of changes in the transport of two fluorescent molecules and in the passive electrical properties. Bioelectrochemistry and Bioenergetics 39(1), 1–12 (1996)

Prausnitz, M.R.: A practical assessment of transdermal drug delivery by skin electroporation. Advanced Drug Delivery Reviews 35, 61–76 (1999)

Puc, M., Kotnik, T., Mir, L.M., Miklavcic, D.: Quantitative model of small molecules uptake after in vitro cell electropermeabilization. Bioelectrochemistry 60, 1–10 (2003)

Pucihar, G., Kotnik, T., Valic, B., Miklavcic, D.: Numerical determination of transmembrane voltage induced on irregularly shaped cells. Annals Biomed. Eng. 34, 642–652 (2006)

Reddy, J.N.: An introduction to the Finite Element Method. McGraw-Hill Science Engineering, New York (2004)

Reilly, J.P.: Applied Bioelectricity, From Electrical Stimulation to Electropathology. Springer, New York (1998)

Rubinsky, B., Onik, G., Mikus, P.: Irreversible electroporation: A new ablation modality - Clinical implications. Technology in Cancer Research & Treatment 6(1), 37–48 (2007)

Schoenbach, K.H., Beebe, S.J., Buescher, E.S.: Intracellular effect of ultrashort electrical pulses. Bioelectromagnetics 22, 440–448 (2001)

Sel, D., Cukjati, D., Batiuskaite, D., Slivnik, T., Mir, L.M., Miklavcic, D.: Sequential finite element model of tissue electropermeabilization. IEEE Trans. Biomed. Eng. 52, 816–827 (2005)

Sel, D., Macek-Lebar, A., Miklavcic, D.: Feasibility of employing model-based optimization of pulse amplitude and electrode distance for effective tumor electropermeabilization. IEEE Trans. Biomed. Eng. 54, 773–781 (2007)

Semrov, D., Miklavcic, D.: Calculation of the electrical parameters in electrochemotherapy of solid tumors in mice. Computers in Biology and Medicine 28, 439–448 (1998)

Sersa, G., Jarm, T., Kotnik, T., Coer, A., Podkrajsek, M., Sentjurc, M., Miklavcic, D., Kadivec, M., Kranjc, S., Secerov, A., Cemazar, M.: Vascular disrupting action of electroporation and electrochemotherapy with bleomycin in murine sarcoma. Br. J. Cancer 98, 388–398 (2008)

Susil, R., Semrov, D., Miklavcic, D.: Electric field induced transmembrane potential depends on cell density and organization. Electro. Magnetobiol. 17, 391–399 (1998)

Tekle, E., Oubrahim, H., Dzekunov, S.M., Kolb, J.F., Schoenbach, K.H., Chock, P.B.: Selective field effects on intracellular vacuoles and vesicle membranes with nanosecond electric pulses. Biophys. J. 89, 274–284 (2005)

Towhidi, L., Kotnik, T., Pucihar, G., Firoozabadi, S.M.P., Mozdarani, H., Miklavcic, D.: Variability of the minimal transmembrane voltage resulting in detectable membrane electroporation. Electromagn. Biol. Med. 27, 372–385 (2008)

Tungjitkusolmun, S., Woo, E.J., Cao, H., Tsai, J.-Z., Vorperian, V.R., Webster, J.G.: Thermal-electrical finite element modelling for radio frequency cardiac ablation: effects of changes in myocardial properties. Medical & Biological Engineering & Computing 38, 562–568 (2000)

Valic, B., Golzio, M., Pavlin, M., Schatz, A., Faurie, C., Gabriel, B., Teissié, J., Rols, M.P., Miklavcic, D.: Effect of electric field induced transmembrane potential on spheroidal cells: theory and experiment. Eur. Biophys. J. 32, 519–528 (2003)

Valic, B., Gajsek, P., Miklavcic, D.: Current density in a model of a human body with a conductive implant exposed to ELF electric and magnetic fields. Bioelectromagnetics (submitted)

Yamamoto, T., Yamamoto, Y.: Electrical properties of the epidermal stratum corneum. Med. Biol. Eng. 14(2), 151–158 (1976)

Yamamoto, T., Yamamoto, Y.: Dielectric constant and resistivity of epidermal stratum corneum. Med. Biol. Eng. 14(5), 494–500 (1976a)

Zupanic, A., Ribaric, S., Miklavcic, D.: Increasing the repetition frequency of electric pulse delivery reduces unpleasant sensations that occur in electrochemotherapy. Neoplasma 54, 246–250 (2007)

Zupanic, A., Corovic, S., Miklavcic, D.: Optimization of electrode position and electric pulse amplitude in electrochemotherapy. Radiol. Oncol. 42, 93–101 (2008)

Optimization and Numerical Modeling in Irreversible Electroporation Treatment Planning

Anže Županič and Damijan Miklavčič*

Faculty of Electrical Engineering, University of Ljubljana, Ljubljana, SI-1000, Slovenia
* Corresponding author E-mail: damijan.miklavcic@fe.uni-lj.si

Introduction

When you overhear someone mention treatment planning, you can bet the conversation is about one of the forms of radiation therapy (RT). In the 1960s, when high powered x-ray delivery systems were readily available to radiation oncologists, the biggest challenge remained to improve the accuracy in locating the tumor and in directing the beams of charged particles. This changed when the first computed tomography (CT) scanners were invented (Lampert et al, 1974). Availability of 3D anatomical data and ever increasing computer processing power gave rise to numerical treatment planning. Together with improved beam generation and delivery technology, treatment planning has enabled RT to better target the tumor and to reduce adverse effects on vital organs (Jaffray et al, 2007). This is exactly what researchers would also like to achieve with irreversible electroporation (IRE): ablate the target tissue and spare as much healthy tissue as possible. Just as treatment planning in RT provides radiation oncologist with the radiation beam intensities and directions that cover the tumor without causing extensive damage to healthy tissue, so can treatment planning in IRE provide physicians with electrode configurations and amplitudes of electric pulses that result in adequate electric field distribution in and around the target tissue.

It might seem a bit early to talk about treatment planning for IRE as it is still in the phase of clinical experimentation. However, RT has been around for more than fifty years and has only become one of the most successful cancer treatments after treatment planning procedures were implemented. And while in RT one had to wait for medical imaging and powerful computers to appear, everything is readily available for IRE. There is also no need to wait for IRE to become a recognized and widely implemented ablation technique, treatment planning is just as important in the experimental stage, as experiences from electrochemotherapy, an application of reversible electroporation that has already made it to the clinic, have shown (Miklavcic et al, 1998; Miklavcic et al, 2000; Semrov & Miklavcic, 1998). Careful experiment planning not only increases the reproducibility of experiments, it also lowers the number of needed experimental animals, and thus improves the overall quality of research.

B. Rubinsky (Ed.): Irreversible Electroporation, BIOMED, pp. 203–222.
springerlink.com

In this chapter we compare features of external beam radiation therapy (and brachytherapy, when probe/electrode insertion questions are analyzed) with IRE features, which are relevant for the treatment planning procedure. By careful analysis we try to determine which features of RT treatment planning can be applied directly to IRE, which could be applied after adjustments and which cannot be applied due to overly different natures of both therapies. At the end we present our recent work in electroporation numerical modeling and genetic algorithm optimization in two illustrative (but hypothetical) examples of treatment planning of tumor ablation by IRE.

A Brief Overview of Irreversible Electroporation Basics

Cell membranes can be permeabilized by exposing them to a high enough electric field, a phenomenon termed electroporation (Neumann et al, 1982). The nature of electrically-induced membrane permeabilization can be predominantly controlled by the amplitude of local electric field. Permeabilization can be either reversible – the membrane stays permeabilized for up to minutes, allowing entrance of molecules that do not normally cross the membrane, and later recovers (Orlowski et al, 1988); or, if electric field strength is increased, irreversible – membranes do not recover and the cells die (Rubinsky et al, 2007). In biomedical research reversible electroporation has become widely used in the last decade, e.g. for electrochemotherapy (Belehradek et al, 1991; Heller et al, 1999; Mir et al, 1998; Serša et al, 2006), gene electrotransfer (Golzio et al, 2004; Hojman et al, 2007; Mir et al, 1999), transdermal drug delivery (Denet et al, 2004; Prausnitz, 1999) and electrofusion of cells (Scott-Taylor et al, 2000; Trontelj et al, 2008), while IRE has gained momentum in the last few years, since Davalos et al showed it can be used to kill cells without considerable thermal effects (Davalos et al, 2005). Further studies on ablation capacity of IRE have confirmed the absence of significant resistive heating during IRE (Al-Sakere et al, 2007; Edd et al, 2006; Miller et al, 2005) and have also demonstrated some additional advantages IRE has over more conventional thermal and chemical ablation techniques. These advantages include: 1) IRE is a non-thermal physical ablation modality, therefore not affected by blood flow (Miller et al, 2005); 2) delineation between treated (ablated) and untreated tissue after IRE is very sharp – only a few cells thick (Lee et al, 2007); 3) IRE affects only cell membranes and leaves extracellular structures intact – preservation of microvasculature is possible (Lee et al, 2007; Maor et al, 2007; Onik et al, 2007); 4) IRE elicits no immune response and can thus be used for treatment of patients with immune system deficiency (Al-Sakere et al, 2007); 5) the procedure is relatively fast compared to other ablation techniques (Lee et al, 2007); 6) IRE allows rapid regeneration of ablated tissue with healthy tissue (Rubinsky et al, 2007); 7) IRE can be accurately numerically modeled – numerical models of reversible electroporation that have been around for quite some time can be easily modified and implemented for IRE modeling (Corovic et al, 2007; Edd & Davalos, 2007; Pavselj & Miklavcic, 2008a).

IRE was tested as an ablation modality in various medical applications, such as ablation of cancer (Onik et al, 2007; Rubinsky et al, 2008), epicardial ablation (Lavee et al, 2007) and prevention of restenosis after angioplasty (Maor et al, 2008). After encouraging primary results of these studies, researchers expressed the need for accurate experimental planning that would: 1) guarantee that thermal effects are indeed negligible; 2) take advantage of the sharp physical delineation between treated and untreated tissue to enable surgically precise ablation and 3) make experimental (and later medical) procedures more reproducible.

Before setting guidelines for proper treatment planning, it seems appropriate to check whether sophisticated and time consuming numerical treatment planning procedures are necessary in all biomedical applications of IRE. If, for example, IRE takes place in homogeneous and isotropic tissues with no vital organ in the vicinity of the treated area, then a simple look-up database of appropriate treatment parameters calculated by simple numerical models validated by experiments would suffice. The same can be said for more complex tissue geometries that do not change much from patient to patient, as is the case in restenosis prevention where electric pulses are applied to blood vessel walls. The look-up database should include electric field distributions generated by the use of different electrode geometries, different electrode configurations and a range of electrical input parameters. Such databases can be constructed using numerical modeling alone, without optimization, and have to provide enough information for the treating physician to choose the proper set of electrodes and electric parameters for each treatment. Optimization becomes important in more complex situations, when the target tissue is located near a vital organ whose function should not be compromised by the treatment. In such situations it is of vital importance to control the magnitude and distribution of the electric field so that as little as possible critical tissue (organ at risk) is compromised by the treatment. This can be effectively accomplished by numerical modeling of IRE and optimization based on anatomical medical imaging, as has been demonstrated recently for electrochemotherapy (Corovic et al, 2008; Zupanic et al, 2008). Numerical modeling may also be necessary for treatment planning in tissues with highly anisotropic properties and highly non-homogeneous tissues, where electric field distribution is otherwise difficult to predict (Pavselj & Miklavcic, 2008b). In all such cases the treatment planning procedure will have to be applied individually for each patient and the electric field distribution will have to be sculpted carefully to ablate all of the target tissue and preserve as much of the critical tissue as possible.

Radiation Therapy vs. Irreversible Electroporation Procedure

Over the past decades progress in imaging technology and computer processors have modernized RT through use of more accurate radiation dose calculation algorithms, more complex dose delivery techniques and modern imaging modalities (Dawson & Sharpe, 2006; Jaffray et al, 2007). Modern radiation therapy techniques, such as 3D conformal RT (Purdy & Starkschall, 1999) and intensity-modulated RT (Boyer et al, 2001) are completely image-guided and computerized, which has enabled more accurate sculpting of radiation doses to clinical volumes (target volumes and critical volumes). RT treatment planning procedure has accordingly become extremely sophisticated, which makes it an ideal benchmark for treatment planning of other biomedical application based on physical agents. A typical RT treatment today generally consists of five major phases: simulation, treatment planning, set-up verification, dose delivery and response assessment (Lecchi et al, 2008). In the simulation phase, data on patients' anatomy are acquired via modern 3D imaging devices. In treatment planning, clinical volumes are first delineated, dose constraints are defined and the treatment plan is determined. Before treatment, imaging is again utilized for control of the clinical set-up. After the radiation dose is delivered the treatment success is periodically evaluated by following tumor response. In this chapter we analyze the first three phases with emphasis on the treatment planning phase.

Similar major phases as for RT can also be defined for IRE treatment: simulation, treatment planning, set-up verification, electric pulse delivery and response assessment. Simulation is probably not necessary in all biomedical applications of IRE; however, it is necessary in all situations that require numerical treatment planning. Treatment planning is currently limited to experienced researchers that are able to "predict" the appropriate electrode configuration and electric pulse amplitude, whether from experience alone or with help of modeling. The appropriate positioning of electrodes can be controlled by real time ultrasound measurements (Lee et al, 2007). Electric pulses are delivered by a clinical electroporator (Bertacchini et al, 2007) and evaluation of treatment success depends on individual IRE application, e.g. when using IRE for tumor ablation, tumor size is periodically measured.

Medical Imaging

The first phase in RT is medical imaging of the whole region surrounding the target tumor tissue. This is usually accomplished with computer tomography (CT), although, if necessary, other imaging modalities, such as magnetic resonance imaging (MRI) and positron emission tomography (PET) are also used as they can provide additional anatomical and physiological information (Lecchi et al, 2008; Newbold et al, 2006; Vanuytsel et al, 2000). If MRI or PET are used, their images are later aligned to the CT images using image-fusion algorithms (Skerl et al, 2007; Slomka, 2004). It is vital for the success of the therapy that the patient's position during imaging is as close as possible to the actual treatment position. In external beam RT this is usually achieved with laser positioning control and patient immobilization (Heinzerling et al, 2008). In brachytherapy, on the other hand, it is very important that imaging is done with probes already implanted inside the body, as their insertion can significantly change the internal organ positions (Potter et al, 2008a; Potter et al, 2008b). During treatment, ultrasound can be used to control internal organ movement (Chandra et al, 2003).

The exact same imaging procedures as in RT can also be used in IRE; however, since IRE is a local treatment, less extensive imaging is necessary. Ultrasound imaging can be very useful in IRE, e.g. for real-time ultrasound guidance of electrode insertion and real-time ablation zone monitoring (Lee et al, 2007). Electrical impedance imaging has also been suggested as a real-time electroporation monitoring modality (Davalos et al, 2004; Davalos et al, 2002; Granot & Rubinsky, 2007; Ivorra & Rubinsky, 2007), as has been current and voltage measurement during pulse delivery (Cukjati et al, 2007). As precise positions of electrodes for IRE are only available after the treatment planning procedure, it would be no use to insert them prior to imaging. Instead, ultrasound imaging will probably have to be relied upon to assure adequate positioning of the electrodes with respect to the target volume and critical volumes.

Delineation of Clinical Volumes and Definition of Dose Constraints

Oncology experts examine the acquired medical images slice by slice and delineate the target volumes and critical volumes for the treatment as defined by the International commission on radiation units and measurements (ICRU) reports 50 and 62 (ICRU-50, 1993; ICRU-62, 1999). If the location of the target tissue is such that target and critical volumes coincide, the expert has to use his/her experience and adjust the volumes with respect to the patient's best interest. Dose constraints for each critical volume and target dose are also defined according to the same ICRU reports. In RT there is no threshold effect: lower doses already cause tissue damage and by increasing the dose the damage increases up to the point where the dose necessary to kill all cancer cells is achieved.

The ICRU reports define clinical volumes mainly according to the probability of error in accurate delineation, according to the probability of microscopic spread of tumor cells outside the main tumor mass and accordingly to the set-up and delivery errors, which were all evaluated by a vast collection of clinically acquired data. As only the set-up and delivery error are radiation specific and are analyzed separately, tumor volumes treated by IRE can also be defined using ICRU reports. In IRE electric field strength is believed to be the main factor controlling the treatment outcome. IRE is apparently, contrary to RT, a threshold phenomenon. Electric fields below the irreversible threshold permeabilize cells reversibly, but do not kill them, while electric fields over the threshold irreversibly permeabilize cells and thus destroy them. This effectively means that tissue damage can theoretically be sculpted to the target tissue and around all critical tissues with great accuracy. IRE thresholds have, however, been found to be tissue specific. Furthermore, several electric pulse parameters affect the threshold values: pulse duration, number of pulses and to some extent also pulse repetition frequency (Edd & Davalos, 2007; Miklavcic & Kotnik, 2004). Prior to any treatment planning, data on thresholds for all target tissues and all critical tissues should be available for a range of electroporation parameters.

Tissue Properties

For radiation therapy data on tissue properties is readily available directly through CT imaging. Namely, CT values correlate well with electron density, the main tissue property used in computation of Compton scattering, which is the most probable interaction of high energy X rays with atomic nuclei in living beings. It is also considered that electron density remains the same during the entire treatment (Ruchala et al, 2000).

Data on human tissue electrical properties (electrical conductivity and electrical permittivity) are harder to come by. Although several reports have been published (Gabriel et al, 1996a; Gabriel et al, 1996b; Miklavcic et al, 2006b; Polk & Postow, 1996), the data between individual studies differ by a factor of 2 or more. A database of electrical properties is desperately needed, preferably with additional data on electrical property variation with age, gender and pathological changes (e.g. different tumors or scarred tissues). Furthermore, because electric tissue properties change dramatically during electroporation (Pavlin & Miklavcic, 2008; Pliquett & Weaver, 1996), these data should be available not only for pre-electroporation, but also for permeabilized tissues. Models of electroporation that take this dynamic change in tissue properties into account, provide us with much more detailed description of electroporation and can describe in vivo phenomena that cannot be explained with models that use constant tissue properties. Data on electrical tissue properties after electroporation are, however, very scarce, although they may prove crucial for accurate treatment planning.

Dose Calculation Algorithms

In modern radiation therapy two main calculation methods are used: convolution-superposition, where the patients dose is computed and lateral transport of radiation, beam energy, beam modifiers and electron density distribution are accounted for (Mackie et al, 1985; Sharpe & Battista, 1993); and Monte-Carlo where the dose is computed directly from first principles (Reynaert et al, 2007). While convolution-superposition is faster, Monte-Carlo is more accurate. Medical physicists in oncology usually choose the method according to necessity (and availability); when time is very important and many calculations are needed, convolution-superposition is used, and when high accuracy is needed, Monte-Carlo based methods are used.

In tissue electroporation modeling several algorithms are used, mostly utilizing the finite element method (for details on finite element modeling in electroporation based applications consult *Finite element modeling of in vivo electroporation* by Pavselj and Miklavcic). The difference between individual models is mostly whether they take into account the changes in tissue electrical properties during electroporation or not. Steady-state models calculate the electric field distribution without incorporating changes in tissue properties. These models give reasonably good results, if the tissues modeled are homogeneous and isotropic. (Edd & Davalos, 2007; Miklavcic et al, 2006a; Miklavcic et al, 2000). When this is not the case, sequential models can be used. These models approximate changes in electrical properties as a function of the magnitude of electric field and thus approximately describe the time course of conductivity increase during electroporation (Sel et al, 2005). Further improvements in accuracy can be achieved by taking into account a certain time-dependency of changes in electrical properties (Pucihar et al, 2008), or even by multiscale modeling that combines single cell electroporation into the bulk tissue models (Esser et al, 2007; Smith & Weaver, 2008; Weaver, 2003). Each of these improvements is computationally more intensive than the previous one, thus taking more time. For treatment planning purposes it is necessary to choose the model that takes the least time to compute, while at the same time maintains an adequate level of accuracy. Since steady-state models cannot accurately simulate IRE in complex anatomies, and multiscale models are so computer intensive that so far they have only been used in 2D, currently the only achievable options are sequential models.

Multiphysics (Applicable to Irreversible Electroporation Only)

One of the main advantages of irreversible electroporation is its non-thermal ablation capacity. When electric pulses are applied to biological tissue, heat is generated in the form of resistive heating and the temperature increases. In order to guarantee non-thermal ablation IRE treatment planning must involve a control for the heat generated by electric pulses. Coupling of electrical and thermal phenomena can provide an estimate of temperature rise and distribution due to IRE (Davalos & Rubinsky, 2008; Edd & Davalos, 2007; Pliquett, 2003); thus it can be used to calculate temperature increases for each individual application of IRE or to generate a conservative range of electric pulse parameters (pulse duration, number of pulses, pulse repetition frequency) that do not elevate tissue temperatures excessively. When the electric field distribution is highly non-homogeneous, with high local peaks, calculation of resistive heating should be included in the treatment planning process.

Forward and Inverse Treatment Planning Procedure

Forward planning is a technique used in RT to produce a treatment plan that consists of a set of physically deliverable modulated beam fluence profiles, which in practice means that, for each beam, a direction, duration and modulated intensity have to be chosen. In forward planning, an initial plan is made by a treatment planner (usually a medical physicist) who uses his/her experience to produce a set of treatment parameters that can deliver sufficient radiation to a tumor while sparing vital organs and also minimizing the dose to other healthy tissues. The treatment dose is then calculated and evaluated by an oncology expert. If necessary, improvements to the plan are made and radiation doses are recalculated. This cycle is repeated until a satisfactory plan is produced. Forward planning is used for the majority of RT treatments.

In more complex cases, when vital organs are near the target volume or the target volume is of complex shape, inverse planning produces better results (Ezzell et al, 2003; Galvin et al, 2004; Lindegaard et al, 2008; Webb, 2003). In inverse planning, which is used in intensity-modulated radiation therapy, a desired dose distribution (dose constraints) in and around the target volume is defined. Then a computer optimization technique is used together with the dose calculation algorithm to determine the optimal set of treatment parameters resulting in a dose distribution that most closely matches the desired one. The optimization algorithm compares the quality of different treatment plans according to an objective function that incorporates dose constraints defined earlier. The constraints are usually implemented using dose-volume criteria: how much target volume and critical volume is covered by the appropriate dose; how much is overdosed and how much underdosed. Biological effects models, such as tumor control probability and normal tissue complication probability are also used: what are the biological consequences of covering a certain volume of the tumor and of critical tissues with a certain dose (Bortfeld, 1999; Lyman & Wolbarst, 1987). Currently, gradient based optimization algorithms and stochastic optimization algorithms are used in radiation therapy (Bortfeld, 2006; Ezzell, 1996). Gradient algorithms are faster, but require a good initial guess to reach a satisfactory solution, while stochastic algorithms, such as simulated annealing and genetic algorithms are slower, but do not require an initial guess.

In IRE, a treatment plan should consist of appropriate electric pulse parameters, i.e. pulse duration, number of pulses, pulse repetition frequency and pulse amplitude, and of appropriate electrode parameters, i.e. electrode geometry, configuration of electrode arrays and sequence of electrode activation in cases when multiple pulses (that would induce different electric field distributions) would have to be delivered. The effect of number of pulses, their duration and repetition frequency to treatment success can be substantial; increasing the number of pulses or pulse duration increases electroporation efficiency, increasing repetition frequency decreases efficiency. Increasing any of the three increases tissue temperature (Macek-Lebar et al, 2002; Pucihar et al, 2002). Since the sequential models cannot currently evaluate the effect of these parameters on electroporation efficiency, all three parameters have to be chosen according to the experimental results and the non-thermal criteria. Numerical treatment planning procedure should therefore only deal with pulse amplitude and electrode related parameters, unless appropriate biological effects models are included.

Forward planning in IRE is possible, but probably not sensible, since inverse planning is not much more time consuming and produces better results. The choice between gradient and stochastic methods is less straight-forward. Gradient methods are faster and more accurate, if a good initial guess is available and if the number of parameters optimized for is not too large. Increasing the number of parameters can result in gradient optimization methods becoming stuck in one of the local optimums a long way away from the global optimum. In such a case, stochastic methods, such as genetic algorithm and simulated annealing, are much more likely to come close to the global optimum (since the methods are stochastic, the probability of reaching the global optimum in a reasonable amount of time are slim, however, they do find an acceptably good solution in a fixed amount of time). An additional advantage of genetic algorithm optimization is its ability to return more than one high-quality suggestion for the optimal parameters, thus giving the treatment planner more than one option of similar quality into consideration. No matter which methods (gradient or stochastic) are used, the IRE treatment planning procedure should follow the RT procedure closely. Target volumes, critical volumes and the appropriate electric field distributions should be defined. Optimization algorithm should then compare the quality of different treatment plans according to an objective function that takes into account that the electric field must be over the IRE threshold in all the target tissue and under the IRE threshold in vital organs and as low as possible elsewhere. If biological effect models ever become available for IRE, they should also be included in the objective function.

Tumor Ablation with Irreversible Electroporation Treatment Planning Examples

We present two examples of treatment planning of ablation by IRE using numerical modeling and a genetic optimization algorithm. In both examples we try to determine the best possible configuration and electric potentials of six electrodes surrounding a subcutaneous tumor - the target tissue. Our goal is to irreversibly electroporate ($E > E_{irr}$) the entire tumor volume, while sparing as much as possible of the hypothetical spherical organ at risk, situated next to the tumor (Figure 1).

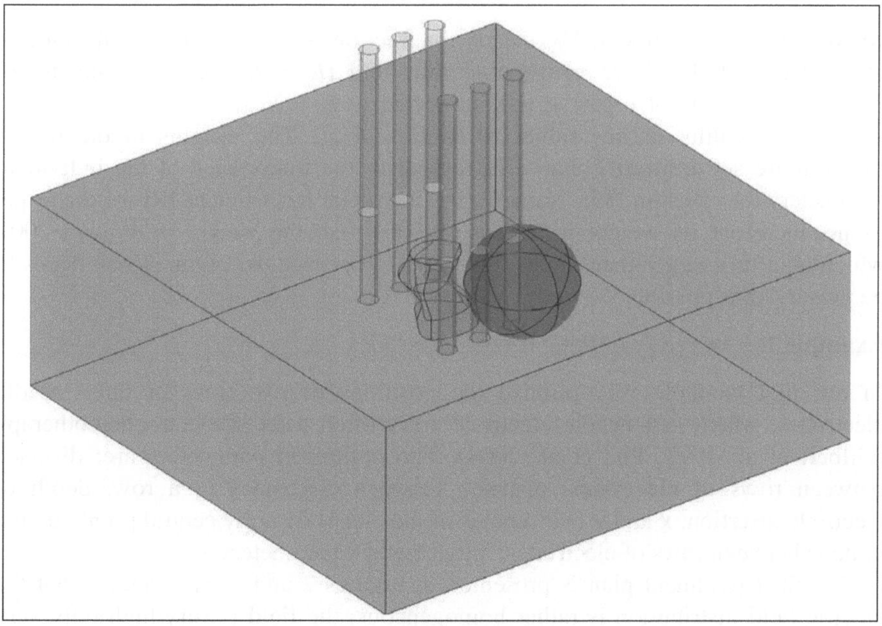

Fig. 1. Model geometry: biological tissue (light blue); tumor (green) – geometry taken from (Sel et al, 2007); organ at risk (dark blue). Needle electrodes (pink – two rows of needle electrodes as in example 1) are inserted into the tissue and appropriate electric potentials are assigned to each electrode so that the entire tumor volume and the least possible volume of the organ at risk is irreversibly electroporated.

In both examples we use the same steady-state numerical model of electroporation; i.e. electric field distribution in the tissue caused by an electric pulse is determined by solving the Laplace equation for static electric currents. All tissues are considered isotropic and homogeneous, the assigned conductivity values being 0.4 S/m for the tumor and 0.2 S/m for healthy tissue and for organ at risk (Cukjati et al, 2007; Pavselj et al, 2005). The IRE threshold is taken to be 800 V/cm, which is the average threshold reported in literature (Davalos et al, 2005). However, this value is only used for demonstration purposes, as the exact threshold is tissue dependent and also depends on electric pulse duration and number.

The genetic algorithm (Holland, 1992) was written in Matlab and was run together with the numerical calculation using the link between Matlab and COMSOL Multiphysics, a finite element software. In both examples the initial population of chromosomes is generated randomly, taking into account the following model constraints: range of distances between electrodes, range of depth of electrode insertion into the tissue and range of electric potential values on individual electrodes. Chromosomes for reproduction are selected proportionally to their fitness, according to the fitness function:

$$F = 10000 \cdot V_{Tir} - 200 \cdot V_{OARir} - 2 \cdot V_{HTir},$$

where F stands for fitness, V_{Tir} stands for fraction of tumor volume subjected to local electric field above irreversible threshold ($E > E_{irrev}$), V_{OARir} stands for fraction of volume of organ at risk subjected to $E > E_{irrev}$ and V_{HTir} stands for volume of healthy tissue subjected to $E > E_{irrev}$. The weights in the fitness function are set arbitrarily, but with respect to the importance of the individual parameters for efficient IRE. Namely, V_{Tir} is crucial for efficient IRE of the target tissue, therefore its weight is largest (10000) than the weight of V_{OARir} (-200), which is in turn larger than the weight of V_{HTir}, because the organ at risk needs to be preserved, if possible.

Example 1. – two rows of three needle electrodes.

In our first example we optimize the positions of two rows of three needle electrodes, which is a needle electrode array often used in electrochemotherapy (Gilbert et al, 1997; Puc et al, 2004). The optimized parameters are: distance between rows of electrodes, distance between electrodes in a row, depth of electrode insertion, x and y coordinates of the electrode array central point and the voltage between rows of electrodes; altogether six parameters.

The final treatment plan is presented in Figures 2 and 3. We can see that the electric field distribution is rather homogeneous; the field is very high only very close to the electrodes and just above the E_{irr} inside the tumor. Electric field is quite high in the organ at risk closest to the tumor as well – all in all E_{irr} is exceeded in 2.43 % of the organ at risk (Table 1).

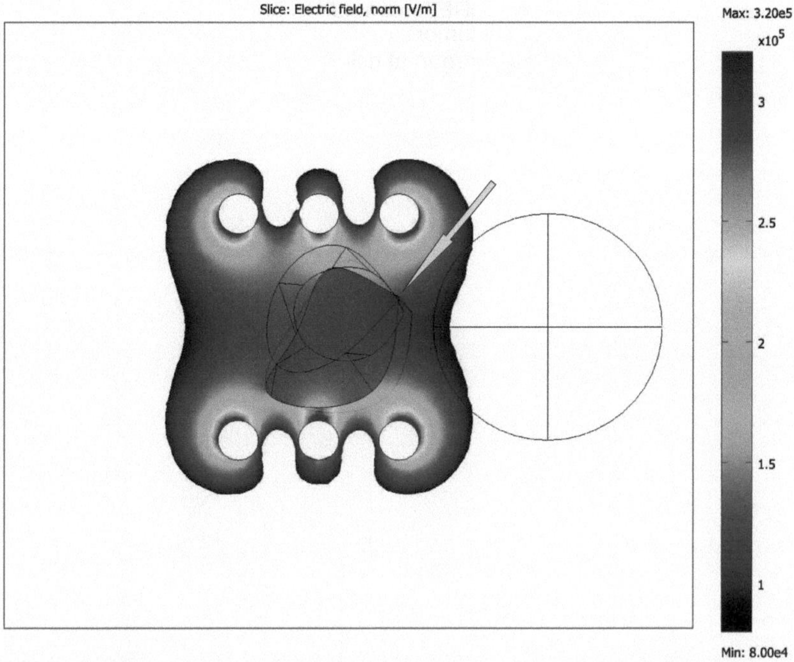

Fig. 2. Local electric field distribution for treatment plan 1 is shown in the XY plane through the center of the tumor. White arrow marks part of the tumor, where electric field is barely over IRE threshold (800 V/cm).

At a first glance it seems that the electric field exceeds E_{irr} in a large volume outside the target tissue and that the treatment planning algorithm should give better results. The obvious change to improve the result would be to put the electrodes more to the left, so that less of the organ at risk gets affected, or perhaps to use only four electrodes instead of six. Actually, none of these two obvious improvements work (data not shown). Moving the electrodes further left causes the electric field on the edge of the tumor (Figure 2) to fall below E_{irr} – as a result the potential on the electrodes has to increase so that the whole tumor volume is covered and this in turn increases the affected volume of organ at risk. Using only four electrodes leads to a similar result.

Example 2. – six needle electrodes.

In our second example we optimize the positions of six individual needle electrodes. Optimized parameters are: x and y coordinates of each of the electrodes separately, electric potential of each electrode and depth of electrode insertion (the same for all electrodes); altogether 19 parameters.

Fig. 3. Target tissue, organ at risk and IRE are presented as contours in the XY plane (left) at three different depths (top: 8 mm; middle: 23 mm; bottom: 38 mm) and in the XZ plane (right) at three different cross-sections (middle: tumor center; top and bottom: 1.2 mm from tumor center).

Table 1. Quality of treatment planning parameters V_{Tir}, V_{OARir}, V_{HT} and treatment planning (computational) time. V_{Tir} and V_{OARir} are normalized by their tissues' respective volumes, V_T and V_{OAR}. All values were calculated using the optimal parameters acquired by the optimization procedure.

	F	V_{Tir}/V_T [%]	V_{OARir}/V_{OAR} [%]	V_{HTir} [mm^3]	No. of parameters	Calculation time [h]
Example 1	9995.0	100	2.4	89	6	1.5
Example 2	9998.1	100	0.8	149	19	4.2

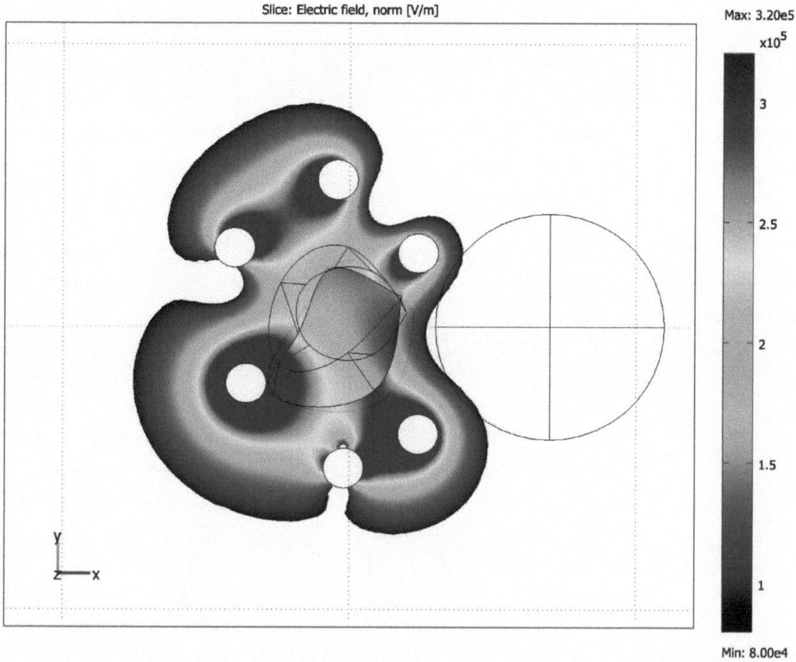

Fig. 4. Electric field distribution for the treatment plan 2 is shown in the XY plane through the center of the tumor. For better comparison, the same color legend (800 V/cm – 3200 V/cm) as in Figure 2 is used. Values over 3200 V/cm are shown in dark red.

The final treatment plan is presented in figures 4 and 5. We can see that the electric field distribution in this case is not at all homogeneous; the field around the electrodes and also in the tumor is much higher than in example 1. Nevertheless, only 0.8 % of the organ at risk is affected (Table 1). We can clearly see that the irregular positioning of the electrodes and different potentials on each electrode result in an electric field distribution "avoiding" high electric fields in the organ at risk on the expense of higher electric field elsewhere.

Both treatment plans provide the treating physician with a set of treatment parameters that successfully ablate the entire target tissue. The treatment plan 2 causes less damage to the organ at risk, however, it takes longer to calculate and causes more damage to the non-critical healthy tissue (Table 1). According to our fitness function *F*, treatment plan 2 is in fact better than treatment plan 1. However, in the clinical environment, the treating physician has control over the fitness function weights, which can be chosen according to expert knowledge and the choice may well be significantly different from ours, which may result in treatment plan 1 winning out over treatment plan 2.

Fig. 5. Target tissue, organ at risk and IRE are presented as contours in the XY plane (left) at three different depths (top: 8 mm; middle: 23 mm; bottom: 38 mm) and in the XZ plane (right) at three different cross-sections (middle: tumor center; top and bottom: 1.2 mm from tumor center)

Conclusions

There are many similarities between radiation therapy and irreversible electroporation. Both treatments depend on the knowledge of the treated anatomy and on medical imaging. Both treatments are based on harmful effects of a physical agent, which can be efficiently and accurately numerically modeled. Both treatments are local – they target a certain volume of tissue and try not to affect the rest. That is why the RT treatment planning methodology can at least partially be translated to IRE treatment planning, while keeping in mind the relevant differences between the two treatments.

RT and IRE treatment planning both rely on medical imaging to provide patient anatomical data. RT also relies on imaging to provide patient tissue properties, while IRE relies on average electrical properties of excised human tissue measured by different researchers. While tissue properties do not change significantly during RT, they do during IRE. Tissue properties and tissue specific IRE thresholds seems to be the biggest challenge that IRE planning has still to address. Before IRE treatment planning takes off, additional research will have to be performed on human tissue electrical properties, before, during and after electroporation. More data on tissue specific IRE thresholds are also needed, at the very least for all target tissues and critical tissues. At the moment IRE thresholds are only available for few tissue types and limited pulse parameters (Miklavcic et al, 2000; Pavselj et al, 2005). Only after these crucial parameters are available will it become possible to accurately model IRE and provide high accuracy treatment plans.

Choosing the appropriate mathematical model and the appropriate optimization algorithm are also very important steps in treatment planning, as is the decisions, which parameters should the algorithm optimize. Currently, the most viable option seems to be the sequential model of electroporation in combination with one of the stochastic optimization algorithms, which generally provide the best results for higher numbers of optimized parameters (Corovic et al, 2008; Sel et al, 2007; Zupanic et al, 2008). When non-homogeneous electric field distributions are expected, electroporation models should also include modeling of the thermal effects to guarantee non-thermal ablation of target tissue (Davalos & Rubinsky, 2008; Edd & Davalos, 2007).

The presented (hypothetical) IRE treatment planning procedure used the steady-state electroporation model, i.e. no changes in tissue conductivity due to electroporation is taken into consideration, and the genetic optimization algorithm to plan IRE treatment of a subcutaneous tumor. Two different numbers of optimized parameters were chosen and two completely different treatment plans resulted in completely covering the tumor with sufficiently high electric field and only minimally affecting the organ at risk. The presented approach represents the basis for developing future IRE treatment planning algorithms.

Acknowledgements

This research was supported by the Slovenian Research Agency. We would like to thank Božidar Casar, Rihard Hudej and Robert Hudej of Institute of Oncology, Ljubljana for fruitful discussion on radiation therapy and brachytherapy treatment planning.

References

Al-Sakere, B., Bernat, C., Andr, F., Connault, E., Opolon, P., Davalos, R.V., Mir, L.M.: A study of the immunological response to tumor ablation with irreversible Electroporation. Technology in Cancer Research & Treatment 6(4), 301–305 (2007)

Belehradek, J., Orlowski, S., Poddevin, B., Paoletti, C., Mir, L.M.: Electrochemotherapy of spontaneous mammary-tumors in mice. European Journal of Cancer 27(1), 73–76 (1991)

Bertacchini, C., Margotti, P.M., Bergamini, E., Lodi, A., Ronchetti, M., Cadossi, R.: Design of an irreversible Electroporation system for clinical use. Technology in Cancer Research & Treatment 6(4), 313–320 (2007)

Bortfeld, T.: Optimized planning using physical objectives and constraints. Seminars in Radiation Oncology 9(1), 20–34 (1999)

Bortfeld, T.: IMRT: a review and preview. Physics in Medicine and Biology 51(13), R363–R379 (2006)

Boyer, A.L., Butler, E.B., DiPetrillo, T.A., Engler, M.J., Fraass, B., Grant, W., Ling, C.C., Low, D.A., Mackie, T.R., Mohan, R., Purdy, J.A., Roach, M., Rosenman, J.G., Verhey, L.J., Wong, J.W., Cumberlin, R.L., Stone, H., Palta, J.R.: Intensity Modulated Radiation T Intensity modulated radiotherapy: Current status and issues of interest. International Journal of Radiation Oncology Biology Physics 51(4), 880–914 (2001)

Chandra, A., Dong, L., Huang, E., Kuban, D.A., O'Neill, L., Rosen, I., Pollack, A.: Experience of ultrasound-based daily prostate localization. International Journal of Radiation Oncology Biology Physics 56(2), 436–447 (2003)

Corovic, S., Pavlin, M., Miklavcic, D.: Analytical and numerical quantification and comparison of the local electric field in the tissue for different electrode configurations. Biomedical Engineering Online 6 (2007)

Corovic, S., Zupanic, A., Miklavcic, D.: Numerical modeling and optimization of electric field distribution in subcutaneous tumor treated with electrochemotherapy using needle electrodes. IEEE Transactions on Plasma Science 36(4), 1665–1672 (2008)

Cukjati, D., Batiuskaite, D., Andre, F., Miklavcic, D., Mir, L.M.: Real time electroporation control for accurate and safe in vivo non-viral gene therapy. Bioelectrochemistry 70(2), 501–507 (2007)

Davalos, R.V., Mir, L.M., Rubinsky, B.: Tissue ablation with irreversible electroporation. Annals of Biomedical Engineering 33(2), 223–231 (2005)

Davalos, R.V., Otten, D.M., Mir, L.M., Rubinsky, B.: Electrical impedance tomography for imaging tissue electroporation. IEEE Transactions on Biomedical Engineering 51(5), 761–767 (2004)

Davalos, R.V., Rubinsky, B.: Temperature considerations during irreversible electroporation. International Journal of Heat and Mass Transfer 51(23-24), 5617–5622 (2008)

Davalos, R.V., Rubinsky, B., Otten, D.M.: A feasibility study for electrical impedance tomography as a means to monitor tissue electroporation for molecular medicine. IEEE Transactions on Biomedical Engineering 49(4), 400–403 (2002)

Dawson, L.A., Sharpe, M.B.: Image-guided radiotherapy: rationale, benefits, and limitations. Lancet Oncology 7(10), 848–858 (2006)

Denet, A.R., Vanbever, R., Preat, V.: Skin electroporation for transdermal and topical delivery. Advanced Drug Delivery Reviews 56(5), 659–674 (2004)

Edd, J.F., Davalos, R.V.: Mathematical Modeling of irreversible Electroporation for treatment planning. Technology in Cancer Research & Treatment 6(4), 275–286 (2007)

Edd, J.F., Horowitz, L., Davalos, R.V., Mir, L.M., Rubinsky, B.: In vivo results of a new focal tissue ablation technique: Irreversible electroporation. IEEE Transactions on Biomedical Engineering 53(7), 1409–1415 (2006)

Esser, A.T., Smith, K.C., Gowrishankar, T.R., Weaver, J.C.: Towards solid tumor treatment by irreversible electroporation: Intrinsic redistribution of fields and currents in tissue. Technology in Cancer Research & Treatment 6(4), 261–273 (2007)

Ezzell, G.A.: Genetic and geometric optimization of three-dimensional radiation therapy treatment planning. Medical Physics 23(3), 293–305 (1996)

Ezzell, G.A., Galvin, J.M., Low, D., Palta, J.R., Rosen, I., Sharpe, M.B., Xia, P., Xiao, Y., Xing, L., Yu, C.X.: Guidance document on delivery, treatment planning, and clinical implementation of IMRT: Report of the IMRT subcommittee of the AAPM radiation therapy committee. Medical Physics 30(8), 2089–2115 (2003)

Gabriel, C., Gabriel, S., Corthout, E.: The dielectric properties of biological tissues.1. Literature survey. Physics in Medicine and Biology 41(11), 2231–2249 (1996a)

Gabriel, S., Lau, R.W., Gabriel, C.: The dielectric properties of biological tissues.2. Measurements in the frequency range 10 Hz to 20 GHz. Physics in Medicine and Biology 41(11), 2251–2269 (1996b)

Galvin, J.M., Ezzell, G., Eisbrauch, A., Yu, C., Butler, B., Xiao, Y., Rosen, I., Rosenman, J., Sharpe, M., Xing, L., Xia, P., Lomax, T., Low, D.A., Palta, J.: Implementing IMRT in clinical practice: A joint document of the American Society for Therapeutic Radiology and Oncology and the American Association of Physicists in Medicine. International Journal of Radiation Oncology Biology Physics 58(5), 1616–1634 (2004)

Gilbert, R.A., Jaroszeski, M.J., Heller, R.: Novel electrode designs for electrochemotherapy. Biochimica Et Biophysica Acta-General Subjects 1334(1), 9–14 (1997)

Golzio, M., Rols, M.P., Teissie, J.: In vitro and in vivo electric field-mediated permeabilization, gene transfer, and expression. Methods 33(2), 126–135 (2004)

Granot, Y., Rubinsky, B.: Methods of optimization of electrical impedance tomography for imaging tissue electroporation. Physiological Measurement 28(10), 1135–1147 (2007)

Heinzerling, J.H., Papiez, L., Chien, S., Anderson, J., Forster, K., Zhang, G., Timmerman, R.: Stereotactic body radiation therapy: Evaluation of setup accuracy and targeting methods for a new couch integrated immobilization system. Technology in Cancer Research & Treatment 7(3), 197–206 (2008)

Heller, R., Gilbert, R., Jaroszeski, M.J.: Clinical applications of electrochemotherapy. Advanced Drug Delivery Reviews 35(1), 119–129 (1999)

Hojman, P., Zibert, J.R., Gissel, H., Eriksen, J., Gehl, J.: Gene expression profiles in skeletal muscle after gene electrotransfer. Bmc. Molecular Biology 8 (2007)

Holland, J.H.: Adaptation in Natural and Artificial Systems: An Introductory Analysis with Applications to Biology, Control, and Artificial Intelligence. MIT Press, Cambridge (1992)

ICRU-50, Prescribing, recording, and reporting photon beam therapy. Bethesda (1993)

ICRU-62, Prescribing, recording, and reporting photon beam therapy (supplement to ICRU Report 50). Bethesda (1999)

Ivorra, A., Rubinsky, B.: In vivo electrical impedance measurements during and after electroporation of rat liver. Bioelectrochemistry 70(2), 287–295 (2007)

Jaffray, D., Kupelian, P., Djemil, T., Macklis, R.M.: Review of image-guided radiation therapy. Expert Review of Anticancer Therapy 7(1), 89–103 (2007)

Lampert, V.L., Zelch, J.V., Cohen, D.N.: Computed tomography of orbits. Radiology 113(2), 351–354 (1974)

Lavee, J., Onik, G., Mikus, P., Rubinsky, B.: A novel nonthermal energy source for surgical epicardial atrial ablation: Irreversible electroporation. Heart Surgery Forum 10(2), E162–E167 (2007)

Lecchi, M., Fossati, P., Elisei, F., Orecchia, R., Lucignani, G.: Current concepts on imaging in radiotherapy. European Journal of Nuclear Medicine and Molecular Imaging 35(4), 821–837 (2008)

Lee, E.W., Loh, C.T., Kee, S.T.: Imaging guided percutaneous irreversible electroporation: Ultrasound and immunohistological correlation. Technology in Cancer Research & Treatment 6(4), 287–293 (2007)

Lindegaard, J.C., Tanderup, K., Nielsen, S.K., Haack, S., Gelineck, J.: MRI-guided 3D optimization significantly improves DVH parameters of pulsed-dose-rate brachytherapy in locally advanced cervical cancer. International Journal of Radiation Oncology Biology Physics 71(3), 756–764 (2008)

Lyman, J.T., Wolbarst, A.B.: Optimization of radiation-therapy.3. A method of assessing complication probabilities from dose-volume histograms. International Journal of Radiation Oncology Biology Physics 13(1), 103–109 (1987)

Macek-Lebar, A., Sersa, G., Kranjc, S., Groselj, A., Miklavcic, D.: Optimisation of pulse parameters in vitro for in vivo electrochemotherapy. Anticancer Research 22(3), 1731–1736 (2002)

Mackie, T.R., Scrimger, J.W., Battista, J.J.: A convolution method of calculating dose for 15-MV X-rays. Medical Physics 12(2), 188–196 (1985)

Maor, E., Ivorra, A., Leor, J., Rubinsky, B.: The effect of irreversible electroporation on blood vessels. Technology in Cancer Research & Treatment 6(4), 307–312 (2007)

Maor, E., Ivorra, A., Leor, J., Rubinsky, B.: Irreversible electroporation attenuates neointimal formation after angioplasty. IEEE Transactions on Biomedical Engineering 55(9), 2268–2274 (2008)

Miklavcic, D., Beravs, K., Semrov, D., Cemazar, M., Demsar, F., Sersa, G.: The importance of electric field distribution for effective in vivo electroporation of tissues. Biophysical Journal 74(5), 2152–2158 (1998)

Miklavcic, D., Corovic, S., Pucihar, G., Pavselj, N.: Importance of tumour coverage by sufficiently high local electric field for effective electrochemotherapy. Ejc Supplements 4(11), 45–51 (2006a)

Miklavcic, D., Kotnik, T.: Electroporation for electrochemotherapy and gene therapy. In: Rosch, P., Markov, M. (eds.) Bioelectromagnetic medicine, pp. 637–656. Marcel Decker, New York (2004)

Miklavcic, D., Pavselj, N., Hart, F.X.: Electric properties of tissues. In: Wiley Encyclopedia of Biomedical Engineering. John Wiley & Sons, New York (2006b)

Miklavcic, D., Semrov, D., Mekid, H., Mir, L.M.: A validated model of in vivo electric field distribution in tissues for electrochemotherapy and for DNA electrotransfer for gene therapy. Biochimica Et Biophysica Acta-General Subjects 1523(1), 73–83 (2000)

Miller, L., Leor, J., Rubinsky, B.: Cancer cells ablation with irreversible electroporation. Technology in Cancer Research & Treatment 4(6), 699–705 (2005)

Mir, L.M., Bureau, M.F., Gehl, J., Rangara, R., Rouy, D., Caillaud, J.M., Delaere, P., Branellec, D., Schwartz, B., Scherman, D.: High-efficiency gene transfer into skeletal muscle mediated by electric pulses. Proceedings of the National Academy of Sciences of the United States of America 96(8), 4262–4267 (1999)

Mir, L.M., Glass, L.F., Sersa, G., Teissie, J., Domenge, C., Miklavcic, D., Jaroszeski, M.J., Orlowski, S., Reintgen, D.S., Rudolf, Z., Belehradek, M., Gilbert, R., Rols, M.P., Belehradek, J., Bachaud, J.M., DeConti, R., Stabuc, B., Cemazar, M., Coninx, P., Heller, R.: Effective treatment of cutaneous and subcutaneous malignant tumours by electrochemotherapy. British Journal of Cancer 77(12), 2336–2342 (1998)

Neumann, E., Schaeferridder, M., Wang, Y., Hofschneider, P.H.: Gene transfer into mouse lyoma cells by electroporation in high electric-fields. Embo Journal 1(7), 841–845 (1982)

Newbold, K., Partridge, M., Cook, G., Sohaib, A., Charles-Edwards, E., Rhys-Evans, P., Harrington, K., Nutting, C.: Advanced imaging applied to radiotherapy planning in head and neck cancer: a clinical review. British Journal of Radiology 79(943), 554–561 (2006)

Onik, G., Mikus, P., Rubinsky, B.: Irreversible electroporation: Implications for prostate ablation. Technology in Cancer Research & Treatment 6(4), 295–300 (2007)

Orlowski, S., Belehradek, J., Paoletti, C., Mir, L.M.: Transient electropermeabilization of cells in culture - increase of the cyto-toxicity of anticancer drugs. Biochemical Pharmacology 37(24), 4727–4733 (1988)

Pavlin, M., Miklavcic, D.: Theoretical and experimental analysis of conductivity, ion diffusion and molecular transport during cell electroporation — Relation between short-lived and long-lived pores. Bioelectrochemistry 74, 38–46 (2008)

Pavselj, N., Bregar, Z., Cukjati, D., Batiuskaite, D., Mir, L.M., Miklavcic, D.: The course of tissue permeabilization studied on a mathematical model of a subcutaneous tumor in small animals. IEEE Transactions on Biomedical Engineering 52(8), 1373–1381 (2005)

Pavselj, N., Miklavcic, D.: Numerical modeling in electroporation-based biomedical applications. Radiology and Oncology 42(3), 159–168 (2008a)

Pavselj, N., Miklavcic, D.: Numerical models of skin electropermeabilization taking into account conductivity changes and the presence of local transport regions. IEEE Transactions on plasma science, 1650–1658 (2008b)

Pliquett, U.: Joule heating during solid tissue electroporation. Medical & Biological Engineering & Computing 41(2), 215–219 (2003)

Pliquett, U., Weaver, J.C.: Electroporation of human skin: Simultaneous measurement of changes in the transport of two fluorescent molecules and in the passive electrical properties. Bioelectrochemistry and Bioenergetics 39(1), 1–12 (1996)

Polk, C., Postow, E.: Handbook of Biological Effects of Electromagnetic Fields. CRC Press, Boca Raton (1996)

Potter, R., Fidarova, E., Kirisits, C., Dirnopoulos, J.: Image-guided adaptive brachytherapy for cervix carcinoma. Clinical Oncology 20(6), 426–432 (2008a)

Potter, R., Kirisits, C., Fidarova, E.F., Dimopoulos, J.C.A., Berger, D., Tanderup, K., Lindegaard, J.C.: Present status and future of high-precision image guided adaptive brachytherapy for cervix carcinoma. Acta Oncologica 47(7), 1325–1336 (2008b)

Prausnitz, M.R.: A practical assessment of transdermal drug delivery by skin electroporation. Advanced Drug Delivery Reviews 35(1), 61–76 (1999)

Puc, M., Corovic, S., Flisar, K., Petkovsek, M., Nastran, J., Miklavcic, D.: Techniques of signal generation required for electropermeabilization. Survey of electropermeabilization devices. Bioelectrochemistry 64(2), 113–124 (2004)

Pucihar, G., Kotnik, T., Miklavcic, D., Teissie, J.: Kinetics of transmembrane transport of small molecules into electropermeabilized cells. Biophysical Journal 95(6), 2837–2848 (2008)

Pucihar, G., Mir, L.M., Miklavcic, D.: The effect of pulse repetition frequency on the uptake into electropermeabilized cells in vitro with possible applications in electrochemotherapy. Bioelectrochemistry 57(2), 167–172 (2002)

Purdy, J., Starkschall, G.: A practical guide to 3-D planning and conformal radiation therapy. Advanced Medical Publishing, Madison (1999)

Reynaert, N., van der Marck, S.C., Schaart, D.R., Van der Zee, W., Van Vliet-Vroegindeweij, C., Tomsej, M., Jansen, J., Heijmen, B., Coghe, M., De Wagter, C.: Monte Carlo treatment planning for photon and electron beams. Radiation Physics and Chemistry 76(4), 643–686 (2007)

Rubinsky, B., Onik, G., Mikus, P.: Irreversible electroporation: A new ablation modality - Clinical implications. Technology in Cancer Research & Treatment 6(1), 37–48 (2007)

Rubinsky, J., Onik, G., Mikus, P., Rubinsky, B.: Optimal Parameters for the Destruction of Prostate Cancer Using Irreversible Electroporation. Journal of Urology 180(6), 2668–2674 (2008)

Ruchala, K.J., Olivera, G.H., Schloesser, E.A., Hinderer, R., Mackie, T.R.: Calibration of a tomotherapeutic MVCT system. Physics in Medicine and Biology 45(4), N27–N36 (2000)

Scott-Taylor, T.H., Pettengell, R., Clarke, I., Stuhler, G., La Barthe, M.C., Walden, P., Dalgleish, A.G.: Human tumour and dendritic cell hybrids generated by electrofusion: potential for cancer vaccines. Biochimica Et Biophysica Acta-Molecular Basis of Disease 1500(3), 265–279 (2000)

Sel, D., Cukjati, D., Batiuskaite, D., Slivnik, T., Mir, L.M., Miklavcic, D.: Sequential finite element model of tissue electropermeabilization. IEEE Transactions on Biomedical Engineering 52(5), 816–827 (2005)

Sel, D., Lebar, A.M., Miklavcic, D.: Feasibility of employing model-based optimization of pulse amplitude and electrode distance for effective tumor electropermeabilization. IEEE Transactions on Biomedical Engineering 54(5), 773–781 (2007)

Semrov, D., Miklavcic, D.: Calculation of the electrical parameters in electrochemotherapy of solid tumours in mice. Computers in Biology and Medicine 28(4), 439–448 (1998)

Serša, G., Čemažar, M., Miklavčič, D., Rudolf, Z.: Electrochemotherapy of tumours. Radiology & Oncology 40, 163–174 (2006)

Sharpe, M.B., Battista, J.J.: Dose calculations using convolution and superposition principles - the orientation of dose spread kernels in divergent X-ray beams. Medical Physics 20(6), 1685–1694 (1993)

Skerl, D., Likar, B., Fitzpatrick, J.M., Pernus, F.: Comparative evaluation of similarity measures for the rigid registration of multi-modal head images. Physics in Medicine and Biology 52(18), 5587-5601 (2007)

Slomka, P.J.: Software approach to merging molecular with anatomic information. Journal of Nuclear Medicine 45, 36S–45S (2004)

Smith, K.C., Weaver, J.C.: Active mechanisms are needed to describe cell responses to submicrosecond, megavolt-per-meter pulses: Cell models for ultrashort pulses. Biophysical Journal 95(4), 1547–1563 (2008)

Trontelj, K., Rebersek, M., Kanduser, M., Curin-Serbec, V., Sprohar, M., Miklavcic, D.: Optimization of bulk cell electrofusion in vitro for production of human–mouse heterohybridoma cells. Bioelectrochemistry 74, 124–129 (2008)

Vanuytsel, L.J., Vansteenkiste, J.F., Stroobants, S.G., De Leyn, P.R., De Wever, W., Verbeken, E.K., Gatti, G.G., Huyskens, D.P., Kutcher, G.J.: The impact of F-18-fluoro-2-deoxy-D-glucose positron emission tomography (FDG-PET) lymph node staging on the radiation treatment volumes in patients with non-small cell lung cancer. Radiotherapy and Oncology 55(3), 317–324 (2000)

Weaver, J.C.: Electroporation of biological membranes from multicellular to nano scales. IEEE Transactions on Dielectrics and Electrical Insulation 10(5), 754–768 (2003)

Webb, S.: The physical basis of IMRT and inverse planning. British Journal of Radiology 76(910), 678–689 (2003)

Zupanic, A., Corovic, S., Miklavcic, D.: Optimization of electrode position and electric pulse amplitude in electrochemotherapy. Radiology and Oncology 42(2), 93–101 (2008)

The Place of the Electroporation-Based Antitumor Therapies in the Electrical Armamentarium against Cancer

Lluis M. Mir

CNRS, UMR 8121, Institut Gustave-Roussy, F-94805 VILLEJUIF Cédex and Univ. Paris-Sud, UMR, Paris 8121, France
Corresponding author E-mail: luismir@igr.fr

Abstract. Irreversible electroporation (IRE) and electrochemotherapy are two electroporation-based antitumor treatments, both relying on the delivery of short electric pulses of a typical duration of 100 microseconds. Constant current, radiofrequency electromagnetic fields and other types of pulses are also used in the clinics or are tested in preclinical trials. In electrochemotherapy, cell reversible electropermeabilization allows the uptake of non-permeant or low-permeant anticancer drugs and the tumor cells killer is the drug. In IRE, the killer is the electric field which irreversibly perturbs membranes structure and cells homeostasis. The characteristics of IRE and electrochemotherapy are compared, showing the differences and the complementarities of these two antitumor approaches.

Abbreviations: IRE: Irreversible electroporation, ECT: electrochemotherapy, SOP: Standard Operating Procedures.

Keywords: Irreversible electroporation, electrochemotherapy, reversible electroporation, electropermeabilization, tumor treatments.

1 Introduction: Reversible and Irreversible « Electroporation »

For at least two decades, the delivery of “permeabilizing” electric pulses to cells and to tissues was concentrated on the use of electrical parameters allowing achieving “reversible” electroporation[1]. The electric pulses, through their effect on the cell membrane, were envisaged as a way to vectorize molecules from the cell environment to the cell inside[2-4]. This approach resulted indeed in the possibility to introduce non-permeant molecules with strictly intracellular targets, those “non-permeant” molecules being devoid of transport mechanisms across the cell membrane and being completely unable to diffuse through the cell membrane. The internalisation of non-permeant molecules such as long DNA (e.g. plasmids, minicircles,…) or RNA molecules (e.g. mRNA) and short DNA (e.g. antisense oligonucleotides) or RNA molecules (e.g. siRNA) has been exploited in basic and applied studies[5, 6].

B. Rubinsky (Ed.): Irreversible Electroporation, BIOMED, pp. 223–233.
springerlink.com

Because *in vitro*, on cells in culture, irreversible electroporation lead to a usually unwanted cell loss, irreversible electroporation was considered as a negative side effect of the electric pulses delivery. Efforts were thus developed to find "optimal" conditions to achieve the electropermeabilization of (if possible) the whole of the exposed cells, with no cell death caused by the electric pulses delivery[4]. *In vitro*, by means of experimental "try and error" optimization procedures, conditions were found where 90% of the cells survived to the procedures (the electric pulses, and also the trypsinization, pipettings, centrifugation, etc) and 98% of the surviving cells were reversibly permeabilized[4]. *In vivo*, aside of the use of the same experimental approach, the development of models to understand electric field distribution in the tissues allowed reaching similar levels of permeabilization and safety[7, 8].

Reversible electropermeabilization has evident applications for example in gene therapy[9-11]. *In vitro*, if viral and chemical approaches are efficient to transfer nucleic acids to cells, there are cells refractory to efficient transformation by these classical approaches. Electrotransfer is then a valid alternative to transduce these cells. *In vivo*, extensive use of virus has been performed, with recent extraordinary clinical achievements [12] but also with many drawbacks (including the death of patients) that have slowed down the progression of this therapeutic approach. Non viral approaches development is nowadays a demand from the end-users, particularly in the frame of the regenerative medicine where safety must be the first point to consider. Indeed, non viral approaches, in particular those derived from the use of physical means, do not seem to introduce long term perturbations to the cells. They are becoming more and more efficient, in particular nucleic acids electrotransfer, which seems to be a reversible electropermeabilization-based technology with an important future[11, 13]. One of its immediate applications could be vaccination, as its easiness and efficacy have been already proved [14, 15].

Interestingly, another reversible electropermeabilization-based technology, electrochemotherapy, is now in the clinical routine in several countries, disseminating to others, in particular in Europe. In the third section of this chapter, we will present this antitumor treatment based on reversible electroporation, and we will compare its characteristics to those of the IRE antitumor approach in the fourth section.

2 Antitumor Treatments Based on the Delivery of Electricity

Before the description of the direct comparison between the IRE and electrochemotherapy antitumor approaches, it is important to present at large the use of electricity to treat solid tumors (until now, no direct treatment of leukaemia has been based on the use of the electricity, except under extracorporeal conditions, in which blood would be treated under continuous flow outside of the body; this approach could have been envisaged from experiments performed about two decades ago, but has not progressed until preclinical trials).

Indeed, electricity is already used for tumor treatment in the clinical routine, under various forms. Some of them are used almost in every place in the world, while the use of others is not yet worldwide extended (at the present time).

2.1 Worldwide Current Technologies

- Ablathermia with radiofrequencies

This is a local treatment of isolated nodules located, for example, in lungs and in liver[16-18]. It is based on the delivery of radiofrequency electromagnetic fields at high power. The local heating of the treated tumor mass provokes the denaturation of the proteins and other biological molecules with the consequent cell killing of the affected cells. The treatment is very efficient and can be followed on line under ultrasounds or CT monitoring because there is a clear frontier between the denatured (heated) area and the surrounding unheated tissue. Moreover treatment planning procedures allow optimizing the treatment. However, the antenna requires a good positioning by the operator. The main drawback is that the coagulated tissue is hardly removed by the body: the residue of the treated tumor mass disappears very slowly, which makes difficult the assessment of the treatment efficacy. This approach is now used in many countries.

- Electrical knife as a main, or adjuvant, tool in surgery

This is a use of the electricity that is very often forgotten in the lists, probably because electrical knifes are currently present in all surgery rooms[19, 20]. It is important to recall that surgery is the first antitumor treatment, and the one that allows the achievement of the largest proportion of cures. Electrical knifes can be used to really cut the tumor mass and remove it from the body, or they can be used as a sort of “adjuvant” technology to interrupt the bleeding caused by the section of small vessels.

2.2 Technologies Used in the Clinical Practice in Various Parts of the World

- Electrochemical treatment

The delivery of a constant current between two electrodes results in a series of electrochemical reactions at the level of the cathode and the anode. Indeed, not only metals can be released form the electrodes, but the injection and the extraction of electrons due to the current flow provoke the generation of reactive species that destroy the biological tissues. Associated to this, there are extreme variations of the pH around the electrodes, for the same reasons[21]. While this approach was initiated in Europe[22, 23], it actually developed in China were several thousands of patients were treated in a few years[24, 25]. Treatment uses constant current at low field amplitude (a few volts per centimetre) for minutes or hours depending on the size of the lesion to be treated. The low field, associated to low current (milliamperes) results in electrochemical destruction without thermal effects[21, 26].

- Hyperthermia with moderate temperature increases

It was shown that moderate increase of the temperature (43 to 45°C) will induce cell suicide and therefore tumor regression without thermal destruction of the tissues[27]. Such a local heating of the tumor can be achieved by means of electromagnetic fields delivery at various frequencies. This method could have been a very soft and safe procedure for tumor eradication[28, 29]. Unfortunately, there is heat transfer to surrounding tissues and through the blood flow which makes difficult to control the local heating, also impaired by the fact that the body thermostat in the hypothalamus reduces the whole body temperature when local heating reaches certain limits. This method, in spite of large research efforts, is almost no longer used.

- Electrochemotherapy

This procedure, based on reversible electroporation, will be described in the next section. It is also a non-thermal treatment.

2.3 Experimental Approaches

Two recent experimental approaches have been described.

- Nanopulses-based treatment

Two groups have recently shown the antitumor possibilities offered by the delivery of "nanopulses" (electric pulses of a submicrosecond duration and very high field amplitude) to treat small tumor nodules. Both 60 ns and 300 ns duration pulses have been used to modify the cells respectively with pulses of 6.5 kV between needles distant of 0.15 cm[30] or fields just greater than 20 kV/cm[31]. The mechanisms are still under analysis as both apoptosis induction and cell electroporation have been invoked. One of these recent papers describes the application of this technology for the treatment of a basal cell carcinoma in a human being, which resulted in a clinical complete regression, histologically confirmed[30].

- Irreversible electroporation

Irreversible electroporation being the subject of this book, and its clinical applications the matter of full chapters, we will limit its description, in section 4, to the comparison with the electrochemotherapy.

3 Electrochemotherapy

3.1 The Electrochemotherapy Concept

We developed electrochemotherapy concept as early as the end of the 80's, based on work on cells in culture. Under optimal reversible electropermeabilization conditions[4], a new pharmacology using transiently permeabilized cells by physical means was then possible. Interestingly, the physical approaches to make the cells transiently permeable do not bring or remove particular cell components (as the chemical vectors do). Therefore, once the physical effects on the cells can be ruled out, the effects observed are indeed those of the substances that have been able to penetrate the cells thanks to the permeabilization of the cell membrane. Under these conditions we found that an increase of thousands of times in bleomycin toxicity could be measured on cells in culture if cells were exposed to the electric pulses at the beginning of their incubation with this anticancer drug[2]. Experiments in mice, with transplanted and spontaneous tumors, demonstrated the antitumor efficacy of this combination, and a increase of about 1000 times of the *in vivo* antitumor capabilities of the bleomycin[32]. Early clinical trials also demonstrated electrochemotherapy feasibility in humans[33, 34]. Later, G. Sersa and colleagues also showed that cytotoxicity of cisplatinum on cells in culture was increased by cells reversible electropermeabilzation, as well as the antitumor efficacy of the cisplatinum *in vivo*, in animal models and in humans[35-37].

The next steps run in two directions, basic science and translational efforts

3.2 Bases of the Electrochemotherapy

Concerning the basic aspects of the electrochemotherapy, we completely analyzed the parameters of the treatment and the reasons of its efficacy. In the mean time we showed its efficacy in different tumor systems, including intracerebral gliomas[38], spontaneous fibrosarcomas in cats[39], transplanted tumors in liver in rabbits[40], etc. In parallel many other groups investigated the possibilities of the ECT, and more than 40 different types of preclinical models were found efficiently treatable by ECT.

We focussed on mechanisms analysis and we found that bleomycin is completely unable to diffuse through the plasma membrane and is devoid of an active transport mechanism at the cell surface (bleomycin actually enters the cells by receptor-mediated endocytosis, a slow and passive mechanism with limited cargo efficiency)[41]. When 500 molecules of bleomycin enter the cell at the time of the treatment (cells remain fully permeable for less than one minute in vitro and for only a few minutes in vivo, as revealed by our experiments), the double-strand breaks (DSB) caused by the bleomycin can only be partially repaired[42]. The unrepaired DSB are the responsible for the cell death: the cells will not be killed immediately as the probability to cut a housekeeping essential gene is extremely low when a few cuts are made in the chromosomes, but division into two daughter cells is impossible as mitosis is blocked due to the presence of broken chromosomes. This explains why the doses of bleomycin used in ECT remain low, and are on purpose kept low, in order to get a differential effect between the normal cells (that do not divide and that remain alive and functional in spite of the few DSB generated in their chromosomes) and the tumor cells that will die at their first division[43]. This differential effect explains, for example, the extraordinary cosmetic effects observed in the treatment of cutaneous or subcutaneous nodules by ECT (as depicted in several clinical papers) and the overall safety of the procedure[44].

We, and others, also showed that a correct coverage of the tumors by an efficient electric field is essential[45] and we have introduced ways to ameliorate the distribution of the electric field in the treated tumor[46, 47]. Another unexpected point that was found in the preclinical and clinical experiments is the absence of bleeding after ECT. Even the removal of needles inserted in liver does not provoke bleeding if appropriate electric pulses have been delivered. In muscle we showed that the electric pulses provoke a physiological vascular lock, histamine-dependent, that only lasts for one to two minutes if muscle fibers reversible electropermeabilisation is achieved (a time insufficient to provoke hypoxia-related side effects)[48]. In tumors, Gregor Sersa and colleagues showed that this vascular lock stands for much longer times, which contributes to the antitumor effects of the treatment[49]. Finally, the cell death provoked by the bleomycin (mitotic cell death) results in a delayed inflammatory process that raises an antitumor response to immunogeneic tumors. This immune response can be largely enhanced by either Toll-like receptors ligands[50] or interleukin-2[51], to an extent in which systemic antitumor effects can be generated by the combination of the ECT and these immunomodulating molecules, which opens particularly interesting perspectives.

3.2 *Electrochemotherapy in the Clinics*

Clinical development followed a three steps process.

First a clinically approved (CE marked) device was developed, the Cliniporator™ (IGEA Inc., Carpi, Italy), within the Cliniporator EU project. Apart from diverse security elements that makes its use (and therefore the electrochemotherapy application) very safe, the Cliniporator™ incorporates a console on which, just after the pulses delivery, a screen displays both the voltage actually applied and the current actually delivered through the electrodes during the pulses. This allows the physician to have a real-time control over the success of the electric pulses application, which very often almost corresponds to a control over the success of the treatment, provide the electrodes cover the tumor entirely (in one or in several applications).

The second step was a clinical trial at the European level, the ESOPE project. It revealed that close to 74% of the cutaneous and subcutaneous nodules treated (in some cases, untreatable using established conventional methods) completely disappeared in 8 to 10 weeks, and 11% more where in partial regression (85% of objective responses)[44]. This trial also allowed the definition and validation of the Standard Operating Procedures (SOP) of the electrochemotherapy for the treatment of cutaneous and subcutaneous tumors[43], which were published in November 2006, together with the results of the trial[44], several very instructive case reports showing the possibilities offered by the electrochemotherapy, as well as the historical, biological and engineering bases of the treatment.

The third step, the present step, is the dissemination of the method. The ESOPE SOP are easily followed by the physicians who have tried the method and continue to apply it, particularly in Europe. In the second half of 2008, 42 centers delivered ECT in Europe, many of them in Italy (where the treatment is reimbursed), but also 5 in Spain, 3 in Sweden, 2 in the UK. Other centers are located in Hungary, Greece, Denmark, Slovenia, France, Portugal, ... If only 62 patients were enrolled in the ESOPE clinical trial between 2004 and 2005, more than 300 patients were treated in 2007, and more than 800 in 2008.

4 The Complementarities of Electrochemotherapy and IRE Approaches

We will first highlight the differences between the two approaches to underline their complementarities in the second part of this section. The comparison is obviously limited to the antitumoral approaches of the IRE, which has other interesting applications like cardiac ablation (in the same way as the reversible electropermeabilization has also other interesting applications like non-viral gene therapy).

From an engineering point of view, the major difference between the two approaches is the number of pulses (8 in electrochemotherapy versus e.g. 80 in IRE) and the voltage (that must be sufficient to generate a ratio between the applied voltage and the distance between the electrodes of 1000 to 1300 V/cm in electrochemotherapy and larger than 1300 V/cm in IRE)[43, 52].

From a biological point of view, the main difference is that the irreversible electroporation and subsequent dissolution of the cells membranes observed after IRE (using electrical parameters actually achieving efficient antitumor effects) seem to block a potential infiltration of immunocompetent cells at least for the 72 hours that follow the treatment[52, 53]. If tumor vasculature destruction is a very important contribution to IRE antitumor effects[54], the absence of capillaries blocks the immune cells movement inside the tumor. In electrochemotherapy, a post-treatment infiltration of the tumors by macrophages and dendritic cells is always observed[50]. In spite of the fact that IRE will not destroy the tumors antigens (as IRE is a non-thermal process), immune reactions as those spontaneously observed with electrochemotherapy (reactions that can be enhanced to the level of the systemic antitumor effects achievement[50, 51, 55]) might be reduced as the tumor antigen source will not be close to the immune system antigen-presenting cells, due to the absence of tumor infiltration. Moreover, the complete treatment of the tumors, which includes the sterilization of the margins, will require also the IRE of the normal cells surrounding the tumor, with their subsequent killing. One can, nevertheless, consider that IRE can then be a secure treatment for immunodepressed patients as no immune response should be necessary in order to get the complete sterilization of the tumor mass[53].

Indeed, in IRE, the killer is the electric field. In ECT, the killer is the drug, bleomycin or cisplatin injected either locally or systemically (i.e. by an intravenous injection). While this difference allows to explain that there is an interesting differential killing effect between the tumor cells and the non-tumor surrounding cells in electrochemotherapy, this difference implies that the tumor mass must be completely covered not only by an appropriate electric field but by both, an appropriate electric field and the drug. Usually drug distribution is not a drawback, but of course one cannot exclude particular situations in which intravenous drug injection could not be acceptable and intratumoral drug injection impossible to perform. Under these circumstances, physicians should apply IRE. Moreover, IRE is also the way to utilize an electroporation-based antitumor therapy if the association of physical means and drugs is not accepted by the regulatory bodies or if the rules impose too high constraints.

Because the principle is based on two levels of the same electric field, and that the same type of square pulses of 100 microseconds duration are used, the complementarities are evident and IRE is benefiting of the work done in the development of the electrochemotherapy.

For example, devices for the production of short and very intense electric pulses (high voltage and high amperage delivered during the pulses) were already developed[43]. They included features discussed with physicians in the frame of the delivery of electric pulses, and the conditions for the acceptance by the physicians of such a novel procedure.

The study of the influence of the needles shape on the field distribution and thus on the anticipated volume of tissue that one can efficiently treat, developed for the electrochemotherapy since almost a decade [7, 56] are directly usable for the IRE approach (both in electrochemotherapy and IRE, the most used type of electrodes are needles, in arrays with different preset geometries, or needles individually inserted).

5 Conclusion

The differences and complementarities between electrochemotherapy and IRE discussed here above are evident if we take into account that both approaches are based on the same bioelectrochemical effect of the electric fields on the cell membranes, however at two different levels of this effect (reversible versus irreversible electroporation). The experimental and clinical development of the electrochemotherapy and IRE will continue to fertilize each other as both technologies present great interest in tumor treatment.

Acknowledgements

The author acknowledges his colleagues for fruitful discussions, and the funding source, mainly CNRS, IGR and EU Commission (CLINIPORATOR project, ESOPE project, EMF-NET coordinated action).

References

1. Mir, L.M.: Therapeutic perspectives of in vivo cell electropermeabilization. Bioelectrochemistry 53, 1–10 (2001)
2. Orlowski, S., Belehradek Jr., J., Paoletti, C., Mir, L.M.: Transient electropermeabilization of cells in culture. Increase of the cytotoxicity of anticancer drugs. Biochem. Pharmacol. 37, 4727–4733 (1988)

3. Gehl, J., Skovsgaard, T., Mir, L.M.: Enhancement of cytotoxicity by electropermeabilization: an improved method for screening drugs. Anticancer Drugs 9, 319–325 (1998)
4. Mir, L.M., Banoun, H., Paoletti, C.: Introduction of definite amounts of nonpermeant molecules into living cells after electropermeabilization: direct access to the cytosol. Exp. Cell. Res. 175, 15–25 (1988)
5. Heller, L.C., Heller, R.: In vivo electroporation for gene therapy. Hum. Gene. Ther. 17, 890–897 (2006)
6. Van Tendeloo, V.F., Ponsaerts, P., Berneman, Z.N.: mRNA-based gene transfer as a tool for gene and cell therapy. Current opinion in molecular therapeutics 9, 423–431 (2007)
7. Miklavcic, D., Semrov, D., Mekid, H., Mir, L.M.: A validated model of in vivo electric field distribution in tissues for electrochemotherapy and for DNA electrotransfer for gene therapy. Biochim. Biophys. Acta 1523, 73–83 (2000)
8. Pavselj, N., Bregar, Z., Cukjati, D., Batiuskaite, D., Mir, L.M., Miklavcic, D.: The course of tissue permeabilization studied on a mathematical model of a subcutaneous tumor in small animals. IEEE Trans. Biomed. Eng. 52, 1373–1381 (2005)
9. Heller, L.C., Ugen, K., Heller, R.: Electroporation for targeted gene transfer. Expert opinion on drug delivery 2, 255–268 (2005)
10. Mir, L.M., et al.: High-efficiency gene transfer into skeletal muscle mediated by electric pulses. Proc. Natl. Acad. Sci. U. S. A. 96, 4262–4267 (1999)
11. Andre, F., et al.: Efficiency of High and Low Voltage Pulse Combinations for Gene Electrotransfer in Muscle, Liver, Tumor and Skin. Hum. Gene. Ther. (2008)
12. Benninghoff, U., et al.: Clinical improvement and normalized Th1 cytokine profile in early and long-term interferon-alpha treatment in a suspected case of hyper-IgE syndrome. Pediatr. Allergy Immunol. 19, 564–568 (2008)
13. Hojman, P., et al.: Physiological Effects of High and Low Voltage Pulse Combinations for Gene Electrotransfer in Muscle. Hum. Gene. Ther. (2008)
14. Bodles-Brakhop, A.M., Draghia-Akli, R.: DNA vaccination and gene therapy: optimization and delivery for cancer therapy. Expert review of vaccines 7, 1085–1101 (2008)
15. Luxembourg, A., Evans, C.F., Hannaman, D.: Electroporation-based DNA immunisation: translation to the clinic. Expert opinion on biological therapy 7, 1647–1664 (2007)
16. de Baere, T., et al.: Usefulness of guiding needles for radiofrequency ablative treatment of liver tumors. Cardiovasc Intervent. Radiol. 29, 650–654 (2006)
17. Zhu, J.C., Yan, T.D., Morris, D.L.: A systematic review of radiofrequency ablation for lung tumors. Annals of surgical oncology 15, 1765–1774 (2008)
18. Hiraki, T., Gobara, H., Mimura, H., Sano, Y., Kanazawa, S.: Percutaneous radiofrequency ablation of lung cancer. The lancet oncology 9, 604–605 (2008)
19. Wang, K., Advincula, A.P.: Current thoughts in electrosurgery. International journal of gynaecology and obstetrics: the official organ of the International Federation of Gynaecology and Obstetrics 97, 245–250 (2007)
20. Advincula, A.P., Wang, K.: The evolutionary state of electrosurgery: where are we now? Current opinion in obstetrics & gynecology 20, 353–358 (2008)
21. Nilsson, E., Fontes, E.: Mathematical modelling of physicochemical reactions and transport processes occurring around a platinum cathode during the electrochemical treatment of tumours. Bioelectrochemistry 53, 213–224 (2001)
22. Azavedo, E., Svane, G., Nordenstrom, B.: Radiological evidence of response to electrochemical treatment of breast cancer. Clinical radiology 43, 84–87 (1991)

23. Nordenstrom, B.E.: Electrochemical treatment of cancer. I: Variable response to anodic and cathodic fields. American journal of clinical oncology 12, 530–536 (1989)
24. Xin, Y., Xue, F., Ge, B., Zhao, F., Shi, B., Zhang, W.: Electrochemical treatment of lung cancer. Bioelectromagnetics 18, 8–13 (1997)
25. Xin, Y., Xue, F., Zhao, F.: Effectiveness of electrochemical therapy in the treatment of lung cancers of middle and late stage. Chinese medical journal 110, 379–383 (1997)
26. Nilsson, E., et al.: Electrochemical treatment of tumours. Bioelectrochemistry 51, 1–11 (2000)
27. Hildebrandt, B., et al.: The cellular and molecular basis of hyperthermia. Critical reviews in oncology/hematology 43, 33–56 (2002)
28. Sreenivasa, G., et al.: Clinical use of the hyperthermia treatment planning system HyperPlan to predict effectiveness and toxicity. Int. J. Radiat. Oncol. Biol. Phys. 55, 407–419 (2003)
29. Wust, P., et al.: Hyperthermia in combined treatment of cancer. The lancet oncology 3, 487–497 (2002)
30. Garon, E.B., et al.: In vitro and in vivo evaluation and a case report of intense nanosecond pulsed electric field as a local therapy for human malignancies. Int. J. Cancer 121, 675–682 (2007)
31. Nuccitelli, R., et al.: Nanosecond pulsed electric fields cause melanomas to self-destruct. Biochem. Biophys. Res. Commun. 343, 351–360 (2006)
32. Mir, L.M., Orlowski, S., Belehradek Jr., J., Paoletti, C.: Electrochemotherapy potentiation of antitumour effect of bleomycin by local electric pulses. Eur. J. Cancer 27, 68–72 (1991)
33. Mir, L.M., et al.: Electrochemotherapy, a new antitumor treatment: first clinical trial. C. R. Acad Sci. III 313, 613–618 (1991)
34. Belehradek, M., Domenge, C., Luboinski, B., Orlowski, S., Belehradek Jr., J., Mir, L.M.: Electrochemotherapy, a new antitumor treatment. First clinical phase I-II trial. Cancer 72, 3694–3700 (1993)
35. Sersa, G., Cemazar, M., Miklavcic, D.: Antitumor effectiveness of electrochemotherapy with cis-diamminedichloroplatinum(II) in mice. Cancer Res. 55, 3450–3455 (1995)
36. Sersa, G., Stabuc, B., Cemazar, M., Jancar, B., Miklavcic, D., Rudolf, Z.: Electrochemotherapy with cisplatin: potentiation of local cisplatin antitumour effectiveness by application of electric pulses in cancer patients. Eur. J. Cancer 34, 1213–1218 (1998)
37. Sersa, G., Stabuc, B., Cemazar, M., Miklavcic, D., Rudolf, Z.: Electrochemotherapy with cisplatin: clinical experience in malignant melanoma patients. Clin. Cancer Res. 6, 863–867 (2000)
38. Salford, L.G., Persson, B.R., Brun, A., Ceberg, C.P., Kongstad, P.C., Mir, L.M.: A new brain tumour therapy combining bleomycin with in vivo electropermeabilization. Biochem. Biophys. Res. Commun. 194, 938–943 (1993)
39. Mir, L.M., et al.: First clinical trial of cat soft-tissue sarcomas treatment by electrochemotherapy. Br. J. Cancer 76, 1617–1622 (1997)
40. Ramirez, L.H., et al.: Electrochemotherapy on liver tumours in rabbits. Br. J. Cancer 77, 2104–2111 (1998)
41. Pron, G., et al.: Internalisation of the bleomycin molecules responsible for bleomycin toxicity: a receptor-mediated endocytosis mechanism. Biochem. Pharmacol. 57, 45–56 (1999)

42. Tounekti, O., Pron, G., Belehradek Jr., J., Mir, L.M.: Bleomycin, an apoptosis-mimetic drug that induces two types of cell death depending on the number of molecules internalized. Cancer Res. 53, 5462–5469 (1993)
43. Mir, L.M., et al.: Standard Operating Procedures of the Electrochemotherapy. Eur. J. Cancer Supplements 4, 14–25 (2006)
44. Marty, M., et al.: Electrochemotherapy - an easy, highly effective and safe treatment of cutaneous and subcutaneous metastases: results of the ESOPE (European Standard Operating Procedures of Electrochemotherapy) study. Eur. J. Cancer Supplements 4 (2006)
45. Miklavcic, D., Corovic, S., Pucihar, G., Pavselj, N.: Importance of tumour coverage by sufficiently high local electric field for effective electrochemotherapy. Eur. J. Cancer Supplements 4, 45–51 (2006)
46. Corovic, S., Al-Sakere, B., Haddad, V., Miklavcic, D., Mir, L.M.: Importance of contact surface between electrodes and treated tissue in electrochemotherapy. Technol. Cancer Res. Treat. 7, 393–400 (2008)
47. Ivorra, A., Al-Sakere, B., Rubinsky, B., Mir, L.M.: Use of conductive gels for electric field homogenization increases the antitumor efficacy of electroporation therapies. Physics in medicine and biology 53, 6605–6618 (2008)
48. Gehl, J., Skovsgaard, T., Mir, L.M.: Vascular reactions to in vivo electroporation: characterization and consequences for drug and gene delivery. Biochim. Biophys. Acta 1569, 51–58 (2002)
49. Sersa, G., Cemazar, M., Miklavcic, D., Chaplin, D.J.: Tumor blood flow modifying effect of electrochemotherapy with bleomycin. Anticancer Res. 19, 4017–4022 (1999)
50. Roux, S., et al.: Tumor destruction using electrochemotherapy followed by CpG oligodeoxynucleotide injection induces distant tumor responses. Cancer Immunol. Immunother 57, 1291–1300 (2008)
51. Mir, L.M., et al.: Systemic antitumor effects of electrochemotherapy combined with histoincompatible cells secreting interleukin-2. J. Immunother Emphasis Tumor Immunol. 17, 30–38 (1995)
52. Al-Sakere, B., et al.: Tumor Ablation with Irreversible Electroporation. PLoS ONE 2, e1135 (2007)
53. Al-Sakere, B., et al.: A study of the immunological response to tumor ablation with irreversible electroporation. Technol. Cancer Res. Treat. 6, 301–306 (2007)
54. Edd, J.F., Horowitz, L., Davalos, R.V., Mir, L.M., Rubinsky, B.: In vivo results of a new focal tissue ablation technique: irreversible electroporation. IEEE Trans. Biomed. Eng. 53, 1409–1415 (2006)
55. Orlowski, S., An, D., Belehradek Jr., J., Mir, L.M.: Antimetastatic effects of electrochemotherapy and of histoincompatible interleukin-2-secreting cells in the murine Lewis lung tumor. Anticancer Drugs 9, 551–556 (1998)
56. Sel, D., Cukjati, D., Batiuskaite, D., Slivnik, T., Mir, L.M., Miklavcic, D.: Sequential finite element model of tissue electropermeabilization. IEEE Trans. Biomed. Eng. 52, 816–827 (2005)

Irreversible Electroporation: First Patient Experience Focal Therapy of Prostate Cancer

Gary Onik[1,*] and Boris Rubinsky[2]

[1] School of Medicine, University of Central Florida, Orlando, Fl 32816 USA
[2] Department of Mechanical Engineering, University of California, Berkeley, CA 94720 USA
and
Center for Bioengineering in the Service of Humanity and Society, School of Computer Science and Engineering, Hebrew University of Jerusalem, Givat Ram Campus, Jerusalem, Israel
* Corresponding author E-mail: onikcryo@gmail.com

Introduction

Irreversible electroporation (IRE) is a new non-thermal ablation modality that uses short pulses of DC electric current to create irreversible pores in the cell membrane thus causing cell death. This method has been shown by Rubinsky et al. to have significant advantages in ablating prostatic tissue, such as rapid lesion creation, rapid lesion resolution, sparing of structures such as vessels, nerves and urethra, and uniform destruction throughout the IRE lesion (1). The underlying principles are well covered in other chapters in this book. This discussion will deal with the first human applications of IRE and whether its theoretical promises of improved clinical outcomes, have been delivered. For a number of reasons, including the ability to carry out extensive post operative biopsies to confirm the adequate ablation of cancer, the first human experience was carried out in the prostate.

The focal therapy of prostate cancer is gaining interest as a potentially new prostate cancer management strategy, falling between watchful waiting on the one hand and morbid whole gland treatments such as radical prostatectomy on the other. Onik et al. introduced the concept of using focal therapy for treating prostate cancer using cryo-ablation in 2002 (2). Other clinical series using cryo-ablation to focally treat prostate cancer have shown excellent cancer control rates with extremely low morbidity (3-5). Cryo-ablative lesions, however, have some distinct limitations, such as variable damage at the cryo lesion's margin, injury to adjacent structures such as rectum, urethra and neurovascular bundle (NVB), and long procedure time due to the need for multiple freeze thaw cycles. These characteristics could limit the widespread acceptance of this modality despite certain demonstrated advantages over the more traditional treatments of radiation and radical prostatectomy (6). Since these disadvantages of cryoablation are the theoretical strengths of IRE it seemed reasonable to explore the first human use of IRE in this setting.

B. Rubinsky (Ed.): Irreversible Electroporation, BIOMED, pp. 235–247.
springerlink.com

Methods

Patient Selection

Patients were considered for cancer targeted IRE ablation, if based on TRUS biopsies their cancer was localized and the maintenance of potency and/or continence was a major concern of the patient. All patients were restaged by a transperienal mapping biopsy biopsied under heavy sedation or general anesthesia. A brachytherapy grid was employed and biopsies were obtained every 5 mm's throughout the volume of the prostate under US guidance according to recently published guidelines for biopsy prior to focal therapy for prostate cancer (7). Each sample was sent separately, labeled as to its location and proximal segment of each core was inked to identify its orientation as to the base and apex of the prostate gland.

Procedure

Patients were placed in the dorsal lithotomy position, under sterile technique and general anesthesia (with paralysis to mitigate muscle contraction associated with the procedure) 18 gauge IRE electrodes were placed under TRUS guidance percutaneously through the perineum. IRE probes were placed to cover the known area of cancer location based on the patients mapping biopsy (Figure 1). Four probes were placed in a roughly square array, 1-1.5 cm apart, with the known area of cancer in the center of the array. Probes were placed to the capsule at the base and insulation was pulled back to expose enough electrode to affect the full length

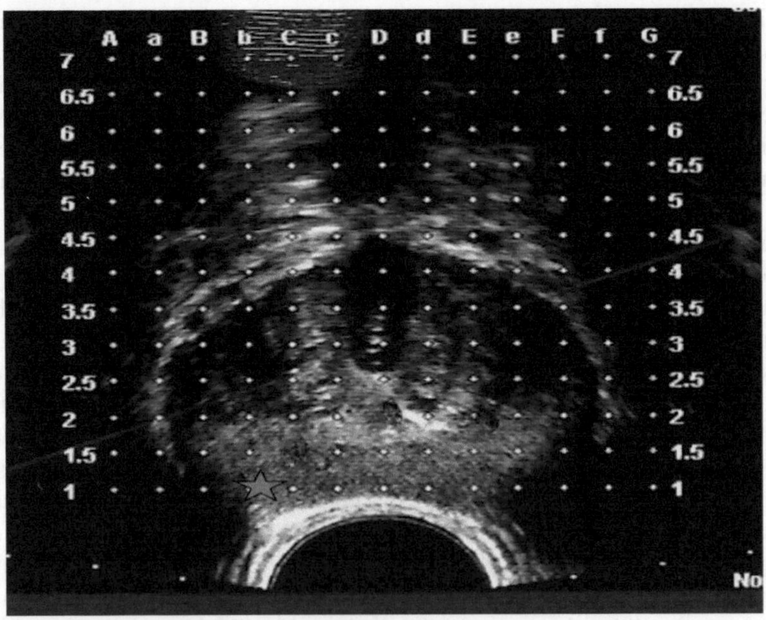

Fig. 1a. Ultrasound showing the brachytherapy grid overlay used during the 3D Prostate Mapping Biopsy. The star represents the area that was positive on this patient. Note that the positiveve area is in the region of the NVB.

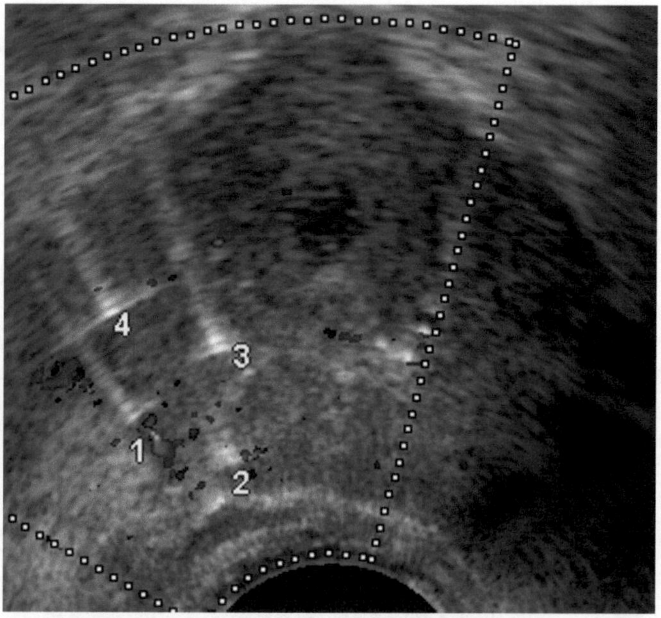

Fig. 1b. Ultrasound showing the IRE probes in the a parallelogram pattern covering the area of he know tumor

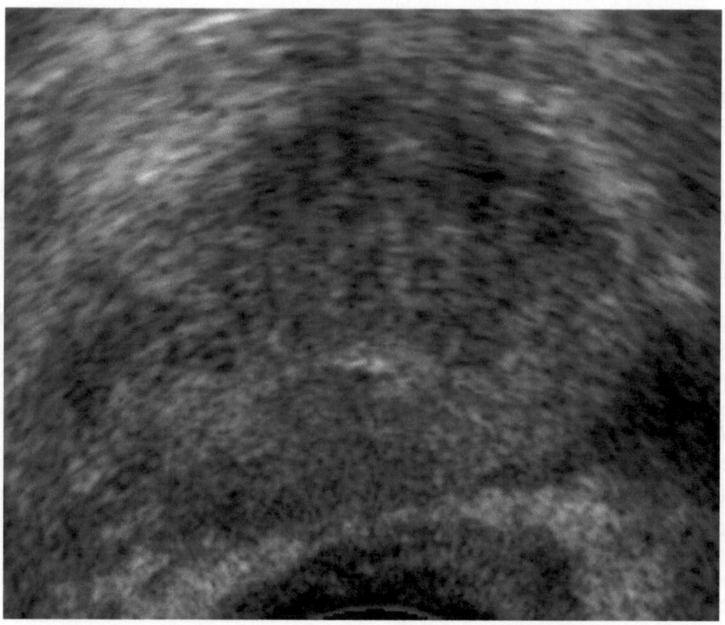

Fig. 1c. Post operative ultrasound at 1 day showing a mixed echogenicity region in the area treated with indistinct capsular margins on the right side of the gland

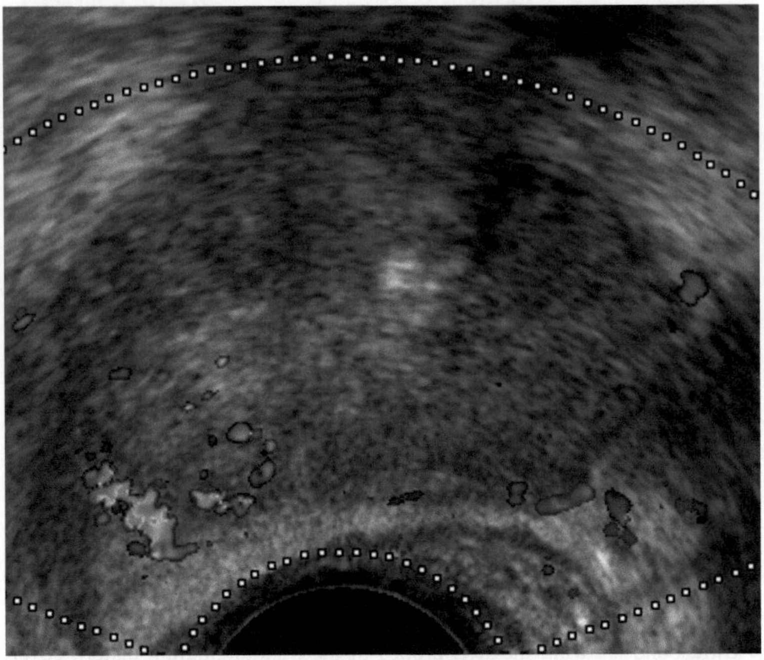

Fig. 1d. Color Doppler US shows increased blood flow in the Neurovascular bundle compared the untreated side. Patient was potent at this time.

of the prostate. Using an IRE generator (Angiodynamics Inc. Queensbury N.Y.) Ninety pulses of 70-100 microseconds in length and 100 millisecond interval between pulses at 1500 volts were delivered between electrode pairs in a bipolar manner until all the permutations for the four probe array were covered (six sets). If the area of the cancer was larger than the 4-probe array or there was a separate focus of cancer an additional array was used to cover the area. When the cancer was adjacent to the neurovascular bundle the array was placed so the ablation would include the bundle itself as well.

When cancer was in the periphery of the gland adjacent to the rectum a 22 gauge spinal needle as placed into Denonvillier's fascia and approximately twenty cc'c of D5W were injected to separate the rectum from the prostate

Patient follow-up

At three weeks post treatment additional transperienal US guided biopsies were conducted in the previously known cancer loci and the immediate surrounding areas areas.

Results

Sixteen patients ranging in age from 40 to 78 were treated using IRE. The other patient characteristics are detailed in Table 1. All patients tolerated the procedure well and all were treated on an outpatient basis. US obtained during the procedure showed a mixed echogenicity in the treated area. Indistinctness of the prostatic capsule was noted when the created lesion included it. Color Doppler ultrasound showed intact flow within the neurovascular bundle immediately after the procedure (Figure 1). The catheter drainage time varied between 0 and 3 days. With the longest time occurring in a patient who had lesions in both sides of the gland treated. All patients were continent immediately and all patients who were potent before the procedure were potent after the procedure. Two patients who had bilateral areas treated required 6 months for full return of potency.

Table 1

Patient	PSA	Gleason	Stage	# tumor foci	Tumor Location	Potent Pre/op	Potent Post/op	Post op biopsy
1	4.1	6	T1c	1	Rt NVB	yes	yes	Neg
2	4.5	7(3+4)	T1c	3	Bilat NVB Lt ant peri-urethral	No	No	Neg
3	7	6	T1c	1	Rt ant peri-urethral	yes	yes	Neg
4	3	6	T1c	1	Rt NVB	yes	yes	Neg
5	5	6		1	Posterior urethral bilateral	yes	yes	Neg
6	3.5	7	T1c	2	Bilat	yes	yes	Refused PSA .2
7	5.1	6	T1c	2	Rt apex, midline	No	No	neg
8	4	7	T1c	1	Left Hemi	yes	yes	Neg
9	4.9	6	T1c	1	Lt NVB	yes	yes	Neg
10	4.99	7	T1c	2	Lt ant, post urethra	No	No	Neg
11	6.1	8	T1c	1	Lt NVB	yes	yes	Neg
12	4.6	6	T1c	1	Lt NVB	yes	yes	Neg treated area, micro focus outside treated area
13	7	7	T1c	2	Bilat, diffuse	yes	yes	Neg
14	2.2	8	Rad Failure	2	Midline, Lt NVB	No	No	Neg
15	2.6	8	Rad Failure	1	Rt NVB	No	No	Neg
16	4.6	7	T2c	1	Midline	Yes	Yes	Neg

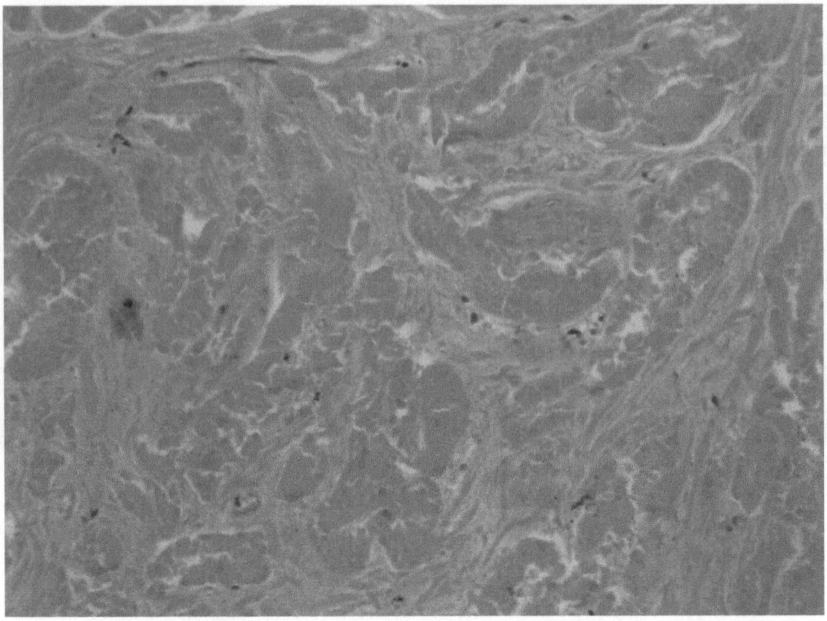

Fig. 2a. Hand E stain of post op biopsy. Ghosts of prostate glands are seen but no viable epithelial cells are identified.

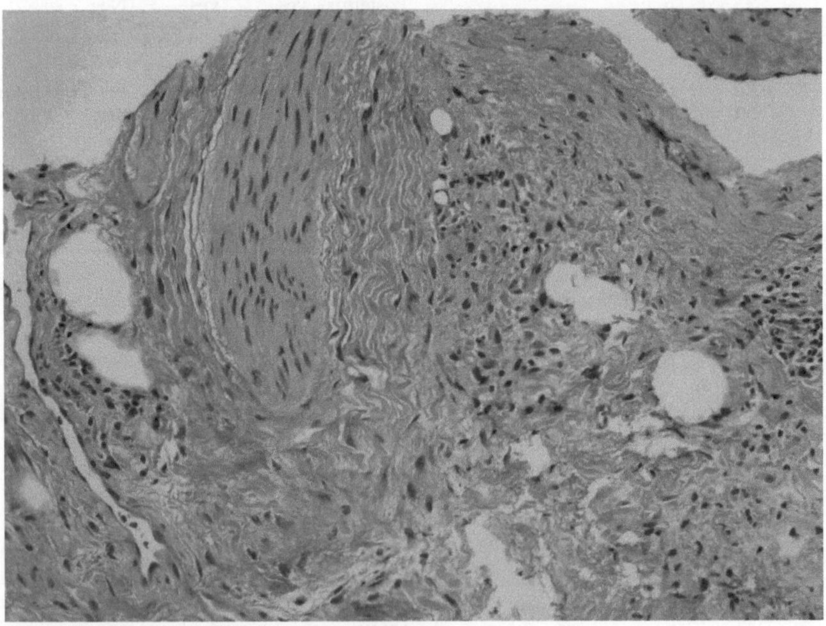

Fig. 2b. Hand E stain of post op biopsy. Post op biopsy showing an intact nerve bundle surrounded by reactive fibrosis.

Postoperative biopsies taken from the area of previously known cancer in 15 patients showed no evidence for cancer. There was one patient with a negligible PSA who refused a post operative biopsy and one in whom a micro focus of Gleason 6 cancer was found outside the treated area. This patient was successfully retreat with focal cryosurgery.

In addition, there was no evidence for any viable glandular tissue in the biopsy specimens. H and E staining showed all epithelial elements gone. There were occasional areas with ghosts of glandular structures without viable cells present. Vascular elements were patent and intact nerve bundles with viable ganglion cells within them were noted, surrounded by necrotic tissue and fibrotic tissue (Figure 2).

Discussion

The use of breast sparing surgery i.e. "lumpectomy" to treat breast cancer revolutionized the local control of that disease. Lumpectomy experience showed that patient quality of life of can successfully be integrated into the equation of cancer treatment without major of treatment efficacy (8). Men with prostate cancer face many of the same issues that breast cancer patients do. A number of recent studies have questioned the efficacy of an aggressive treatment approach to prostate cancer. Current management of prostate cancer covers both ends of the treatment spectrum. Patients can elect no treatment at all i.e. "watchful waiting" (9) or aggressive whole gland treatments such as radical prostatectomy. Focal therapy, in which just the known area of cancer is destroyed, appears to be a logical extension of the watchful waiting concept. Focal therapy minimizes the risks associated with expectant management since the clinically threatening index cancer has been treated. Minimizing prostate trauma, by treating only a portion of the prostate, has been shown to decrease the risk of lifestyle altering complications associated with morbid whole gland treatments such as impotence and incontinence (3-5).

The main theoretical objection to a focal therapy of prostate cancer is that prostate cancer is often a multi-focal disease. As with breast cancer, however, prostate cancer is a spectrum of diseases, some of which are may be amenable to focal therapy. The prostate cancer pathology literature clearly shows that a significant number of patients have a single focus of prostate cancer and that many others have additional cancer foci that may not be clinically significant (10-13) Until now however, little attention has been paid in trying to differentiate those patients with uni-focal, from multi-focal disease, since all treatments were aimed at either total gland removal or destruction.

In a study examining radical prostatectomy specimens Djvan et al. (10) showed that patients with unifocal disease constituted one third of the cases. In addition, Villars et al (11), showed that 80% of multi-focal tumors are less than .5 cc's indicating they may not be of clinical significance. This study was confirmed by Rukstalis et al (12) and Noguchi et al (13) in which pathologic examination showed that uni-focal tumors were present in 20% and 25% of patients respectively and using the size criteria of .5 cc's or less as an insignificant tumor, an additional 60% and 39% of patients might be a candidate for a focal treatment approach. Clearly then based on the known pathology of prostate cancer an opportunity exists to investigate a focal treatment approach.

Newer biopsy techniques now being used, in which the gland is biopsied transperineally every 5 mm's using a brachytherapy type grid, could also have an impact on excluding patients with significant multi focal disease. A recent paper by Crawford et al, (14) using a computer simulations on RP and autopsy specimens, demonstrated that trans-perineal prostate biopsies, spaced at 5mm intervals through the volume of a patient's prostate had a sensitivity of 95% in finding clinically significant tumors. Results by Onik and Barzell et al (15-16) in patients undergoing 3D mapping biopsies for further staging, showed that cancer could be demonstrated in 50% of patients who previously had negative biopsies in the supposedly uninvolved prostate lobe.

Since the anatomy of the prostate gland does not make it amenable to partial removal or lumpectomy, tumor destruction by another modality is needed to realize a "lumpectomy" in a male. Cryoablation was the obvious choice since it has a long history of effective tumor treatment in various parts of the body. Approximately 8 years ago prostate cryoablation was approved by Medicare as a treatment for primary prostate cancer (removing it from the investigational category) (17). Long term data is now available that confirms cryoablation is a treatment competitive to both surgery and radiation in treating prostate cancer. A recent prospective randomized study by Donnelly et al has shown cryoablation equivalent to eternal beam radiation in treating prostate cancer (18).

Using cryoablation as the method of treatment Onik et al (2) published the first initial series of patients focally treated for prostate cancer. Further recent studies have shown focal therapy to be efficacious for treating prostate cancer while minimizing morbidity such as incontinence and impotence (3-5). Based on this preliminary work focal therapy is now gaining significant interest as an alternative management strategy in prostate cancer patients particularly those patients in whom watchful waiting might be considered (19-21). The search for the ideal focal therapy treatment modality is now underway.

Cryoablation and other methods of thermal destruction of tissue do have certain inherent limitations which include, non-uniform destruction particularly at the margin of the lesion, protection of tumor by the heat sink effect next to large vessels, destruction of tissue collagen with associated destruction of normal structures such as small vessels, nerves, urethra, ureters and bowel. In the treatment of prostate cancer these limitations translate into possible recurrences of cancer adjacent to neurovascular bundles or the opposite complete destruction of the NVB with the subsequent result of impotence. Other complications can result from tissue destruction of normal structures such as incontinence, if the sphincter is effected or urethro-rectal fistula if the rectum is effected.

IRE as described in chapter has a number of demonstrated advantages over the well known thermal based ablation methods. Many of these advantages stem from the mechanism of cell destruction used by IRE. IRE rather than causing destruction through coagulative necrosis such as in RFA (radiofrequency ablation) or through cellular and micro-vascular disruption as in cryo-ablation, causes cell death by the mechanism of irreversible electroporation. (IRE). When cells are exposed to a pulsed electrical field with the proper parameters, holes are created in the cell membrane increasing it's permeability. Depending mainly on the magnitude of the trans-membrane potential the cell is exposed to, the holes may close, ie reversible electroporation, or remain open, irreversible electroporation (IRE).

Reversible electroporation occurs within a narrow range of parameters roughly between 300 V/cm2 and 600 V/cm2 although cell size, pulse duration and number of pulses can affect these ranges. The application of reversible electroporation to tissues has found important applications in biotechnology and medicine. Electrogenetherapy (EGT) is the in-vivo insertion of genes into cells in tissue through reversible electroporation and presents an alternative to viral vectors (22). Electrochemotherapy is accomplished by injecting drugs or macromolecules into a targeted area. Electrodes are placed into or around that area to generate a reversible permeabilizing electric field in the tissue, thereby introducing the drugs or macromolecules into the cells of the affected area (23), In addition, potent but normally impermeable anti-cancer drugs such as bleomycin are used with electroporation to ablate tissue.

In the above applications the electroporation has to be reversible and irreversible electroporation is consciously avoided. Therefore, the electrical parameters that induce irreversible electroporation where studied only as an upper limit to the range of electrical parameters that induce reversible electroporation. Irreversible electroporation, however, has been studied extensively in in-vitro cellular systems. For instance IRE has been considered an effective means for killing both gram negative and positive bacteria responsible for water contamination. Davalos and Rubnsky (24) were the first to suggest that irreversible electroporation could be used for surgical tissue ablation without the inherent limitations of chemotherapeutic injection or the narrow window of parameters associated with reversible electroporation. They speculated that this could result in significant advantages over the currently used thermal ablation methods. This was later confirmed with the first animal study creating liver lesions and prostate lesions using IRE (6,25).

Our initial patient experience in treating prostate cancer, confirms many of the advantages seen in the animal studies using IRE. In terms of tissue destruction, previous studies showed that IRE lesions show uniform necrosis throughout, with a very narrow zone of transition to unaffected tissue. Post-operative biopsies in our patients showed complete epithelial cell ablation in all the cores taken from all patients. No viable glandular elements were identified. Due to the apoptotic nature of cell death associated with IRE, ghosts of glandular structures were still identifiable (Figure 2). Of interest. is that these results were independent of Gleason score with successful treatment of patients with Gleason score of 7 and 8. We believe that part of the initial success of our successful tumor ablation was that the dosimetry of the IRE protocol, ie voltage, pulse length, and pulse number was determined with in vitro testing using human cancer cells (26).

Despite the uniform destruction of glandular cellular elements evidenced in our biopsies, sparing of normal structures was also demonstrated. Post op pathology and color Doppler US demonstrated intact and functioning micro and macro vasculature particularly in the prostate neurovascular bundle. This characteristic appears to be unique to IRE lesions. While IRE does cause endothelial cell death it does not cause vessel occlusion.

Also noted on our biopsies was the preservation of nerves and ganglion cells within the IRE lesions despite them being included in the IRE lesion. As far as we know this phenomena is unique to IRE and this along with vasculature preservation, has major implications for the preservation of potency associated with IRE treatment of prostate cancer. All of our patients who were potent prior to treatment were immediately potent after treatment, while it did take approximately 6 months for two patients who were treated bilaterally to regain their potency. While excellent potency rates have been associated with focal cryoablation of prostate cancer, like nerve sparing radical prostatectomy full return to function can often take many months (27).

Previous animal studies also demonstrated preservation of the urethra without sloughing or major damage. This is quite different than cryoablation where active warming of he urethra is required to prevent major urethral complications. It has been suggested in the past that this limited the use of cryoablation when tumors were adjacent to the urethra. This does not appear to be a limitation with IRE since a number of our patients were successfully treated despite tumor being adjacent to the urethra. Of note is that two of our patients were also radiation failures. So far neither of these patients has demonstrated any local complications of tissue sloughing, a problem that is very common with cryoablation in this patient population.

This structure sparing characteristic of IRE had one interesting although unanticipated benefit. In one of our patients the tumor involved the midline ejaculatory duct region. Despite very aggressive treatment in this area the patient's ejaculatory ducts remained intact allowing for normal ejaculation post operatively. Again we believe this is unique to IRE ablation.

Based on our animal studies, the structural integrity of bowel does not appear to be effected by IRE. This should add a margin of safety to IRE that is not available with cryoablation or heat ablation such as RF or HIFU, in which bowel injury can lead to the devastating complication of urethro-rectal fistula. Despite the apparent safety of IRE in this regard we still created additional space between the rectum and the prostate by injection of D5W into Denonvillier's fascia to maintain a extra margin of safety.

Despite these major advantages our animal studies indicate that IRE will ablate both smooth and striated muscle (28). Although all patients in this initial experience were continent immediately, care still needs to be taken not to destroy both the internal and external sphincter which would result in incontinence.

From a technical point of view IRE is similar to other transperineal ultrasound guided procedures such as cryosurgery and brachytherapy. General anesthesia with patient paralysis is needed however due to the muscle contraction associated with the electrical pulsing. The speed of the IRE procedure is however impressive compared to cryoablation and HIFU. The multiple pulses for the treatment are delivered in a few minutes and rather than long process of freezing and thawing associated with cryoablation or the numerous small ablation zones of HIFU.

Despite our patients being treated in a focal manner it remains possible that IRE could be used to treat the whole prostate gland. Whether the complication profile for whole gland IRE would remain the same as in our initial focal experience remains to be seen. Our previous experience with focal therapy however has lead us to believe that, particularly with medium and high risk patients, knowing the location of the tumor and targeting it directly, with its associated areas of possible extra-capsular spread, is a superior strategy. Whole gland treatment inevitably leads to a compromise in the placement of probes to destroy portions of the gland uninvolved with cancer leading to a local recurrence rate, which appears to be higher than the 10% retreatment rate associated with focal therapy (29). This is local failure rate is certainly acceptable since, as with other ablation technologies retreatment with IRE remains a possibility when a local recurrence is detected.

Lastly, while our first patient experience with IRE was carried out in the prostate, many of the advantages of IRE over thermal ablation technologies would extend to other organs. For instance, IRE's ability to treat tumors next to vessels without vessel damage or tumors protection by a heat sink, along with the ability to treat near bowel without damage could make IRE a viable ablation modality for pancreatic carcinoma.

In conclusion, IRE is a new non-thermal ablation modality with significant advantages over heat or cold, based tumor destruction. It's ability to spare nerves and vessels apparently results in minimal effect on potency making it particularly suited to focal therapy of prostate cancer.

References

1. Rubinsky, B., Onik, G., Mikus, P.: Irreversible electroporation: a new ablation modality–clinical implications. Technol. Cancer Res. Treat. 6(1), 37–48 (2007)
2. Onik, G., Narayan, P., Vaughan, D., et al.: Focal nerve sparing cryoablation for the treatment of primary prostate cancer: A new approach to preserving potency. Urology 60(1), 109–114 (2002)
3. Onik, G., Vaughan, D., Lotenfoe, R., Dineen, M., Brady, J.: Male lumpectomy: focal therapy for prostate cancer using cryoablation. Urology 70(suppl. 6), 16–21 (2007)
4. Bahn, D.K., Silverman, P., Lee, F., Badalament, R., Bahn, E.D., Rewcastle, J.C.: Focal prostate cryoablation: initial results show cancer control and potency preservation. J. Endourol. 20(9), 688–692 (2006)
5. Lambert, E.H., Bolte, K., Masson, P., Katz, A.E.: Focal cryosurgery: encouraging health outcomes for uni-focal prostate cancer. Urology 69(6), 1117–1120 (2007)
6. Bahn, D.K., Lee, F., Bandalament, R., et al.: 7-year outcomes in the primary treatment of prostate cancer. Urology 60(suppl. 1, 2), 3–11 (2002)
7. Bostwick, D.G., Waters, D.J., Farley, E.R., Meiers, I., Rukstalis, D., Cavanaugh, W.A., Ragde, H., Dineen, M.K., Bahn, D., Scionti, S., Babian, R., Ellis, D.S., Rewcastle, J.C., Burke, H.B., Andriole, G.L., Onik, G., Barqawi, A.E., Maksem, J., Barzell, W.E.: Group consensus reports from the Consensus Conference on Focal Treatment of Prostatic Carcinoma, Celebration, Florida (February 24, 2006); Urology 70(suppl. 6), 42–44 (December 2007)

8. Santiago, R.J., Wu, L., Harris, E., Fox, K., Schultz, D., Glick, J., Solin, L.J.: Fifteen-year results of breast-conserving surgery and definitive irradiation for Stage I and II breast carcinoma: the University of Pennsylvania experience. Int. J. Radiat. Oncol. Biol. Phys. 58(1), 233–240 (2004)
9. Messing, E.M., Thompson, I.: Follow-up of conservatively managed prostate cancer: watchful waiting and primary hormonal therapy. Urol. Clin. North Am. 30(4), 687–702 (2003)
10. Djvan, B., Susani, M., Bursa, B., et al.: Predictability and significance of multifocal prostate cancer in the radical prostatectomy specimen. Techniques in Urology 5(3), 139–142 (1999)
11. Villars, A., McNeal, J.E., Freiha, F.S., et al.: Multiple cancers in the prostate. Morphological features of clinically recognized vs. incidental tumors. Urology 60(2A), 19–25 (2002)
12. Noguchi, M., Stamey, T.A., McNeal, J.E., et al.: Prognostic factors for multifocal prostate cancer in radical prostatectomy specimens: Lack of significance of secondary tumors. J. Uril. 170(2pt), 459–463 (2003)
13. Rukstalis, D.B., Goldknopf, J.L., Crowley, E.M., et al.: Prostate cryoablation: A scientific rationale for future modifications. Urology 60(2A), 19–25 (2002)
14. Crawford, E.D., Wilson, S.S., Torkko, K.C., et al.: Clinical staging of prostate cancer: A computer-simulated study of transperineal prostate biopsy. BJU Int. 96(7), 999–1004 (2005)
15. Gary Onik, M.D., Winston Barzell, M.D.: Transperineal 3D Mapping Biopsy of the Prostate: An Essential Tool in Selecting Patients for Focal Prostate Cancer Therapy. Urologic Oncology: Seminars and Original Investigations (in Press)
16. Barzell, W.E., Melamed, M.R.: Appropriate patient selection in the focal treatment of prostate cancer: the role of transperineal 3-dimensional pathologic mapping of the prostate–a 4-year experience. Urology 70(suppl. 6), 27–35 (2007)
17. Health Care Financing Administration. Medicare Coverage Policy: Decisions. Bagley GP. Cryosurgery ablation of the prostate (#CAG-00031), `http://www.hcfa.gov/coverage/8b3-f1`.htm (accessed November 7, 2001)
18. Donnelly, B.J., Saliken, J.C., Brasher, P., et al.: Randomized controlled trial comparing external beam radiation and cryoablation in localized prostate cancer. In: Proceedings of the American Urological Association of 2007 (2007), Abstract #1141
19. Eggener, S.E., Scardino, P.T., Carroll, P.R., Zelefsky, M.J., Sartor, O., Hricak, H., Wheeler, T.M., Fine, S.W., Trachtenberg, J., Rubin, M.A., Ohori, M., Kuroiwa, K., Rossignol, M., Abenhaim, L.: International Task Force on Prostate Cancer and the Focal Lesion Paradigm. Focal therapy for localized prostate cancer: a critical appraisal of rationale and modalities. J. Urol. 178(6), 2260–2267 (2007) (Epub October 15, 2007)
20. Ahmed, H.U., Pendse, D., Illing, R., Allen, C., van der Meulen, J.H., Emberton, M.: Will focal therapy become a standard of care for men with localized prostate cancer? Nat. Clin. Pract. Oncol. 4(11), 632–642 (2007)
21. Mouraviev, V., Mayes, J.M., Sun, L., Madden, J.F., Moul, J.W., Polascik, T.J.: Prostate cancer laterality as a rationale of focal ablative therapy for the treatment of clinically localized prostate cancer. Cancer 110(4), 906–910 (2007)
22. Neumann, E., Schaefer-Ridder, M., Wang, Y., Hofschneider, P.H., et al.: Gene transfer into mouse lyoma cells by electroporation in high electric fields. J. EMBO 1, 841–845 (1982)

23. Mir, L.M., Belehradek, M., Domenge, C., Luboniski, B., Orlowski, S., Belehradek, J., Schwaab, B., Luboniski, B., Paoletti, C.: Electrochemotherapy, a novel antitumor treatment: first clinical trial. C. R. Acad. Sci. Ser. III 313, 613–618 (1991)
24. Davalos, R.V., Mir, L.M., Rubinsky, B.: Tissue ablation with irreversible electroporation. Annals of Biomedical Engineering 33(2), 223–231 (2005)
25. Onik, G., Mikus, P., Rubinsky, B.: Irreversible electroporation: implications for prostate ablation. Technol. Cancer Res. Treat. 6(4), 295–300 (2007)
26. Miller, L., Leor, J., Rubinsky, B.: Cancer cells ablation with irreversible electroporation. Technology in Cancer Research and Treatment 4(6), 1–6 (2005)
27. Ellis, D., Jones, J., Pisters, L., et al.: Subtotal/Partial gland prostate cryoablation: Results of 341 patients from multiple centers tracked with the cold registry. Presented AUA 2008, Orlando FL (2008), Abstract Number:08-AB-95484-AUA
28. Lavee, J., Onik, G., Mikus, P., Rubinsky, B.: A novel non-thermal energy source for surgical epicardial atrial ablation: irreversible electroporation. Heart Surg. Forum 10(2), E162–E167 (2007)
29. Donnelly, B.J., Salikan, J.C., Ernst, D.S., et al.: Prospective trial of cryosurgical ablation of the prostate: Five year results. Urology 60(4), 645 (2002)

Human Experience with Irreversible Electroporation

Kenneth Thomson

The Alfred Hospital, Melbourne, and Alfred Health, 55 Commercial Road,
P.O. Box 315 Prahran, Victoria 3181, Australia
Corresponding author E-mail: K.Thomson@alfred.org.au

Considering the remarkable opportunities and safety profile demonstrated in the animal studies using irreversible electroporation, it would seem intuitive that this method would provide a safer, more effective and more widely applicable treatment for solid tumours in humans.

The first patients in Australia were recruited in October and November 2008 and the first procedure performed on an elderly Chinese lady with recurrent hepatocellular carcinoma. The procedures were performed in a phase 1 safety trial framework and a number of patients with incurable and widespread malignancies were treated using irreversible electroporation applied by a commercial device (the NanoKnife™). Because of the constraints of the trial, it was decided to limit the application to only liver, lung and kidney tumours. Prior to this trial irreversible electroporation had been performed in humans only in the prostate gland. Unfortunately the prostate gland remains off-limits in our local area because of the multiple competing methods of treatment of prostate cancer localised to the gland.

Once a suitable cohort of volunteers for this new treatment had been obtained, the challenge became how to apply the energy safely and how to ensure that the whole target zone had been treated adequately.

Our experience with interventional radiology and percutaneous procedures over the past 30 years have provided us with sufficient expertise to be able to place either the bipolar or uni-polar electrodes in the desired locations without difficulty.

Just like the animal experiments, we used general anaesthesia with muscular paralysis to ensure that the energy applied to the electrodes did not cause severe muscle contraction. Even with the patient fully paralysed, the energy delivered by the NanoKnife™ is sufficient to cause contraction of a muscle in the immediate vicinity of the electrodes. Additional patient monitoring using BIS monitors and a direct arterial pressure monitor was used as during the application of the electroporation, the electrocardiogram tracing was significantly distorted by the electrical energy. In a few patients, the electrical energy generated extrasystoles and in one patient, a series of contractions which did not provide adequate cardiac output for several seconds.

As a result of these cardiac arrhythmias, an ECG synchronising device was used to deliver the NanoKnife™ energy 50 μs following the peak of the R wave. While this device prevented the arrhythmias, the delivery of the energy was markedly delayed as in most patients only one or two pulses could be delivered per heartbeat. In practice however, the time of delivery of the energy is not a rate limiting factor for the procedure.

B. Rubinsky (Ed.): Irreversible Electroporation, BIOMED, pp. 249–254.
springerlink.com

In the animal experiments, ultrasound was the primary method of localisation for placement of the electrodes. However in humans, especially in the case of metastases from colorectal carcinoma, ultrasound visualisation solutions can be difficult and for this reason, computed tomography was used for our image guidance. This was even more necessary in the lungs and in most of the renal tumours that were treated.

In those patients in whom ultrasound could be used, similar findings in terms of immediate loss of ultrasound echogenicity with electroporation were observed. An unexpected finding was the generation of relatively large amounts of gas immediately adjacent to the electrodes. This gas dissipated quite rapidly into the venous system and did not impact on the imaging, in fact it aided the computed tomography guidance. Without contrast enhancement, and in the absence of gas bubbles, there was no change in tissue attenuation to indicate the treated zone on computed tomography.

The other complicating factor in the group of patients we chose to treat was the inability to reliably place a grid pattern of electrodes over the tumour. This was because of the overlying ribs, scapula and other vital organs. Unlike the prostate, where a rectangular grid could be used to accurately space the electrodes throughout the gland, in the liver, the spacing of the electrodes was usually performed in an oblique manner from a limited access point in the intercostal space. This aspect of the treatment remains the most difficult in terms of planning and execution. As our experience has grown, we have moved from planning a "slab to slab" delivery to planning a "point to point" delivery of electroporation. We have also reduced the electrode exposure from 40 mm to 20 to 30 mm. This has had the additional effect of increasing the resistance of the electrodes and limiting "overcurrent" episodes. Positioning the electrodes through the tumour mass remains the most time-consuming facet of the procedure and in some respects, the more sophisticated electrodes available for competing technologies such as radio frequency ablation make those procedures quicker to perform as the electrode placement is very straightforward.

Early in our experience this resulted in "skip lesions" as the entire tumour had not been completely electroporated. Even now it is still difficult to be sure with electroporation that the entire treatment zone is "dead" at the time of completion. The available current display is simply an indication rather than a guarantee of electroporation and current flow across the tissue. The display gives an indication of the current throughout the length of the pulse and ideally, the current should increase in both across the pulse and across the whole train of pulses.

The promise of preservation of the structural integrity of the tissue was achieved and as a result of this we have been able to place the electrodes in an extremely aggressive manner with respect to vital organs. Where a tumour lies adjacent to a large bile duct, blood vessel or other vital structure, with imaging guidance it is a simple matter to position the electrodes in such a way as not to puncture the vessel or structure yet provide a zone of electroporation which involves the region of three vessel structure. Likewise lesions adjacent to the gall bladder, stomach, diaphragm and right atrium have been accessed effectively without evidence of damage to these adjacent vital structures.

In terms of complications, apart from the cardiac arrhythmias mentioned earlier, there have been no complications directly related to the irreversible electroporation. Cases of haematuria have occurred when the collecting system has been punctured by an electrode and in one patient the left adrenal gland was unintentionally electroporated. The most remarkable feature of recovery following irreversible electroporation with the NanoKnife™ is the almost complete absence of post-ablation pain. In this group of patients who have been subjected to most other alternatives including chemotherapy, surgery and thermal ablation, this feature of the NanoKnife™ is most remarkable. From a histological point of view, tissue biopsies taken one month after the procedure demonstrated "coagulative necrosis" with preservation of tissue structure. On CT follow-up at periods between one and eight months, there has been no evidence of residual damage to blood blood vessels or bile ducts. Since the biliary endothelium would have suffered the same fate as the tumour cells with electroporation, it is surprising that we have not seen evidence of bile duct stricture. Vascular endothelium and smooth muscle should also be ablated with irreversible electroporation but we have not been able to detect any deleterious effect to blood vessels in our patients. Biochemistry shows marked elevation of the transaminases which resolves to pretreatment levels within 10 to 14 days depending on the volume of tissue treated. In most patients there has been persistent mild elevation of alkaline phosphatase.

Comparison with thermal ablative technologies has been performed indirectly. Where the underlying state of the liver permits, there is a more rapid resolution of irreversible electroporation lesions compared to lesions created by radio frequency ablation. However in patients who have had other treatment modalities and have underlying cirrhosis, resolution of the irreversible electroporation lesions has been slow. Most patients will still show evidence of a zone of ablation on the one-month follow-up computed tomography scan unlike the animal studies where resolution was almost complete at 14 days.

In the kidney, successful irreversible electroporation of an area of tumour results in a focal scar which is almost indistinguishable from the lesions produced by other ablative technologies. The major difference with irreversible electroporation is the ability to treat right into the central portion of the kidney and collecting system without causing thermal damage and late strictures. In the kidney, there has been no postprocedure pain and none of the patients treated have developed a "post-ablation syndrome".

In the lung our experience is still very limited. Pneumothorax is almost inevitable but not unexpected and there is rapid development of airspace consolidation in the treated region. It can be difficult to obtain the same type of current flow across a lung lesion as we have seen in the liver and kidney and to date our clinical response has been disappointing. We have however demonstrated that the treatment can be applied both to the mediastinum and to the hilum of the lung with a high degree of safety. There is a significant application for patients with a limited number of lung metastases provided the technology of electroporation can be improved with respect to the lungs.

Going forward we shall be directing our activities to the liver and kidney for focal tumours that could be potentially cured with one NanoKnife™ procedure. Particularly for those patients who have focal disease, the almost complete lack of toxicity and absence of post-procedure pain from irreversible electroporation with the NanoKnife™ make this procedure most enticing.

At other sites researchers are beginning to apply this technology in humans in these and other areas including the prostate, pancreas and even the brain. The challenge will be to develop easier electrodes to use and more secure method of measuring the immediate efficacy of the procedure.

Illustrations

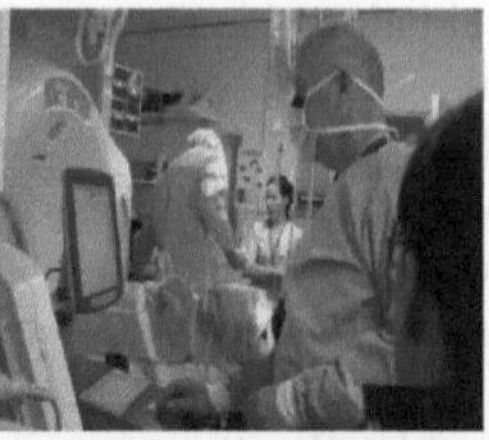

The procedures were done in a CT suite under general anaesthesia. Electrodes were placed with CT and or ultrasound guidance. The author is operating the NanoKnife^{TM}

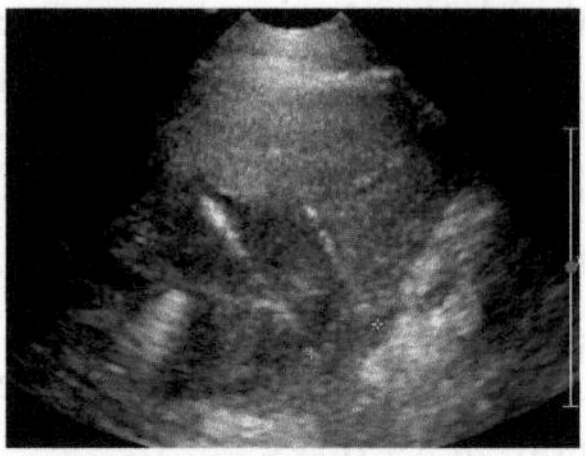

Ultrasound image with two electrodes and their gas bubbles in the plane of view. Echogenic structures to the left and below the electrodes are due to prior electrode activation. It is assumed that the bubbles are generated from electrolytic processes on the electrodes.

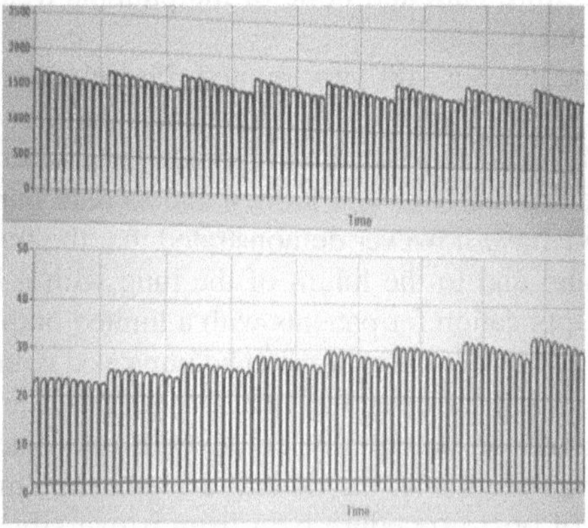

Post electroporation display. Top panel is voltage which falls across each series of pulses as the capacitor discharges. The lower panel shows increasing current flow which indicates electroporation has been achieved. Current is determined by voltage and electrode length (resistance).

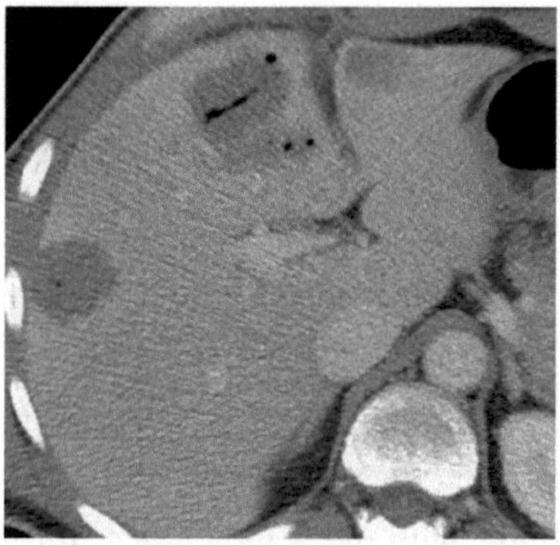

Contrast enhanced (delayed portal venous phase) CT post NanoKnife^{TM}.. Electroporated areas show lack of enhancement even though major vessels are still patent.

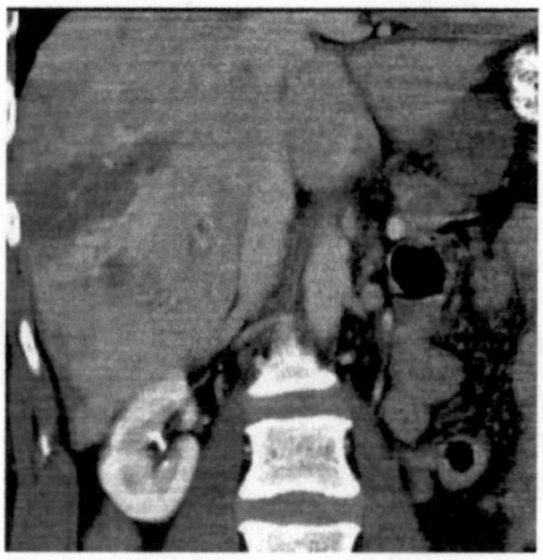

Patient with neuro-endocrine tumours. A large slab of electroporation has been performed. The oblique shape results from an inability to access a grid pattern through rib and intercostal spaces. Note the patent vessels within and at the inferior border of the treated zone on the portal phase CT scan.

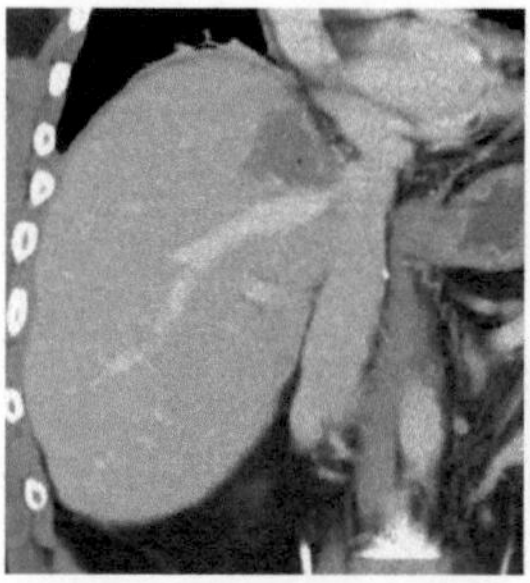

Colorectal carcinoma near right atrium, diaphragm and hepatic vein IVC confluence. Successful procedure without damage to these structures and no post procedure pain.

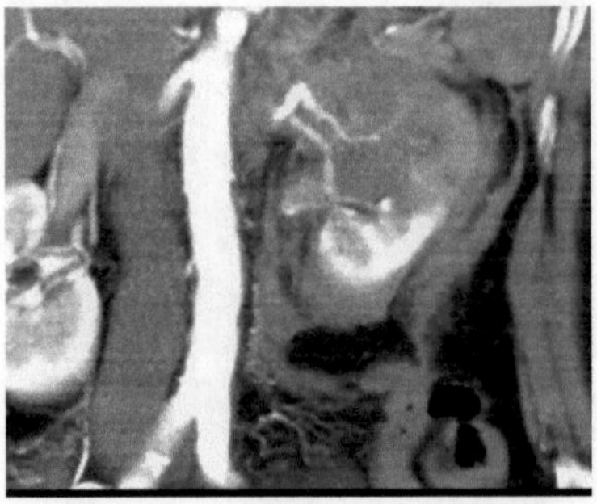

Upper pole "nephrectomy" with the NanoKnife™ for a renal metastasis. Upper pole major arteries are patent at the end of the procedure. Minor haematuria 24 hours. No post procedure pain in spite of minor haemorrhage in the perirenal space.

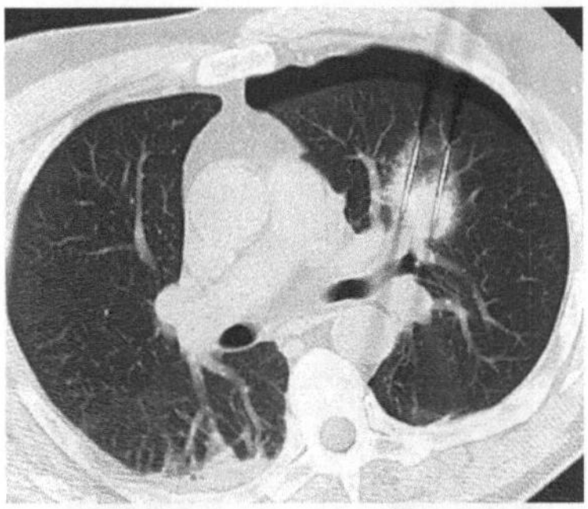

Solitary hilar metastasis with electrodes in position astride the upper pole pulmonary artery. No haemoptysis after the NanoKnife™ procedure. Pneumothorax resolved on drainage in 24 hours.

Irreversible Electroporation Systems for Clinical Use

Claudio Bertacchini, Pier Mauro Margotti, Enrico Bergamini, Mattia Ronchetti, and Ruggero Cadossi*

IGEA S.p.A., Via Parmenide, 10A, I 41012 Carpi (MO), Italy
* Corresponding author E-mail: r.cadossi@igeamedical.com

Introduction

Electroporation is a technique that uses micro to milliseconds electric pulses to create pores in the cell membrane, thus allowing molecules that, due to their physical and/or chemical properties, would normally not be able to cross the cell membrane, to enter the cell (1-5). Electroporation finds applications in many fields in particular for gene insertion in cells (electrogenetherapy) (6,7) and for the treatment of cancer (electrochemotherapy). In electrochemotherapy, the combination of chemotherapy and electroporation of tumour cells, the effects of drugs that usually show little cytotoxicity are greatly increased (8). The opening of pores in the cell membrane allows the chemotherapeutic agent to enter the cell at greater, more effective concentration and exert its cytotoxic action by killing the target cell (9-11).

However, if cells are unable to seal the pores formed because the electric field is maintained for longer times (i.e. when using a high number of pulses) cell death due to the loss of homeostatic mechanisms occurs. This phenomenon is termed "irreversible electroporation". The threshold for irreversible electroporation strongly depends on the type of tissue under treatment (13).

The main advantage of irreversible electroporation is the possibility to avoid the use of drugs, as it relies only on the effect of the applied electric field to kill the target cells, for this reason irreversible electroporation can be used as an ablation modality also when treating low proliferating lesions such as benign tumours. To obtain irreversible electroporation of the target tissue high electric fields need to be generated, the electrical pulses are applied by means of electrodes inserted in the lesion to be treated. Since the amplitude and the gradient of the field depend on the applied voltage as well as on the distance between the electrodes, high values of the electric field can be achieved arranging the electrodes close to each other or increasing the potential applied between electrodes. However, when treating deep seeded tumours, the goal is to introduce the electrodes through the skin both to minimize the procedure's impact and avoid open surgery; thus, introducing a large number of closely arranged electrodes to effectively porate the target tissue is not desirable. In this scenario the goal is to treat a volume of approximately 50 to 70 cm^3 with up to 6 electrodes. This requires applied voltages in the order of 3000 V to obtain irreversible electroporation. The specifications of the device ensure that the minimum value of the electrical field gradient in a volume 40 cm^3 is

B. Rubinsky (Ed.): Irreversible Electroporation, BIOMED, pp. 255–272.
springerlink.com

at least 800 V/cm, this is considered to be the threshold for irreversible electroporation in most cells (13). Electrode design is extremely important, and both monopolar or bi-polar electrodes are used. Typically, electrodes used with the device are 15 cm long stainless steel needles, partially insulated, with a diameter of 1 mm. The conductive, non insulated, distal end of the electrodes can be up to 4 cm long, it should be noted that increasing conductive surface of the electrodes increases the current absorbed by the tissue, possibly exceeding irreversible electroporation system specifications; this can be avoided by using electrodes certified as compatible with the device. The electrodes may be inserted under ultrasound guidance to verify their positions and ensure coverage of all the target tissue.

The high voltages and currents (up to 50 A) needed represent the main problems to be taken into account when designing a device for irreversible electroporation. Safety issues arise because of the high energy involved and because of the variable working conditions present in the operating theatre. In fact, the delivered current depends on the ohmic characteristics of the tissue under treatment and can differ from point to point, particularly in the typical heterogeneous tissue that constitutes tumours. Moreover, the tissue can have its electric characteristics altered during the treatment as a consequence of the deep modifications caused to the cells and the extra cellular environment when high currents are applied. However, irreversible electroporation does not cause significant heating of the tissue: its activity is not due to thermal effects on cells and protein or extracellular matrix denaturation does not occur during treatment.

Consequently in the design of an irreversible electroporator, one has to keep under control the patient leakage current, to design a sturdy and fast high voltage pulse generator, with an high reactivity in case of failure and provide a user-friendly interface for the operators to minimize the chance of user error.

This paper describes the solutions adopted to solve all these challenging issues in the design and implementation of a device for irreversible electroporation to be used in the clinical practice.

General Safety Remarks

In addition to general safety requirements for medical devices stated by regulation and standards (14,15,16), specific safety issues characterise electroporation devices. The principal hazards derive from the high energy that is accumulated on capacitors and from the delivery of high electrical current to the patient, which involves the risk of electrocution for both the patient and the operator.

Energy release to the patient must be reliably controlled and limited: unintended or incorrect release has to be avoided. Strictly related to this issue is device ruggedness: in the absence of adequate protective measures, a failure of some critical component may lead to a complete, unwanted discharge of the accumulated energy.

Reliable energy delivery control characterises the normal device operation and can be ensured by specific safety measures implemented in the software/firmware. This goal is achieved by carrying out risk analysis in the earliest phases of the device architecture design, in order to identify those hardware parts that must support software safety features; ensuring that critical device control is carried out by reliable programmable systems, and implementing best software/firmware development practices (17).

Energy delivery limitation refers to fault conditions. It is a strategic choice: risk control may be obtained either by limiting the probability of a failure, in particular for those critical components whose failure may lead to uncontrolled energy delivery to the patient, or by implementing an independent system that prevents energy delivery above the maximum normal-condition value; both solutions have pros and cons. In any case the system must be sufficiently rugged against short circuits and sparks, likely to occur between electrodes due to the unpredictability of the resistive load of biological tissues, the presence of conductive solutions or human error, to confine such failure to a remotely likely event.

The likelihood of an electroporation treatment to lead to electrocution depends on several factors: the applied pulse voltage, the length of the pulses, the number of pulses, the pulse repetition rate and, of course, the distance between the hearth and the electrodes (18). Generally speaking, synchronisation of treatment delivery with the refractory period of the cardiac cycle (ECG synchronisation) is always advisable when there is not enough confidence that the electroporation treatment cannot determine a current density lower than fibrillation threshold of the myocardium.

General Device Structure

The electroporation device is composed of two main parts: the user interface (UI), that calculates the treatment parameters based on data inserted by the operator, shows and elaborates data and signals measured during the treatment, and a power unit (PU) which actually generates the pulses by using the parameters provided by the interface and records signals.

The UI is a medical grade personal computer suitable for use in a operating room and it guarantees compliance with standards for medical electrical devices. A standard Operating System (OS) is used to provide an intuitive graphical user interface, allowing the operator to easily set treatment parameters and to process, download and analyse measured treatment data. Once the parameters have been set, the PU should be able to perform the treatment independently from the non real-time OS and the unpredictable delays associated with it. This approach guarantees the execution of the operations involved in the treatment delivery with an exact timing sequence and to react immediately to the events being monitored. The described configuration allows a fine control over the charge on the capacitors, the pulse length, the pause between pulses and to immediately react to safety-related alarms.

Strict control of timing delays and signals can be better provided by an FPGA (Field Programmable Gate Array) based platform than a microcontroller. The FPGA processes the treatment parameters, controls if the capacitors store enough energy to deliver a high-voltage, high-current pulse, then handles the generation of pulses and the measurement of the delivered treatment.

A schematic representation of an electroporator device is shown in Figure 1. The PU is composed of high energy parts (highlighted in yellow in Figure 1) and a digital control part. The control block mainly includes the FPGA and an interface for communicating with the UI that directly supervises and handles the operations of the other blocks. The high voltage capacitors store all the energy for a sequence of pulses with the desired voltage, as commanded by the FPGA. Once the capacitors have charged, the FPGA waits for a signal to proceed with the treatment. The signal is provided directly by the operator by means of a double pedal with IPX-8

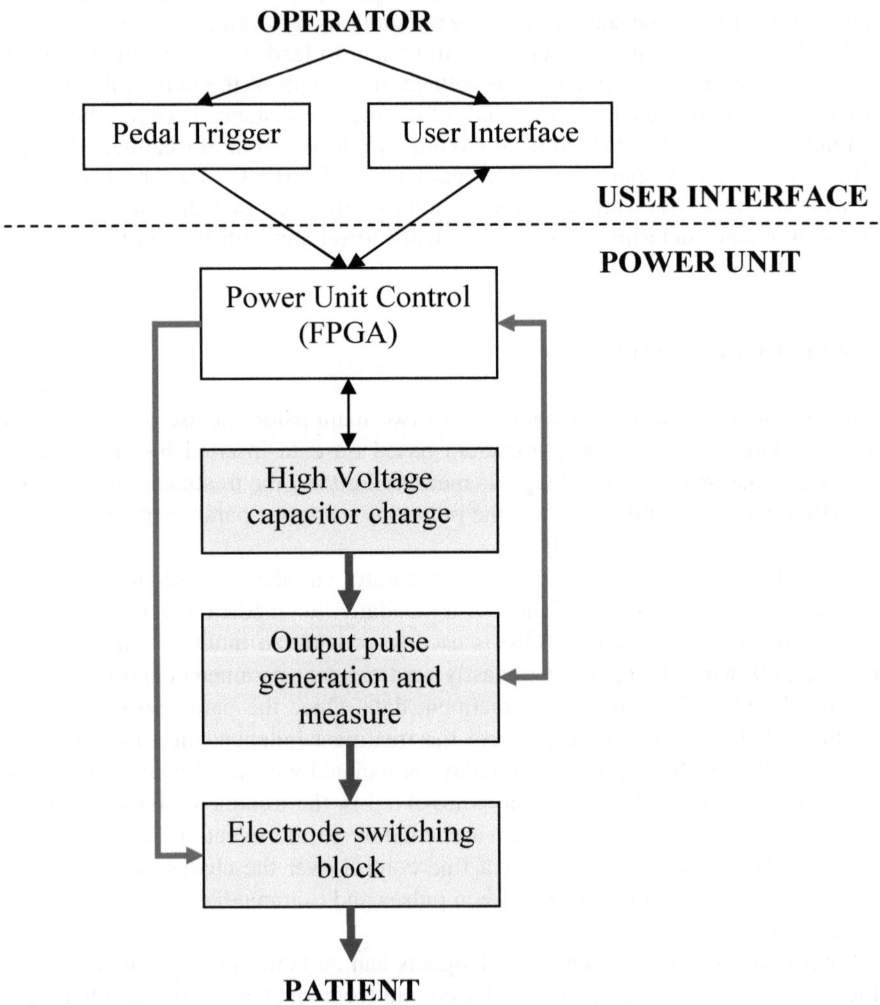

Fig. 1. Chart of the main blocks composing the device. The power unit is composed of high energy parts, highlighted in yellow, and a digital control part.

degree of protection from liquids (as imposed by the standards for the use in operating theatres). The first pedal is used by the operator to enable the second pedal that actually starts the treatment.

Once triggered, the FPGA is in charge of two main tasks: it controls the output pulse block and acquires the measure of voltage and current actually delivered to the patient through the electrodes. The pulse length can be suitably controlled by using two high voltage IGBT (Insulated Gate Bipolar Transistor): one to control the pulse length and the second dedicated to avoid the delivery of uncontrolled pulses under failure conditions.

The electroporator device is able to use up to six independent electrodes to apply an electric field with the most appropriate shape and intensity to homogenously cover the target tissue. Several standard electrode configurations are preprogrammed in the device to allow treatment of simple, regularly shaped lesions, however the operator is presented with the possibility of programming custom treatments and define all relevant parameters: number of electrodes used, potential applied between each couple of electrodes, number of pulses to be applied, and polarity (figure 2). Once the desired treatment parameters are defined, the device delivers a single low intensity pulse between each couple of electrodes. The

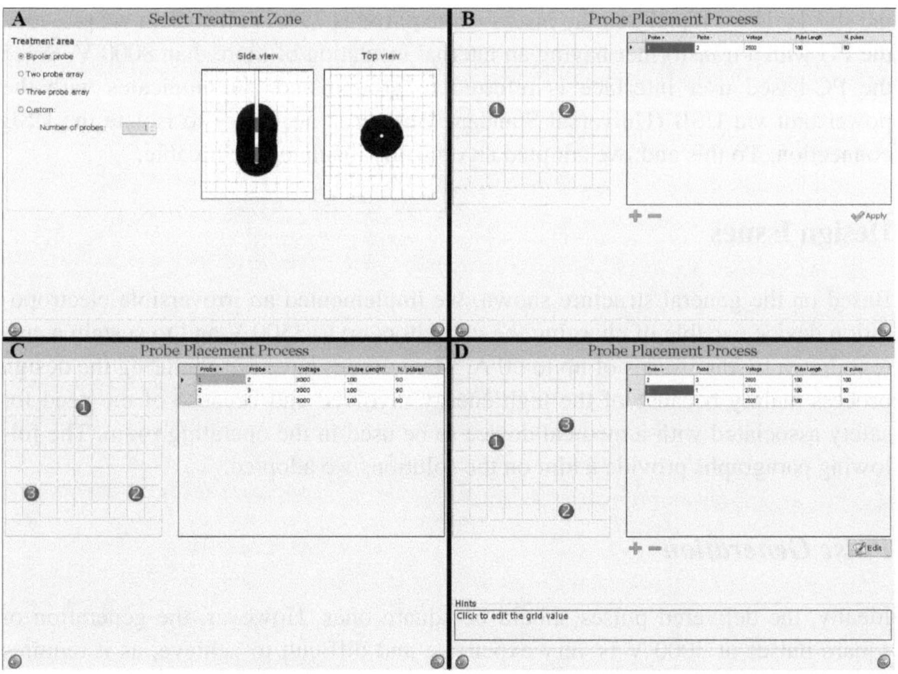

Fig. 2. Examples of treatment parameters: standard single bipolar electrode (A), standard two probes array (B) standard three probes array (C), custom programmed three probe array, applied potential between each electrode is manually selected by the operator (D).

"pre-pulse" allows the device to determine if the treatment can be completed as programmed and allows the detection of errors due to erroneous positioning of the probes such as two probes touching (short circuit); in the event of a problem in a pair of probes the device signals which electrodes are involved to the operator.

A switching block is needed to route the treatment to the electrodes in sequenced fashion. In order to implement such capability, high voltage SPDT (Single Pole Double Throw) relays are used. For each electrode two relays are involved: one to select the polarity of the pulse and the second to connect the electrode. The second relay also has the additional purpose of performing an initial test of the device efficiency conducted using an internal resistive load, which is part of a complete diagnostic self test that the device undergoes when powered on to identify any accidental failure before treatment begins.

Finally the FPGA directly controls the switching activity of the relays of the switching block by connecting each electrode in turn to the output pulse generation block.

In order to reduce design complexity, we chose to consider the whole PU as the patient circuit, defined as any electrical circuit containing conductive parts which are not insulated from the patient (16). This means that for sake of safety not only the high voltage part is completely insulated from ground but the whole PU. Based on the standards we found that the insulation required is 7500 V, therefore we provided the PU with a transformer having an internal insulation of more than 8000 V. Since the PC-based user interface is referred to ground and communicates with the power unit via USB (Universal Standard Bus), it is necessary to isolate the USB connection. To this end, we adopted an optically insulated USB cable.

Design Issues

Based on the general structure shown, we implemented an irreversible electroporation device capable of charging the capacitors up to 3500 V and to sustain a current drawn by the tissues of up to 50 A. Many issues have raised during the design process mainly because of the high energy involved and because of the need for safety associated with a medical device to be used in the operating room. The following paragraphs provide a hint on the solutions we adopted.

Pulse Generation

Ideally, the delivered pulses should be square ones. However, the generation of square pulses at 3000 V is very expensive and difficult to achieve, as it requires the design of a very-high-voltage amplifier. We investigated two alternative approaches for the generation of pulses that approximate the ideal square shape, without requiring the use of an amplifier: the pulse transformer and the high voltage generator.

Pulse Transformer

When using a pulse transformer, pulses at lower voltage are applied to the primary coil of a step-up pulse transformer, which generates the required high voltage pulses on the secondary coil. The energy required is supplied by the charge stored on capacitors. The output energy to be delivered is stored on capacitors connected in parallel to reach the desired capacity. The value of the required maximum charged voltage on the capacitors is dependent on the transformer ratio.

Since the pulse transformer is less efficient at low frequencies than at higher frequencies, a bigger transformer is required to generate longer pulses. Another issue with the use of a transformer is the major difference between the achieved pulse shape and the ideal square one, especially in presence of low-resistance loads.

Tests have been carried out using a toroidal pulse transformer having 148 mm outer diameter; 56 mm height; 3.8 kg weight. The outcome was that the pulse transformer did not allow pulses longer than 30 μs and pulse repetition frequency had to be lower than 81 Hz (1/12300 μs). Figure 3 shows the effect of transformer's core saturation that occurs at higher pulse repetition frequencies.

The load effect became significant with loads lower than 250 Ω, as shown in Figure 4: as load impedance decreases, the voltage pulse amplitude on the secondary coil of the transformer becomes significantly lower and the pulse shape is also affected. If subjected to low impedance loads the transformer is not able to deliver the expected pulses in terms of applied voltage and current; for this reason a device employing the pulse transformer is not suitable for our purposes.

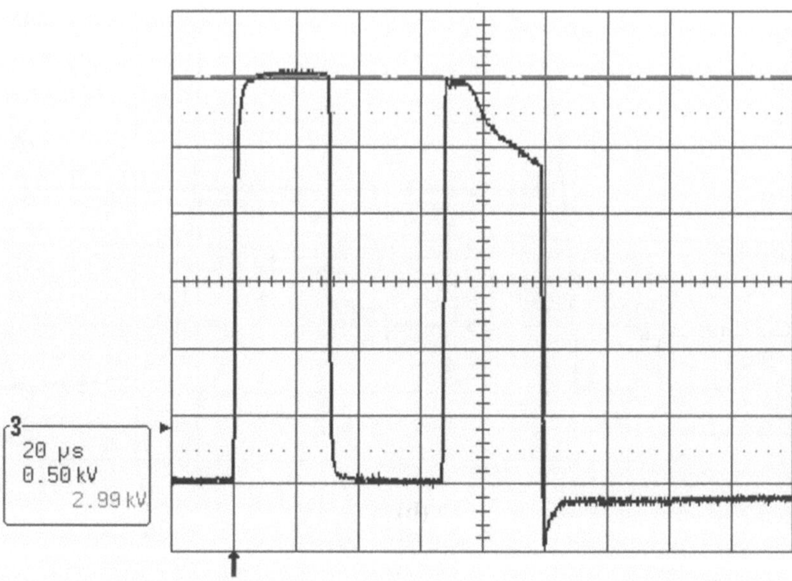

Fig. 3. Two Pulses at 3000 V, 30 μs long, 30 μs pause, with a 500 Ω load

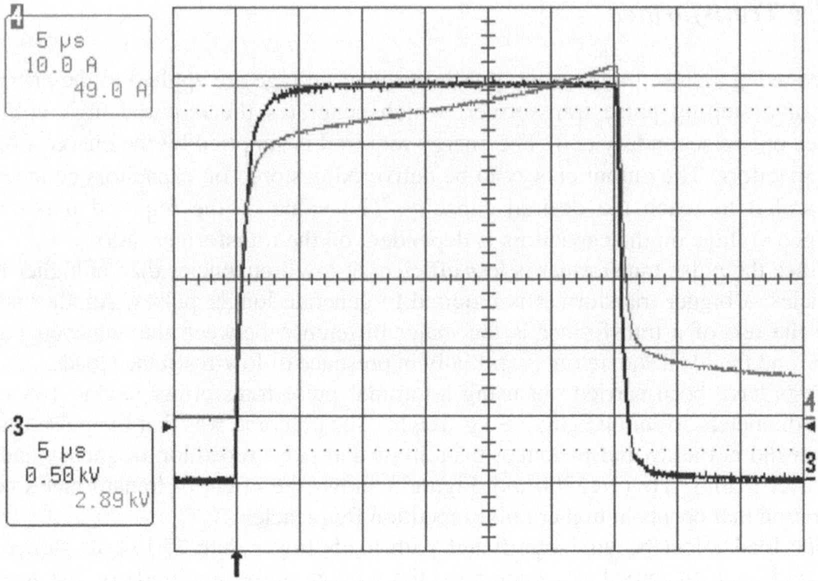

(a)

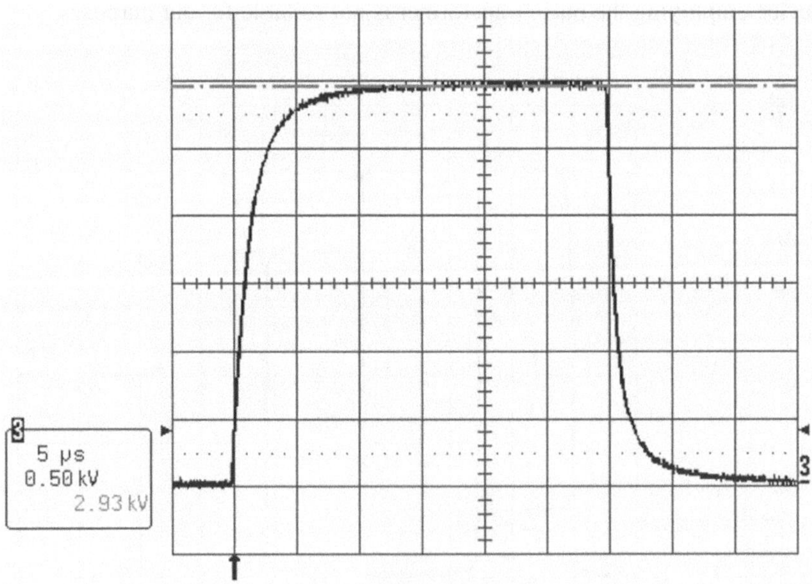

(b)

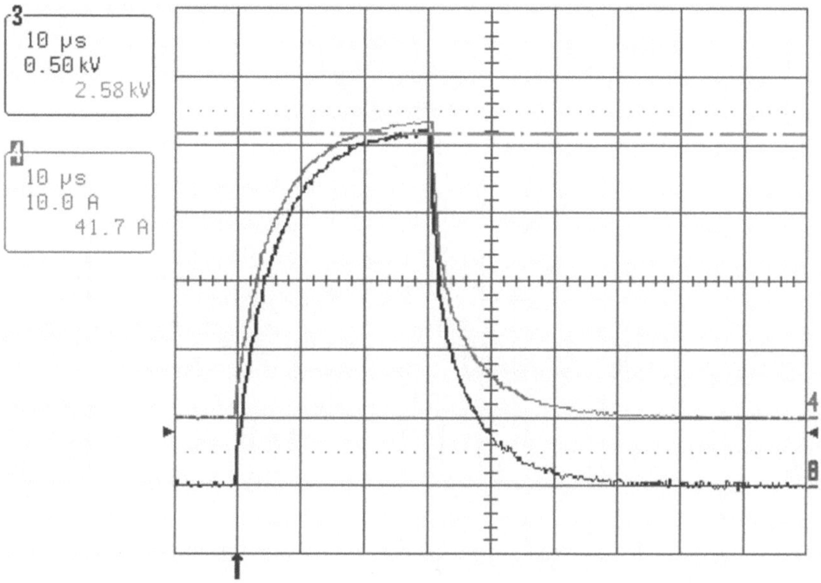

(c)

Fig. 4. Pulse voltage (green) and current (blue) dependence from load with a 3000 V nominal amplitude, 30 μs long pulse: (a) 500 Ω, (b) 250 Ω, (c) 60 Ω.

High Voltage Generator

When using a high voltage generator the capacitors are directly connected to the load by means of switches in order to generate the pulses. With this approach the ideal square pulses are approximated to a time-controlled free discharge of a capacitor (see Figure 5).

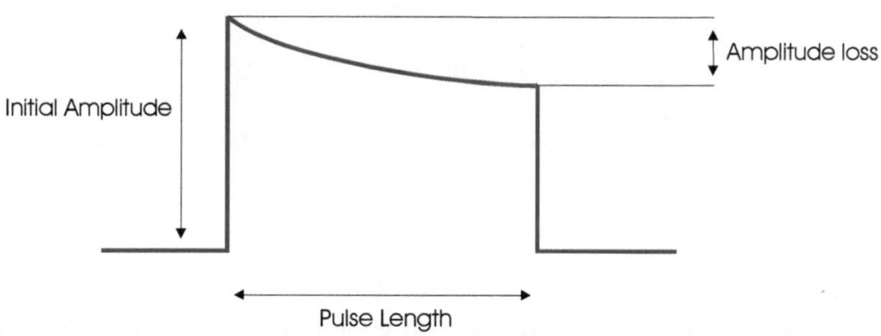

Fig. 5. Truncated slow-decay exponential. The square wave pulse is produced by a partial discharge of a large capacitor, which requires the interruption of high currents against high voltages.

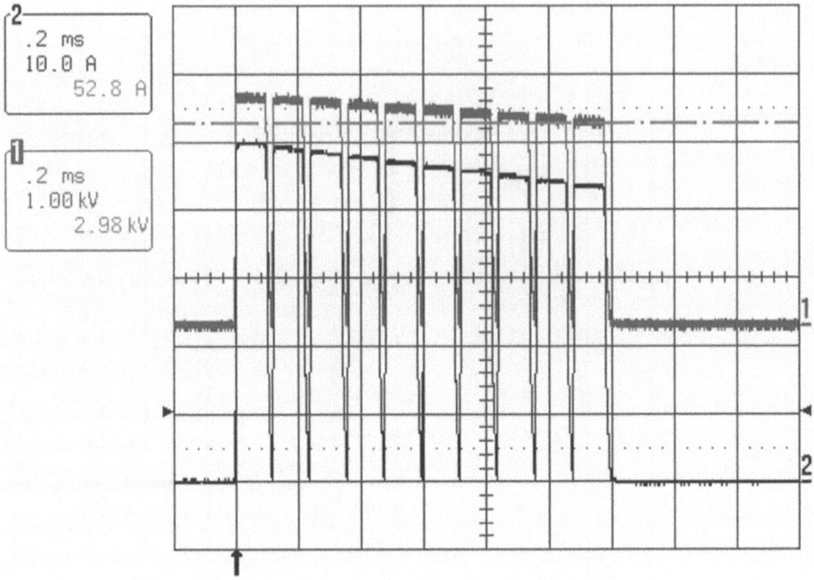

(a)

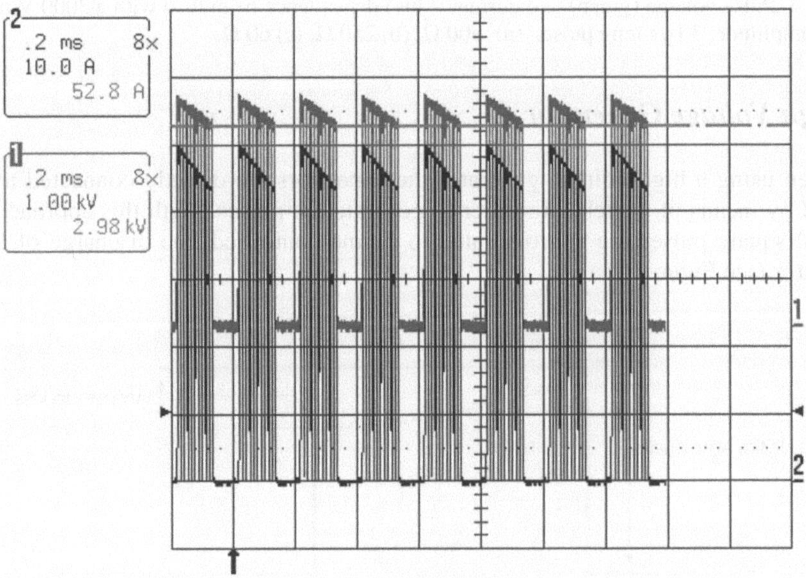

(b)

Fig. 6. Voltage and Current wave forms at 3000 V (at the end of ten 100 μs, 20 μs spaced pulses) on a 68 Ω load (approx. 50 A peak current), obtained using the high voltage generator approach. A single pulse sequence in (a). The repetition of 8 sequences in (b). The dwell between sequences is not shown.

The quality of this approximation depends on the load. This means that such kind of pulse could be well approximated to a square pulse only if the current drawn by the load is low. This is because the current flowing on the load discharges the capacitors, thus causing the amplitude loss shown in Figure 5. A small discharge current causes a little drop and the amplitude at the end of the pulse is only slightly lower than the initial capacitor voltage. On the other hand if the load has a particularly low resistance it will draw high current, and the pulse shape will have a considerable drop, although starting from the required voltage. Control of pulse parameters, such as amplitude and duration, is critical to ensure complete electroporation of the target tissue, for this reason the capacitor should be dimensioned so that the time constant of the voltage decay is much longer than pulse duration also in the least favourable load conditions. Our system was designed to ensure that voltage decay is limited to 10% of the initial capacitor voltage when tested at the maximum current draw allowed of 50 A, this condition presents when a 3000 V pulse is applied to an impedance load of 60 Ω.

The high voltage generator approach requires a high voltage power supply that charges the capacitor at the desired initial pulse voltage and high voltage capacitor of sufficient capacitance. The required high voltage may be obtained connecting capacitors in series, in spite of the total capacitance, which results in a fraction of the capacitance of each single capacitor.

In order to generate the pulses a high voltage Mosfet switch or IGBT can be used to connect the load with the capacitors only for the required time of the pulse length.

After some evaluation about dimensions, costs and considering the specifications, the high voltage generator approach generally has shown to be the best solution.

Speed Issue

Another main issue is the device responsiveness. The pulses delivered can have a length between 20 μs and 1 ms and require a fast control mechanism. Due to the frequency of the pulses and the high currents involved, we found that a safety protection using fuses is not reliable enough, since it is hardly possible to have fuses with a reaction time of few microseconds. For this reason we decided to implement a dedicated fast redundant hardware solutions to protect both the patient and the device.

Variable Load

A peculiarity of the electroporation process is that the resistive load of a biological tissue varies greatly and is therefore unknown at the time of treatment. Moreover, the resistance depends on the physical properties of the electrodes used and there is a decrease of the resistance during the treatment due to the changes induced by the electrical field in the target tissue.

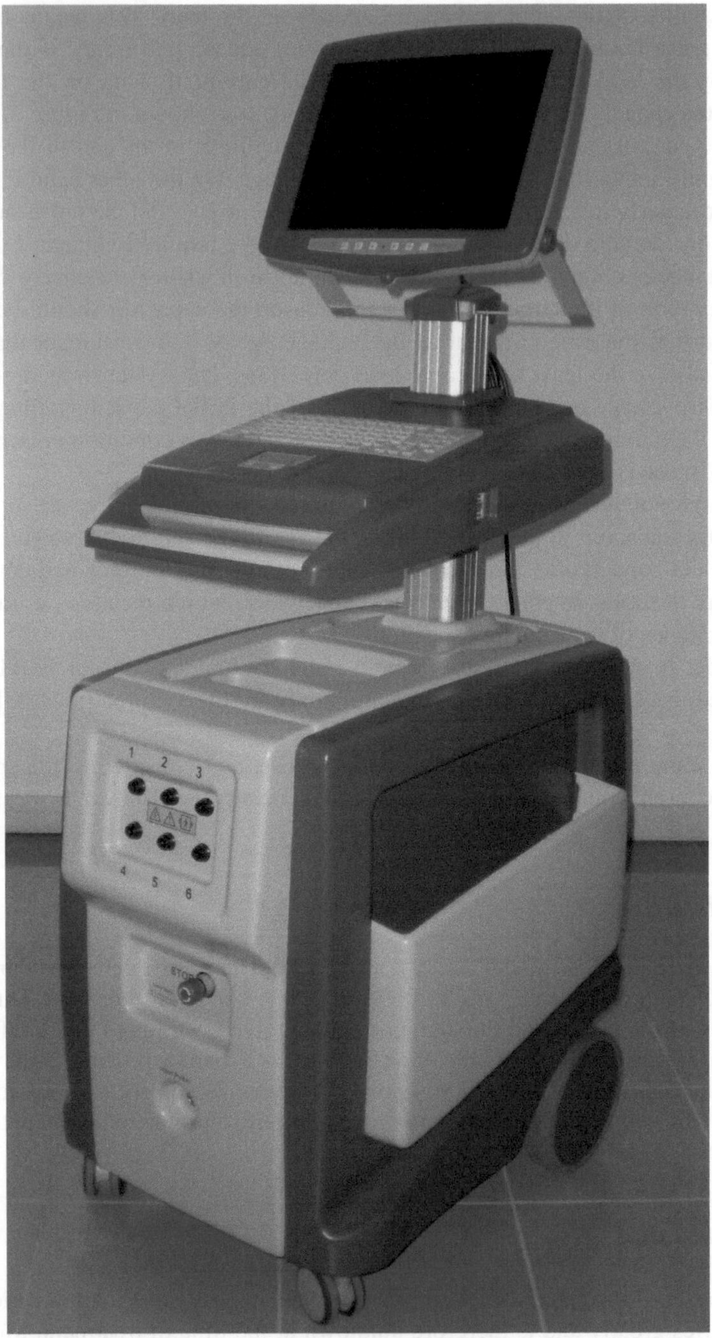

Fig. 7. Trolley containing the high voltage power unit at the bottom and the PC with the user interface (keyboard, touchpad and display) on top

As a consequence, it is almost impossible to make an accurate prediction of the load without a preliminary test and it is safer to consider the least favourable scenario. For this reason we designed the device considering the event of sparks and short circuits and introduced a current limitation to avoid injuries to the patient and failure of the device. Simulations allowed to estimate that in the worse non-short-circuit condition, the load should not draw more than 50 A, therefore it is assumed that if during the treatment the drawn current exceeds this threshold, it is probably due to a short circuit or spark occurring at the electrodes. In this case the device interrupts the pulse sequence. Furthermore, due to the heterogeneous electrical properties of biological tissues and tumours in particular, considerable effort was made to provide feedback to the operator after the procedure to allow the assessment of electroporation efficacy. The device was designed to report "low current warnings" that imply a partial tissue electroporation in the event that the current between any given couple of electrodes does not reach a pre determined threshold (1.5 A). The electrode that generated the problem are reported so that the operator can take the necessary steps to succefully complete the treatment.

Size Issue

The size of the device can vary depending on the choices made during the design phase. We chose to build a device provided with a trolley (figure 7), allowing us to have larger space for the electronics. The advantages of this choice are: the required electrical insulation is achieved more easily and it is possible to increase the robustness to noise induced by the high voltage parts. We decided to put the PC for the UI into a separate part close to the LCD display and the keyboard. The PU is located at the bottom of the trolley.

Software Architecture

As shown in the device structure, a double level of control is desirable. The software consists of two main parts that interact with each other but are logically and functionally independent: a high level part implements the software graphical user interface and a low level control, implemented in hardware and in the FPGA firmware, that is responsible for all functions that directly control the power part of the device.

The Graphical User Interface

The Graphical User Interface (GUI) was developed thinking of ease of use and enable also an inexperienced user to focus on the therapy rather than on the device usage.

The main structure of the GUI is a typical wizard application with the 'next' and 'back' buttons that allow navigation between pages.

The inner task of the GUI is to monitor the status of the PU at fixed time intervals and display the main information regarding the capacitors charge and the treatment delivery process.

The GUI is based on the following structure:

COMM_LAYER: a layer that performs the basic communication with the power device, supervises the USB communication and performs basic read/write operations;

CHECK_LAYER: checks the input data inserted by the operator to prevent loading of inconsistent parameters on the power unit;

TRANSLATION_LAYER: elaborates all data received from the power unit in a format such that the user can quickly analyze the information on the treatment delivered.

The COMM_LAYER and TRANSLATION_LAYER run in the background at all times during the treatment delivery and ensure a strict timing synchronization of the different tasks and between the GUI and PU control firmware. This feature is implemented by means of a handshake protocol that enables the operator to have full monitoring of the ongoing operations. In this way a treatment can be aborted by the operator at any time via software and both the PU and the GUI return in a safe state, waiting for new inputs. On the other hand, if an error occurs during an operation, the GUI immediately notifies it to the user who can decide how to proceed.

In order to improve safety in the use this device tests are performed at start-up before the GUI actually starts operating. These tests check the main functionality of the device such as a live communication between UI and PU, the correct firmware version on the power unit, complete check of the RAM chips of the PU control hardware, a complete charge and discharge cycle of the capacitors, a test on internal test loads, etc (Figure 8).

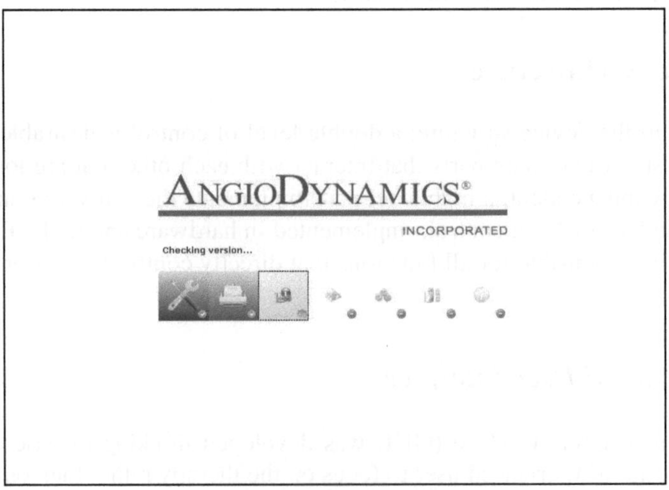

Fig. 8. Screenshot of device tests at start up

Power Unit Control

A FPGA has been chosen to control the whole PU. By using a dedicated board we allow the PC-based UI to represent the FPGA as a series of addresses where the treatment parameters are stored. The FPGA firmware has the structure shown in Figure 9 and it is divided into several parts, each one with a well defined task. The main parts are: the interface block used to communicate with the PC; the charge control part to handle the charge of the capacitors; the generation part to control the pulse generation; the measurement part to allow the measure of voltage and current during the treatment and the switching part to control the relay configurations.

The system controls the pulses by means of two high voltage IGBTs. These two components have different purposes: the first is directly driven by the FPGA to

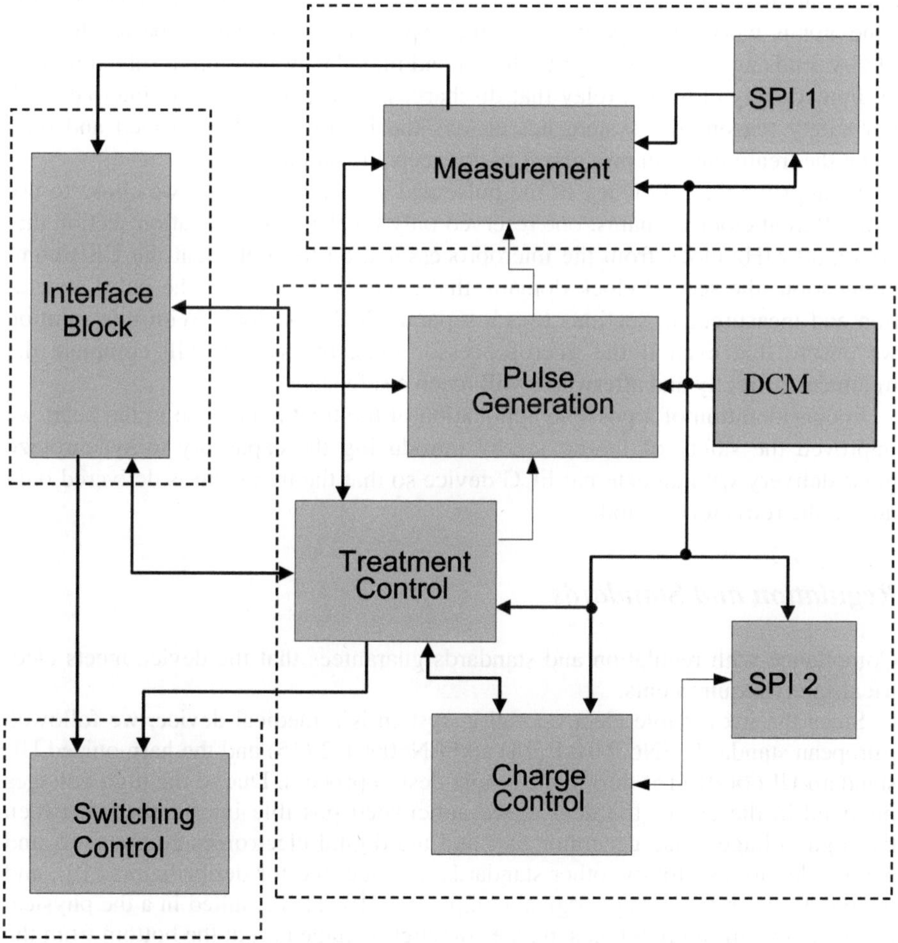

Fig. 9. FPGA firmware architecture

determine the desired pulse length, and the second IGBT is used to limit the maximum length of each pulse to 1ms, for safety reasons in the event of failure. Since the specifications allow a maximum current of 50 A, we used the IGBTs to control the current as well. We improved a standard driver with a circuit capable of readily reacting to an over-current, by causing the pulse to stop within 10 to 20 μs.

Since both the capacitor voltage and the current and voltage of the pulses are not required to be measured at a high rate, the use of analogue-to-digital converters with a serial interface has been preferred. In this way a compact SPI (Serial Peripheral Interface) communication system can be used between the measure blocks and the FPGA.

The charging of the capacitors is operated by a high voltage generator with a maximum voltage over 3500 V. The charge control block enables the generator when a charge phase is required and disables it both during the pulse delivery and during any discharge process. Furthermore, the charge control block checks if the generator is working properly: when the expected voltage cannot be reached the FPGA sends an alarm message to the UI and puts the system into a safe condition by immediately closing a relay that discharges the capacitors on an internal load. For safety reasons the system has always the internal load connected and only when the treatment is in progress it re-connects the output section.

To improve the efficiency of the pulse and generation control we chose to use two different clock domains: one reserved only for the communication section that uses a 30 MHz clock from the microprocessor used to implement the USB communication; the second clock domain, that controls the charge, the pulse generation and measurement sections uses a separate 20 MHz clock. With this solution we ensure that even if the microprocessor fails, the system will complete the treatment correctly and afterwards will reach a safe state.

In consideration of a possible application of the treatment close to the heart we improved the safety of the device by introducing the capability to synchronize pulse delivery with an external ECG device so that the treatment is delivered only during the refractory period.

Regulation and Standards

Compliance with regulation and standards guarantees that the device meets electrical safety requirements.

Since the irreversible electroporation system is a medical device, we followed European standards EN60601-1 (14) and EN60601-2 (15) and the harmonized UL standard UL60601 (16) during the whole design process. Due to the high voltages involved in the use of the device, we anticipated possible interferences between the high voltage pulse generator part and the digital electronics components and we decided to also follow other standards, like the one for defibrillators (19), and the one for high frequency surgical equipment (20). This resulted in a the physical layout shown in figure 9 that separates the high voltage part at the bottom from the digital components.

Fig. 10. Intenal layout of the power unit of a irreversible electroporation system

Conclusions

The development of an irreversible electroporation system approved for the use in the clinical practice, allows a novel approach to tissue ablation. Tumour ablation based on irreversible electroporation relies solely on the application of an intense enough electric field to the target tissue to cause cell death. The device has been developed to allow the use of up to six independent electrodes. The electrodes are not restricted to a fixed geometry, rather they can be independently positioned based on the tumour size, shape and position, to ensure that the target is entirely enclosed within the applied electric field, thus ensuring complete tissue ablation. The result of the development process is a reliable, safe for the operating theatre, high performing medical device for irreversible electroporation based tissue ablation.

References

1. Neumann, E., Rosenheck, K.: Permeability Changes Induced by Electric Impulses in vescicular membrane. J. Membrane Biol. 10, 279–290 (1972)
2. Crowley, J.M.: Electrical Breakdown of Biomolecular Lipid Membranes as an Electromechanical Instability. Biophysical Journal 13, 711–724 (1973)

3. Zimmermann, U., Vienken, J., Pilwat, G.: Dielectric Breakdown of Cell Membranes. Biophysical Journal 14, 881–899 (1974)
4. Weaver, J.C.: Electroporation of Cells and Tissues. IEEE Transactions on Plasma Science 28, 24–33 (2000)
5. Puc, M., Corovic, S., Flisar, K., Petkovsek, M., Nastran, J., Miklavcic, D.: Techniques of Signal Generation Required for Electropermeabilization. Survey of Electropermeabilization Devices. Bioelectrochemistry 64, 113–124 (2004)
6. Neumann, E., Schaefer-Ridder, M., Wang, Y., Hofschneider, P.H., et al.: Gene Transfer into Mouse Lyoma Cells by Electroporation in High Electric Fields. J. EMBO 1, 841–845 (1982)
7. Gehl, J.: Electroporation: Theory and Methods, Perspectives for Drug Delivery, Gene Therapy and Research. Acta Physiol. Scand. 177, 437–447 (2003)
8. Mir, L.M., Belehradek, M., Domenge, C., Luboniski, B., Orlowski, S., Belehradek, J., Schwaab, B., Luboniski, B., Paoletti, C.: Electrohemotherapy, A Novel Antitumor Treatment: First Clinical Trial. C. R. Acad. Sci. Ser. III 313, 613–618 (1991)
9. Mir, L.M.: Therapeutic perspective of in vivo cell electropermeabilization. Bioelectrochemistry 53, 1–10 (2000)
10. Marty, M., Sersa, G., Garbay, J.R., Gehl, J., Collins, C.G., Snoj, M., Billard, V., Geertsen, P.F., Larkin, J.O., Miklavcic, D., et al.: Electrochemotherapy – An Easy Highly effective and Safe Treatment of Cutaneous and Subcutaneous Metastases: Results of ESOPE (European Standard Operating Procedures of Electrochemotherapy) Study. Eur. J. Cancer Supp. 4, 3–13 (2006)
11. Mir, L.M., Gehl, J., Sersa, G., Collins, G.C., Garbay, J.R., Billard, V., Geersten, P.F., Rudolf, Z., O'Sullivan, G.C., Marty, M.: Standard Operating Procedures of the Electrochemotherapy: Instructions for the Use of Bleomycin or Cisplatin Administered either Sistemically or Locally and Electric Pulses Delivered by the Cliniporator^{TM} by Means of Invasive or Non-invasive electrodes. Eur. J. Cancer Supp. 4, 14–25 (2006)
12. Rubinsky, B., Onik, G., Mikus, P.: Irreversible Electroporation: A New Ablation Modality – Clinical Implications. Technology in Cancer Research and Treatment 6, 37–48 (2007)
13. Davalos, R.V., Mir, L.M., Rubinsky, B.: Tissue Ablation with Irreversible Electroporation. Annals of Biomedical Engineering 33, 223–231 (2005)
14. European Standard EN60601-1 Medical electrical equipment - Part 1: General Requirements for safety, 2nd edn. (1998)
15. European Collateral Standard EN60601-1-2; 2. Collateral Standard: Electromagnetic Compatibility Requirements and tests
16. UL Standard for Safety for Medical Electrical Equipment, Part 1: General Requirements for Safety, UL; 60601-1, 1st edn. (2003)
17. General Principles of Software Validation, Final Guidance for Industry and FDA Staff, CDRH (2002)
18. Patrick, R.J.: Applied Bioelectricity: from electrical stimulation to electropathology. Springer, Heidelberg (1998)
19. IEC 601-2-4:1983-01; Medical electrical equipment - Particular Requirements for the safety of cardiac defibrillators and cardiac defibrillator-monitors
20. IEC 601-2-2:1998-09; Medical electrical equipment - Part 2: Particular Requirements for the safety of high frequency surgical equipment

The Use of Irreversible Electroporation in Food Preservation

Alex Golberg[1], Judith Fischer[2], and Boris Rubinsky[1,*]

[1] Center for Bioengineering in the Service of Humanity and Society,
School of Computer Science and Engineering, Hebrew University of Jerusalem,
Givat Ram Campus, Jerusalem, Israel
[2] Stern College for Women, Yeshiva University, New York, NY, 10016, USA
* Corresponding author E-mail: rubinsky@cs.huji.ac.il

Abstract. The need for food preservation has faced mankind since ancient times. Food preservation technologies have evolved as a result of the progression of human knowledge. The more knowledge we have about the environment we live in, the more sophisticated tools we develop to control it. Any significant breakthrough in physics and engineering has resulted in the improvement of food preservation methods. For example, the invention of electricity in 1600 undoubtedly led to major breakthroughs in fresh food storage. Primarily,various thermal methods were invented; thermal sterilization, refrigeration, and cooled storage are among them. However, as years passed, the additional method of pulsed electric field (PEF) treatment was developed based on the observation of the effect of certain electric fields on cell membranes when delivered as high amplitude short length pulses. Stimulated by customer demand for high quality products displaying the same properties as untreated food, this method is currently in the latest stages of research progress before being implemented industrially. The exact molecular mechanism by which PEF inactivates cells is not yet known. One proposition is that PEF causes the formation of nanoscale pores in the cell membrane, a phenomenon termed electroporation. In the particular case when the process causes cell death, it is called irreversible electroporation (IRE). In the last four decades, IRE food disinfection has been successfully performed on numerous products and bacteria types. Particular contamination problems concerned with specific products have been evaluated, resulting in the investigation of various process parameters and protocol development. This will allow the implementation of IRE technology in the food industry.

The Need for Food Preservation

The need to preserve food has faced mankind from the very beginning. Food preservation was required even in the times before food production started, when society functioned as a hunter-gatherer construct. The main reason for this was the fact that food consumption is constant while production is seasonable, and it was therefore necessary to stockpile food for as long as possible.

B. Rubinsky (Ed.): Irreversible Electroporation, BIOMED, pp. 273–312.
springerlink.com

Centralized food production, which began in 10-11 BC, fostered the economical and technological development of the societies[1]. The need arose to store food which was as fresh as possible until it was delivered to the customers. The primary technologies developed early on were concerned with food production and management, as food supply is a basic need for growing and evolving societies. The discovery of electricity by William Gilbert in the 1600 led to monumental improvements in food preservation technologies throughout the next 400 years. Before the utilization of electricity, most of European countries had a simple diet of bread, dried peas, and salted meat and fish during long winters [2]. After its invention, electricity became the basis of the food supply chain, starting with production and continuing through processing storage and distribution. However, some basic preservation methods such as sun drying are still used even in the most technologically developed countries. Preserving food by heating it and sealing it in jars or cans began in the early nineteenth century, followed by pasteurization of wine and later milk in order to kill spoilage organisms [2]. With the development of steamships and refrigeration in the nineteenth century, international food trade was transformed. Beef could be shipped from Argentina to England and bananas from Central America to New York. Worldwide food exports went from 4 million tons in the 1850s to 18 million tons thirty years later, and 40 million tons by 1914[3].

Before continuing any discussion of food preservation, it is necessary to define the term “spoilage.” In most cases the quality of a food product is determined by the acceptance of specific product traits by a customer. A product may be rejected because of contamination with chemical agents, insects, rodents, physical factors or microorganisms.

Recent reports show that up to 25% of the world's food supply is lost due microbiological spoilage[4]. In 1983, a joint commission of the FAO and Word Health Organization (WHO) concluded that "illness due to contaminated food is perhaps the most widespread health problem in the contemporary world" [5]. Developing countries relying heavily on food import(up to 60%) are the main targets of food spoilage damage because they do not have access to modern methods of food storage after food arrives. Each year, up to 30% of people in developed countries suffer from food-borne diseases, resulting in many hospitalizations and deaths [6], while in developing countries the cases of food borne diseases are not properly documented. WHO estimates that food borne diseases caused by contaminated food and water cause up to approximately 700,000 deaths in Africa alone. This amounts to one third of world global deaths from food illnesses [7].

Major Preservation Technologies

Several major techniques of food preservation have evolved throughout the history of mankind and are currently accepted worldwide as trusted and time verified technologies. These methods often serve as standards in the development of novel technologies. The following paragraphs will present a summary of the main existing food preservation technologies. A detailed review about food preservation technologies can be found in the handbook of preservation [8].

Freezing. Reducing temperature to -18ºC is one of the most effective food storage methods. Biological reactions rates depend on temperature; thus, almost all reactions are stopped and microorganisms cease functioning in low temperatures used for food storage. However, this method does not kill existing bacteria and post thawing contamination may be caused by an inner existing microbial load.

Chilling. Low temperature reduces the growth rate of most microorganisms, and spoilage can be delayed this way. Chilling to 0°C allows for a relatively longer storage time for various food products, without significant influence on their primary physical shape and color which may change due to water phase transition in lower temperatures.

Fermentation and brining: In the processes of fermentation and brining, microorganisms are used to produce new products. For example, wine is made from grape juice via fermentation. New products usually show a long shelf life due to the dominating culture in the product. In addition, the fermentation process changes the chemical composition of the product – acid level increases while sugar is consumed by the yeast-dominant culture, which is carefully chosen for the process.

Acidification: Most existing pathogens proliferate normal human physiological conditions. Reducing pH value of foods to the levels of 4.2-4.5 is therefore lethal for most of them. The growth of spoilage microorganisms is also decreased in this condition, with the exception of some acid tolerant bacteria and molds.

Dehydration. Microbial growth strongly depends on water activity. Therefore, a long shelf life can be achieved by removing the water from product.

Chemical preservatives. Chemical preservation is one of the most effective and widely used methods which not only eliminates microbial population during initial treatment but also restricts recontamination likely to occur during packaging, transportation and storage. Different inorganic (e.g. sulphite and nitrite) and organic (e.g. benzoic and sorbic acids) are approved by regulatory institutions for use in food products. Instead of acting solely as efficient preservation agents, chemical preservatives may add a non-desirable taste to the product. In addition, part of the world population is allergic to particular preservatives. Furthermore, long debates concern whether or not various agents are carcinogenic. Regulations obligate producers to label food products with exactly the kind and amount of preservatives used. Recent years have seen a clear customer trend to look for food products which are free of chemical preservatives.

Canning, vacuum and modified atmosphere packaging (MAP): Modifying the atmosphere within the food package strongly influences microorganism population growth. Vacuum packaging almost eliminates the growth of microbes which demand oxygen. Besides microbial control, removal of oxygen also reduces the rate of chemical reactions demanding oxygen. Oxidation of fatty acids is a significant example of prolonging product shelf life by reducing the level of oxygen in the environment. Introduction of a gas with inhibitory effect on bacteria, such as CO_2, is a cheaper manufacturing method which impacts different microbial populations other than those dependent on oxygen.

Thermal methods: All live organisms consist of basic biological elements including proteins, lipids and carbohydrates. In different high temperature ranges, these basic biological elements begin changing their natural forms, altering their initial functions. In 70°C and higher temperatures, structural proteins and enzymes change their initial forms and functions. Most proteins undergo a degradation process at 90°C. High temperature treatments preserve food by taking advantage of this property of structural proteins, and effectively kill almost all microorganisms.. The efficiency of this process in microbial growth reduction comes at the cost of impacting the product itself, which consists of same basic biological elements as the microorganisms do. Treatment side effects can affect taste and flavor, and may induce color changes. A variety thermal methods are used in industry, including boiling, heating (pasteurization and steam sterilization), microwave and ohmic heating.

Safety and Quality Loss

The numerous factors characterizing food quality must be quantified in order to determine a clear cutoff point for what constitutes unacceptable food. Color, texture, flavor (smell and taste), shape and absence of abnormalities are considered the most important qualities in this realm.

Changes in a product begin from the moment of its production. Contamination may occur in all links of the product's supply chain, starting with manufacturing, continuing to packaging and transportation, and extending to customer distribution. The goal of preservation technology is to keep food products safe and close to their original quality. The role of regulation institutions in various countries is to provide clear quality demands of manufacturers and distributors of food products.

The overriding priority in food preservation and treatments processes is to minimize the presence and growth of microorganisms which may cause food poisoning or food-borne infections [9]. The incorporation of preservation technology into the supply chain allows for mild preservation treatment, thus avoiding product quality loss which may be caused by processing treatment or packaging.

Modern Need and Customer Demand

As their standards of living improve, customers become more and more concerned about the food they consume. In recent years, there has been a wide demand in developed countries for high-quality, convenient food products [10]. Natural taste, freshness and flavor are critical components of this demand [9]. Lack of chemical preservatives and flavor changes caused by treatment processes can significantly contribute to the attractiveness of a particular product for purchase. Customers are often willing to pay higher prices for natural, chemical-free and safe products [11] [12]. According to this demand trend, industry is shifting to milder processing and preservation technologies. However, such a compromise may result in a dangerous laxity in food safety. An increasing number of food-borne disease outbreaks has been attributed to fresh fruits and vegetables and related fresh, ready-to-eat products (e.g. freshly squeezed juices, ready-to-eat salads) [13]. The above analysis leads to the conclusion that there is a need for new food preservation technologies which will produce foods of the high natural quality demanded by customers but are also safe for consumption and have an acceptable shelf life. One of the technologies which is currently being widely investigated is the implementation of irreversible electroporation as a means of causing microbial inactivation in liquid food products [14]. IRE treatment maintains the flavor, color, taste, and nutritional value of food while destroying microorganisms [14] [15].

However, the implementation of novel technologies such as IRE in the food industry has been slow due to many factors, starting from process safety, continuing to products' commercial value and ultimately to customers acceptance. Food industry is considered one of the most conservative industries, often using historically developed and proved processes. Thus, penetration of new technologies demands significant proof of their value. Guidelines for process equivalency and international approaches to novel safety management have been previously discussed in the literature [10].

Introduction to Irreversible Electroporation Theory and Applications

When certain electrical fields are applied across a cell, they can have the singular effect of the permeabilization of the cell membrane, presumably through the formation of nanoscale defects in the cell membrane [16]. The process of cell membrane permeabilization was coined electroporation in the early 80's [17]. Sometimes this process leads to cell death, a process called irreversible electroporation. This occurs primarily when the electrical fields cause permanent permeabilization of the membrane and the consequent loss of cell homeostasis[18]. The history of irreversible electroporation can be traced back to 1898, when Fuller reported on the first use of irreversible electroporation to sterilize water [18]. First reports on the use IRE for sterilization of liquid foods were given for milk at the beginning of the 20th century, when a process called "Electropure" involved attempted use of direct electricity for pasteurization [19] [20]. However, despite the fact that the process utilized electricity, the nature of microbial inactivation that occurred was due to thermal factors. "Electropure" was less energy efficient than a method introduced later called Ultra High Temperature Short Time sterilization (UTH). Later, when interest in using electricity for food sterilization reappeared, the process evolved into the ohmic heating process.

Perhaps the most fundamental study on the use of IRE for sterilization of liquid solutions can be found in a series of papers by Sale and Hamilton from the early 1960’s [21] [22] [23]. These studies followed the pioneering work of Doevenspeck, who published the first report of implementing pulsed electric fields in food processing [24].

The empirical studies of the 1960's and 1970's were followed by attempts to establish a theory of electric field influence on the cell membrane. First models , which proposed an electromechanical membrane breakage of a membrane [25], could not explain the pulse-length dependence of the critical voltage [26] or the dependence of the membrane lifetime on its total area [9] [27]. Current theory proposes that the process of electroporation is concerned with the physiological effect of pore formation [28]. Briefly, it is assumed that hydrophobic pores are spontaneously and randomly generated due to thermal motion of phospholipid molecules. The location and size of these pores vary randomly. If the radius of hydrophobic pore exceeds a critical radius r_c, it overcomes an energy barrier and turns to a hydrophilic pore, which is far more stable and assumed to allow cell content lyses. The hydrophilic pore energy formation is reduced by the combined effects of thermal flocculation and the electric field across the membrane [28]. In spite of the fact that electroporation was shown to be a "all or nothing event", a claim which was recently argued [29], and the possibility of electroporating 100% of cells during the application of the electric pulse, it is not promised that all cells will die [9]. Studies point to a clear difference between electroporated and dead cell curves in the irreversible electroporation voltage range [9]. It is assumed that the main reason for cell death is the release of an intracellular substance caused by increased permeability of the membrane, structural membrane changes, or osmotic swelling caused by the application of pulses [9]. The detailed molecular process of the phenomenon is not yet understood. Cell survival is explained by pores resealing after the application of the electric field ceases [30] [31] [29] [32]. This ability of the membrane to rebuild its structure allows a given cell to survive and proliferate after the treatment, causing the treatment process to be incompletely efficient. A pore resealing model based on temperature was developed by Saulis, which shows the impact of post-treatment temperature on irreversible electroporation effectiveness [31]. It was observed experimentally that moderately elevated temperature after treatment (37°C) prevented complete inactivation, while incubation of cells at low temperature range (4-10°) stabilizes the pores, increasing the treatment effectiveness [30] [26].

Due to the lack of established theory relating to molecular processes, the kinetic models for microbial inactivation are mostly empirical and are based on specific experiments with particular products and microorganism strains[9]. This is the reason that in almost all cases it is necessary to empirically adopt the electroporation process to a specific product and its inherent safety concerns [9] [33]. Hülsheger and Niemann were the first to propose a mathematical model for inactivation of microorganisms using pulsed electric fields [34]. Their model was based on the dependence of the survival ratio on the electric field intensity alone. Later, Hülsheger proposed a new model incorporating a treatment time dependence into the model [35]. In 1995, Peleg proposed a mathematical description of survival rate based on a sigmoid curve; he used electric field strength, treatment time and number of pulses as input parameters [36]. Examples of studies implementing the Hülsheger and Peleg model appear in the report prepared for FDA [33]. The Weibull kinetic model has gained an interest for its potential to describe the survival curves [37]; however, this model does not incorporate the physical parameters of the system. Numerous studies have been done in order to establish the influence of different process parameters. Empirically derived curves can aid in the predictability of various processes; however, it seems that only a complete understanding of the molecular mechanism behind electroporation can provide a general kinetic model. In spite of the lack of established molecular mechanism models underlying the process, IRE has made its way into academia in recent years as efficient non-thermal method for cell disruption in water purification, food sterilization, pharmaceutical treatment and various medical applications [9] [16] [38] [39]. The remarkable effectiveness of this technique primes it for industrial implementation in the coming years.

Applications of Irreversible Electroporation in Food Industry

Technology

IRE will be used for pumping products as part of continuous flow treatment, as described by Bushnell [40]. It is clear that the treatment efficiency depends on the flow and chamber design. Different chamber designs such as parallel plates, coaxial [41] cylinders, and co-linear configurations have been evaluated for liquid food processing [42]. It has been experimentally shown that from all proposed chamber configurations, a co-flow chamber allowed the highest temperature set at a specific pressure without partial discharge, and therefore the highest pulse energy density, with the same applied voltage [9]. While showing the same sterilization efficiency, co-flow reactors allow for the lowest exit temperature [9], which contributes to pore resealing delay and does not bear negative effects on product quality.

In order to be incorporated into high volume industry, this process should be energy efficient. Several recent works address this concern by proposing specific devices[43] or process designs[44] which demonstrate the feasibility of this process as energetically efficient in industrial application. A pilot water sterilization process demanded a total energy consumption of about 40 J/ml [44].

For further information relating to design of pulsed electric field device, it is recommended to start with a recently published technology review of pulsed electric field equipment. More detailed material appears in the literature relating to pulsed power systems and switching devices [45].

Specific attention should be paid to the electrode material used in the system design, because different electrode materials have different pulse length limitations due to variation in the electrochemical double layer capacitance. The preferred material in the food industry is stainless steel, because it is cheap, easy to shape, and protected against wear. It has been used in the food industry for decades [46]. However, problems associated with electrolyses, bubble formation, electrode corrosion and metal release have been observed when stainless steel electrodes were used in food processing [46] [47] [48] [49]. In order to avoid these undesirable processes, alternative electrode materials such as platinum (Pt) [40] or polymer coating [50] were proposed. However, their use seems unpractical at this time because Pt is expensive, and the use of polymer coating in industry has not yet been fully developed. Toepfl and others performed a comparative study of stainless steel and carbon electrodes [51]. A significant increase in treatment efficiency when using graphite electrodes was observed at electric field strength of 16 kV/cm at different initial treatment temperatures. The authors explained their findings by reasoning that the electric field distribution in the treatment zone experienced improved homogeneity caused by reduced bubble formation. Gas bubbles, formed by electrolysis, cavity effects, or release of dissolved gases (formed by heating) have a lower dielectric breakdown strength than the liquid media; their presence will lead to perturbations of the electric field distribution [9] [51] . Detailed electrochemical research is needed in order to evaluate different electrode materials and their impact on treatment efficiency and safety.

Process Factors

Electric Field Intensity

Based on electroporation theory and numerous published reports, it is clear that the treatment efficacy mainly depends on the intensity of the applied electric field. The field amplitude is the main parameter which impacts the process efficacy. The critical electric field E_c (electric field intensity below which inactivation does not occur) increases with the transmembrane potential of the cell. It has been shown that a transmembrane potential in the order of 1 V is sufficient to induce cell IRE [52] [53]. However, E_c does not only depend on the cell structure size and orientation in the field [28] [54], but also on the surrounding environment properties (pH, conductivity, ionic strength) [33].

Heiz et al. presented an equation based on previous work by Neumann [55] which helps predict critical field strength as a function of cell shape [54]. Using these calculations, *Listeria monocytogenes* (short rods sized at about 0.4-0.5X0.5-2μm) require a critical field strength which is only one tenth the strength of that required by *Saccharomyces cerevisiae* (exhibiting an ellipsoidal shape of about 3-15X2-8 μm) [54]. However, the equations and model are used as an estimate only, due to large diversity in cell size and shape of microorganisms even within the same family. In the final application, two different impacts of the electric field are investigated. The first and more common method involves increasing the field's amplitude to tens, even hundreds, of kV/cm [54], while reducing the pulse length to several μsec or tens of nsec [56] [57]. This approach is based on existing kinetic models which show that amplitude has a stronger effect than pulse length or number [35]. Creating such a high pulse amplitude while keeping pulse lengths low demands specific and cumbersome equipmentwhich will hardly be available to industry in the near future. An additional challenge is the creation of a rectangular shape pulse whose length is in the nanosecond range. The second approach concentrates on electric field intensity in the range of critical potential E_c, but changes other treatment parameters such as pulse length and number of pulses, or divides the total pulse number to several groups with significant time intervals between groups [58]. This approach demands a relatively longer time for treatment due to increased length of pulses and pulse numbers. However, it uses currently available pulse power technology; this will allow the commercial penetration of the technology to proceed much more easily.

Treatment Time

Treatment time is defined as the product of the number pulses and the pulse duration [33]. Currently implemented pulse lengths vary from nsec to msec in length, and depend on other system parameters. Experimental results [21] [22] , supported by the theoretical models of Hülsheger and others [34] show that pulse width influences microbial reduction by affecting E_c : longer widths decrease $E_{c.}$ The dependence of pore resealing (leading to bacteria survival) on time between pulses was also investigated [59]. A higher number of longer pulses may result in an undesirable product temperature increase; adverse reactions on electrodes may also begin to occur. A model for calculating electrochemical treatment limitation due to electrode reaction was presented by Roodenburg [46]; it shows dependence of maximum pulse duration as a function of electrode material. This parameter's dependence has certain limit. It was reported that in certain cases, the number of pulses applied results in a saturation level [39], and additional pulses do not influence the inactivation level. In addition, it was reported that for an electric field strength 1.5 times higher than E_c, the critical treatment time would remain constant [39]. Recent studies also show that implementing different pause periods between pulses and dividing pulses to several groups allow the researcher to work with lower fields and longer treatment times. However, this must be done without significantly increasing sample temperature, and the treatment efficiency must be maintained [58].

Pulse Shape

The most efficient pulse shape for food sterilization has not been decided yet. Rectangular, exponentially decaying, damped oscillating , bipolar rectangular and bursts of pulses have all been introduced and tested [9]. However most reports conclude that arectangular shape should be used [33] [9]. Energy comparison between different pulse shapes support the trend of using rectangular pulses, as they were showed to be the most efficient [33] [60] [61].

Treatment Temperature

One aim of novel, non thermal technologies is to eliminate the side effects of thermal treatments while keeping the product safe. Product temperature during all phases of treatment, which can be divided into inlet temperature, chamber temperature and outlet temperature, is a critical factor in process quality and efficiency [9] [39] [62]. Although attempts have been made to reduce the side effects of temperature, it has been clear from outset that temperature has a high synergistic effect on treatment efficacy due to a significant influence on cell membrane fluidity and stability [62]. Cell membrane stability is highly affected by a temperature-dependent phase shift from gel to a liquid crystalline structure [63]. Increasing treatment temperature from 40 to 50°C increased inactivation of *S.*

dublin in milk from 1 to 4 log-cycles [64]. Similarly increasing treatment temperature from 24 to 60°C enhanced inactivation of *L. brevis* [65]. The temperature increase from 3°C to 30°C enhanced *E. coli* inactivation in an electric field of 35 kV/cm [66] [67]. One of recent studies showed significant damage to *E.coli* at 50°C while applying only a 5 kV/cm field [68]. Several other studies also demonstrated increased IRE efficiency with increased temperatures [69] [70] [71]. It was also found that increasing treatment temperature to a range of 35 to 55°C can reduce the electric energy required for a 6 log-cycle inactivation of *Escherichia coli*, from far above 100 kJ/kg to 40 kJ/kg, when operating at an initial treatment temperature of 55°C [72]. A more comprehensive analysis of energy consumption by IRE in food preservation was presented in detail in recent revie ws[9] [73]. A first order model which described both electric field strength and temperature dependency was proposed by Sensoy et al. [74] and Fox et al.[75].

Since in most applications several pulses are applied, attention should be paid to conductivity change of the product in the chamber due to the temperature increase. This conductivity/temperature change will cause a drop in the peak field strength between the electrodes and the specific energy input W. A control system is needed in order to compensate for this field drop [54]. In addition, product temperature will not be homogeneous after treatment, with higher temperatures occurring near the chamber wall. This temperature increase may also cause the field to be inhomogeneous, leading to some degree of inefficiency [9]. A finite element model [76] and an analytical model for specific geometries were introduced [9] in order to optimize the process. A computer simulation for temperature behavior in a continuous flow IRE treatment system was developed and evaluated with the goal of assisting in the development of a more homological chamber [77].

The final stage of temperature influence on the process efficiency is the outlet. It was shown that even if all cells are electroporated, some of them may remain viable due to the capability of pores to reseal [78] [30]. Saulis developed a mathematical model for predicting the fraction of cells which have been irreversibly damaged during electroporation [31]. The model proposes that the pore resealing phenonmenon is strongly dependent on temperature, a claim which is supported by previous experiments [30] [26].These works show that a moderate temperature increase to 30-37°C should facilitate a complete sealing of pores and thus allow some microorganisms to survive, while incubation at low temperatures of 4-10°C should stabilize pores, thereby increasing the inactivation rate. This shows that a cooling system should be implemented at the end of the treatment process in order to stabilize pores, while a warming device should precede the process in order to increase the percentage of larger pore formation.

Influence of a Kind of Microbial Contamination

Due to the huge variability in the microbial population, significant differences were observed in the critical factors needed for inactivation of different microorganism varieties and strains [51] . Among the identified parameters are cell size [54] [79], geometry/shape [54] [79], orientation relative to electrodes [51] [54], membrane structure [21] [22] [79] [35] [29], cell concentration [39], and growth stage [35[[67] [80] [81] [32]. A difference of field impact between different strains and mutants of *E.coli* was observed [32]. This is of great interest when selecting the appropriate target strains for testing the efficacy of novel processes, since an indicator organism must be defined [33].

Cell Size and Geometry

It was shown that the critical field strength sharply increases when the characteristic dimension of the cell (i.e. the radius of sperical cells or the shorter half-axis of elliptical cells) is shifted to smaller values. Variations in cell shape also require a considerable raise of field strength. For example, a rod-shaped cell requires an electrical field more than five times stronger than that required by a spherical cell with the same characteristic dimension [54]. A comparative work of Aronsson and others quantitatively showed that *Saccharomyces cerevisiae*, which has ellipsoidal shape of 3–15X2-8 μm, was the most sensitive organism, followed by *E. coli* straight rods possessing a geometry of 1–1.5X2–6 μm [82].The most resistant organisms were *L. innocua,* which is a regular short rod with dimensions of 0.4-0.5X0.5-2 μm [82] and *L. mesenteroides*, a spherical cell measured at 0.5-0.7X0.7-1.2 μm [82] [79]. However, Garcıa et al. showed that these general assumptions do not always work and in many cases, the required field strength depends on environmental factors such as the pH of the medium [83]. *Y. enterocolitica*, which was one of the smallest bacteria investigated, was more IRE sensitive than larger cells such as *L. plantarum*. General assumptions about the influence of the cell size or shape, which had been previously proposed, were experimentally proven in the same study within each tested group at pH 7 [83].

Membrane Structure

Only few comparative systematic studies were done in the area of membrane structure. Hülsheger et al. found that Gram-positive bacteria and yeasts are less sensitive to electric pulse treatment than Gram-negative bacteria when a low number of pulses was applied; treatment with higher number of pulses revealed same reduction rate across all tested microorganisms [35]. However, Garcıa et al. demonstrated that the aforementioned general assumptions work only under particular conditions. This was based on comparative research carried out with 8 bacterial strains, 4 gram-positive and 4 gram- negative [83]. For instance, *L. monocytogenes*, a gram-positive bacterium which showed the highest PEF resistance at pH 7.0, was one of the most sensitive at pH 4.0. Bacterial spores and mold ascospores are far more resistant to PEF treatment than the organisms discussed above [84]. The differences are due to varying morphological and biochemical properties of various cell parts, membranes in particular [85]. However, the molecular mechanism which can explain the differences and help to predict the needed field is still unclear [9]. Studies comparing energy demand for microbial inactivation in milk found that gram-positive *L. innocua* required a longer treatment time than gram-negative *P.flourescens.* Gram-negative bacteria underwent the same log reduction as gram positive bacteria with the same electric field intensities but only half of the number of pulses [86].

Cell Concentration

The initial concentration of microorganisms may influence the inactivation efficiency of IRE treatment [33]. Inactivation of *E. coli* in a model food system of simulated milk ultrafiltrate was not affected when the concentration of microorganisms was varied from 10^3 to 10^8 CFU/mL after being subjected to 70 kV/cm, 16 pulses, and a pulse width of 2 μsec [39]. Increasing *S. cerevisiae* inoculation from 10^4 to 10^6 CFU/mL resulted in a slight increase of the survival fraction when applying one pulse at 25 kV/cm for 25 μsec [87]. This effect was explained by cluster formation of yeast cells and possible hidden microorganisms in low electric field sites. Molinari et al. showed that in spite of growth phase of the cells, the reduction of *S. cerevisiae* population using 3 pulses at 8kV/cm with a 2 msec pulse width was linearly enhanced by increasing the inoculum concentration. When the inoculum's concentration was $5*10^6$ CFU/mL, a 2.2 log cycle reduction was achieved, and this was increased to a 5.5 log-cycle when the inoculum's concentration was increased to 10^{11} CFU/mL .The authors concluded that the cell concentration is a relevant parameter in determining the inactivation of *S. cerevisiae* by IRE treatment [81]. However, recently, the effect of initial *E.coli* concentration in collagen gels was investigated, and it was concluded that the *E.coli* population was reduced by same 2.5 log cycle or more after 100 pulses applied at 45kV/cm, irrespective of the initial cell densities used for the gel inoculum [88].

The discussed experimental results are not consistent with conclusions, and systematic research using several pathogens should be carried out.

Culture Growth Stage

A general claim is that logarithmic phase cells are more sensitive to stress than lag and stationary phase cells [33]. Microbial growth in the logarithmic phase is characterized by a high proportion of cells undergoing division, during which the cell membrane is more susceptible to an applied electric field. Hülsheger et al. performed a comparative study and concluded that cells in the logarithmic growth phase are more sensitive to IRE than those in the stationary growth phase [35] [89] [32]. *E. coli* cells in the logarithmic phase were more sensitive to IRE treatment when compared to cells in the stationary and lag phases [67], when growing on non selective media [32]. Wouters et al. reported that *L. innocua* cells which had been incubated for a prolonged period of time in the stationary phase were more resistant to IRE treatment, indicating that the physiological state of the microorganism plays a role in inactivation by IRE [90]. They also demonstrated that younger *L. innocua* cultures were more sensitive to IRE inactivation. Some studies with *S. cerevisiae* have shown that higher susceptibility of actively growing cells to IRE also occurs with yeast cells [80] [81]. Gaskova et al. reported that the killing effect of IRE in the logarithmic phase is 30% greater than in the stationary phase of growth [80]. One suggested theory proposes that the area between the mother and daughter cells are more susceptible to an applied electric field than the regular part of cell envelope, and the number of cells in the state of proliferation or shortly afterward is higher in the logarithmic growth phase than in the stationary phase. This makes these "tender" cells more sensitive to the electric field [91]. Molinari et al. performed a study on yeasts investigating the differential influence of growing stage and inoculum concentration on IRE efficiency, since the physiological state of treated cells and a higher yeast concentration may contribute to the previously observed results as the culture time evolves [81]. They concluded that yeast inactivation by pulsed electric fields depends on the inoculum concentration, but is independent of the growth phase [81]. This behavior was explained in the following way: a high yeast concentration would increase the probability of the formation of "equivalent cells" and the IRE treatment efficiency would consequently increase [81].

The molecular mechanism of microbial resistance to IRE in the stationary phase of growth is a matter of increased interest. Stationary phase sigma factor (RpoS) changes the specificity of RNA polymerase. This activates the expression of many genes, some of which are considered responsible for the generalized increase in stress resistance upon entry into stationary phase [92]. Robey et al. proved that the lack of the stationary phase inducible sigma factor RpoS resulted in decreased resistance to high hydrostatic pressure in *E. coli* [92]. Using selective media, Somolinos et al. showed that in spite of primary results on non selective media, which suggested that that resistance to IRE is controlled by mechanisms used to survive stationary phase stresses, the intrinsic resistance of both logarithmic and stationary strains was very similar [32]. This leads to the proposal that the intrinsic resistance does not depend on the growth phase, but the rather is related to the varying capacity of cells to perform sublethal damage repair. Based on these results, the authors suggest that the proteins synthesized under the control of the sigma factor RpoS are, among other functions, responsible for repairing cell membranes which were sublethally injured by IRE treatment [32]. These recent observations, coupled with the studies on the impact of environmental factors on repair mechanisms, suggest that a molecular mechanism involved in the stationary phase combined with external factors is partially responsible for microbial resistance to pulsed electric fields. As such, Saulis' pore resealing theory [30] [31], which is based on temperature, can be extended from physical factors to intrinsic biologically controlled ones. Current data suggests that pore resealing after electroporation depends on temperature, medium and proteins synthesized under the control of the sigma factor RpoS. In addition, it has been shown that IRE alters the expression of molecular chaperone, which is responsible for membrane reconstruction, in *Listeria monocytogenes* [93].

The direction of IRE influence on proteins and protein expression is only beginning. Ensuring the highest inactivation rate of microorganisms by electroporation will demand intense research in order to define the parameters involved as well as evaluate the methods which impact them.

Product Parameters

As discussed previously, particular environmental factors can dramatically impact the process efficiency. Since food products are complex materials, they affect both the physical parts of the system (real developed electric field strength and the pulse parameters of length, shape and raising time, temperature changes) and the biological ones (because food is a microorganism growing medium, which impacts growth rate and pore resealing efficiency), as described previously. The following factors were investigated and identified as critical for a correct evaluation of the process.

Conductivity

The dielectric property of a food is closely related to its physical structure and chemical composition [33]. The treatment chamber configuration determines the resistive load and therefore the properties of the discharging circuit. The load resistance of a chamber is dependent on the conductivity of the treated media; this results in limitations in the range of media conductivity to which the electric pulses can be applied [51]. Food products without the addition of salt have conductivity in the range of 0.1 to 0.5 S/m. Products with high electrical conductivity reduce the resistance of the chamber and consequently require more energy to achieve a specific electrical field strength. Therefore, when processing high salt products, the salt can be added after processing in order to increase the process efficiency [33]. The conductivity of the liquid media determines the voltage drop across the treatment chamber. If the resistance of connectors and protective resistors are in the same range as that of the treatment chamber, there will be a high voltage drop, leading to a reduced peak voltage across the electrodes [42] [51]. Adaptation of the limiting protective resistors to low resistance systems might be difficult, as most of the available switching systems, particularly semiconductor switches, require a maximum current limitation in case of short circuit [51]. During the electrical discharges, temperature is increasingly dependent on energy input in the respective volume element. This leads to changes in media conductivity and electric field distribution and increases the needed capacity of the system, leading to energy consumption increase [42] [54].

Homogeneous liquids with low electrical conductivity provide ideal conditions for continuous treatment with the IRE method [33]. Inactivation of *Lactobacillus brevis* with IRE showed that as the conductivity of the fluid increased, the resistance of the treatment chamber was reduced. This in turn reduced the pulse width and decreased the rate of inactivation [69]. Treatment of *E. coli* suspended in distilled water (conductivity of the distilled water was 0.2 mS/ cm) led to a total bacterial inactivation within 35 pulses, whereas in fat-free milk and in phosphate buffer surviving organisms were still present after 63 pulses [94]. The reduced inactivation rate at elevated ionic strengths was explained by the stability of the cell membrane when exposed to a medium with several ions [94] [95]. A 2-log cycle reduction of *L. innocua* was observed in a medium with a conductivity of 0.79 S/m , while a reduction of almost 4.5 log-cycles was observed in a medium with a conductivity of 0.27 S/m at a 40°C medium outlet temperature [90]. However, a study using a model system showed that resistivity had no effect on PEF effectiveness on *E. coli* and *L. innocua* [96]. These apparent controversial results may be due to the specific microorganisms or media used [33].

pH

Since pH is one the most important factors affecting biological systems, its level both outside and inside the cell influences IRE efficiency on specific microorganisms [89] [97]. The first investigation of this parameter showed no effect of pH on IRE efficiency [34]. However, the subsequent studies showed that pH has a great impact on the treatment efficiency. A synergistic killing effect between the high electric pulse and organic acid was observed at pH 3.4, but not at pH 6 [98]. When the cell suspension of *E. coli* 0157:H7 was treated with five electric pulses in the presence of benzoic or sorbic acid at pH 3.4, the count decreased by 5.6 and 4.2 log-cycles, respectively, while a log cycle reduction of only 1.2 was observed at pH 6.4. The investigators proposed that treatment at pH 3.4 may indicate that the electric treatment enhanced the entry of the undissociated acids into the bacterial cells [98]. Vega-Mercado et al. studied the effect of pH and ionic strength of the medium during IRE treatment [99]. The inactivation ratio increased from undetectable to 2.5-log cycles when ionic strength solutions were adjusted from 168 to 28mM. At 55 kV/cm (8 pulses), as the pH was reduced from 6.8 to 5.7, the inactivation ratio increased from 1.45- to 2.22-log cycles [99]. The IRE treatment and ionic strength were jointly responsible for electroporation and compression of the cell membrane, whereas the pH of the medium affected the cytoplasm when the electroporation was complete [33]. Wouters et al. investigated the influence of pH on IRE efficiency for the inactivation of different microorganisms. *S. cerevisiae* SU 51, a representative eukaryotic microorganism, *E. coli* NCTC 9001, a representative gram-negative microorganism, and the representative gram-positive microorganisms of *L. plantarum* LA 10-11, *L. innocua* NCTC 11289, and *L. innocua* NCTC 11289. Inactivation was studied in pH 4.0, 4.4, 5.0, and 6.0 phosphate buffers. A clear trend was observed: a lower pH made cells more vulnerable to IRE treatment [90]. Different species and even different strains of the same microorganisms showed different resistance. The *L. innocua* strain, VBLLi 02-53, was much more resistant and its survival rate was 2.5 log-cycles greater when it was tested under the same conditions as *L. innocua* strain. The authors concluded that efficiency is dependent on type, strain and lower pH values of the medium [90]. Garcıa et al. evaluated and compared the IRE resistance of four Gram-positive (*Bacillus subtilis*, *L. monocytogenes*, *L. plantarum*, *Staphylococcus aureus*) and four Gram-negative (*E. coli*, *E. coli* O157:H7; *Salmonella serotype Senftenberg* 775W, *Y. enterocolitica*) bacterial strains, all under the same experimental conditions [83]. The conclusion was that the influence of other microbe characteristics on IRE effectiveness, such as gram type or geometrical form, depend on the pH of the treatment medium as a major parameter. The target (marker) microorganism may also vary as a function of the treated medium pH. The most IRE resistant bacterial strain at pH 7.0 was the pathogen *L. monocytogenes*. In contrast, the pathogens *S. Senftenberg* 775W and *E. coli* O157:H7 were the most IRE resistant strains at pH 4.0. The combination of IRE and subsequent holding under acidic conditions has been demonstrated to be an effective method in to the pursuit of achieving a higher level of microbial inactivation. This phenomenon was explained by the inability of sublethally injured cells to repair the membrane in the presence of acid [29] [32] [100]. Another study by Somolinos et al. proposed that that the cytoplasmic membrane is a target directly involved in inactivation by IRE, but the final mechanism might vary depending on the pH of the treatment medium. *S. cerevisiae* cells were IRE-treated at 12.0 kV /cm with 50 pulses in a citrate–phosphate buffer of pH 7.0 and pH 4.0. They were then allowed to recover in both non-selective and selective media, which allowed the evaluation of the recovery and thus total treatment efficiency [29]. The number of inactivated cells was the same for treatment media of pH 7.0 and 4.0. As indicated by the difference in the number of survivors recovered in the non-selective and selective media, the occurrence and extent of sublethal injury in *S. cerevisiae* did not depend on the treatment medium pH. There was a large proportion of sublethally injured cells , about 95%, when *S. cerevisiae* cells were IRE-treated in a citrate–phosphate buffer at a pH of both 4.0 and 7.0 [29]. These results disagree from those previously published, where most gram-positive bacteria studied were significantly more resistant to PEF treatments at pH 7.0 in comparison to pH 4.0, whereas Gram-negative bacteria were more resistant at pH 4.0 [100]. In addition, the occurrence of sublethal injury also appeared to be dependent on the treatment medium pH [100]. Subsequent to IRE treatment, incubation of the survivors in different recovery liquid media showed that sublethal injuries in *S. cerevisiae* were reparable as a function of the pH of the treatment medium and the pH and composition of the recovery liquid media. For instance, whatever the treatment medium pH used, cells were unable to repair their damages in a citrate–phosphate buffer of pH 7.0, while citrate–phosphate buffer of pH 4.0 allowed cell repair. The same observations were made for other recovery media. It was concluded that for different types of bacteria and yeast cells, the ability to survive after IRE seemed to be favored by acidic pH during subsequent holding. Even a nutrient-free medium such as a citrate–phosphate buffer allowed cell repair at an acidic pH. In addition, the novel proposal was made that a different kind of lesions and, therefore, a different mechanism of inactivation, takes place during IRE treatment of yeasts, depending on the treatment medium pH [29].

The relationship between the shape parameter of the Weibull model and the pH of the treatment medium is described by the Gompertz equation. It was developed in order to predict the influence of the pH of the treatment medium on the inactivation of *L. plantarum* by IRE treatments at different electric field strengths [101]. The model was validated in orange and apple juice.

Water Activity

Changes in the acid content or a_w of a food dramatically influence the kind of microbes that grow in the food. Generally, a decrease in water activity represses the killing effect of the IRE [102]. For *E. coli*, the effect of lowering the water activity was more pronounced at lower pH values. At 30°C and pH 5.0, a 4.6 log-cycle reduction was observed for an a_w of 1.00 [102]. When the a_w value was lowered to 0.94, a decrease in inactivation level of the order of 1 log-cycle reduction was observed. Furthermore, raising the pH to 7.0 entirely eliminated the a_w effect for IRE inactivation of *E. coli*, and no significant difference in inactivation level was observed [102]. The outcome of adjusting the a_w value was less pronounced for *S. cerevisiae* [102]. The microbial resistance of *L.*

monocytogenes increased with a reduction in the a_w of the medium [97]. After 800 µsec at 28 kV/cm 2.26, 1.55 and 0.21 log reduction was obtained using a McIlvaine buffer of pH 7.0 and $a_w > 0.99$, 0.97 and 0.93, respectively [97]. The significant inactivation of *E. cloacae* by IRE was reported with increasing water activity of a chocolate liqueur [103] . As the a_w of chocolate liqueur increased from 0.48 to 0.89, the log reduction of *E. cloacae* increased from 0.1 to 1.3 [103]. Dry products such as spices [104] and flour [105] were treated with very limited success (less than 1-log cycle reduction), purportedly due to the low water content in these products [106].

Neidhardt et al. suggested that a reduction in the membrane permeability and fluidity due to a thickening in the cell membrane resulting from reduced a_w values [107]. In high pressure processing, the resistance to inhibition at reduced a_w values was proposed to be the result of cell shrinkage during water loss, caused by osmotic pressure [108]. Aronsson and Rönner proposed the same explanation for reduced IRE efficiency at lower water activity levels [102]. In addition, a reduced cell size due to shrinkage was proposed as an additional reason for increased cell tolerance to IRE [102].

The influence of water activity (a_w) on microbial growth was investigated in depth and summarized by Jay [109].

Hurdle Technology Implementation of IRE in Food Treatment

Microbial growth occurs due to the capability of microorganisms to maintain the stability and balance of their internal environment [9]. The internal environment is separated from the external environment by a phospholipid membrane which acts as a semi permeable barrier. In other words, it controls the passage of nutrients into the cell and the passage of metabolic products out of the cell [23]. Food preservation is based on disturbing the cell homeostatic mechanisms, leading to growth inhibition and cell death [9] due to the cell's inability to respond to the stress [110]. By placing a number of sublethal stresses, or hurdles, on a microbial cell, the organism expends energy to resist the hostile environment. This can potentially lead to metabolic exhaustion and cell death [111]. Hurdles targeting the same elements within the cell have an additive inhibitory effect only, whereas synergistic effects may result from disturbing several functions of the cell [112] [113]. This multi-target effect is the essence of hurdle technology, which is based on the idea that attacking various cellular targets will make the organism apply every possible repair mechanism simultaneously, and the activation of stress shock proteins and other defending mechanisms will also become more difficult. However, in some cases, antagonistic effects of combined technologies were observed [9] and applied stresses in combination led to a reduced lethality as compared to that observed for each stress independently . This multi-target approach may also help to counter stress adaptation associated with the sublethal treatments [114] discussed previously. A detailed review of various hurdle effects of novel non-thermal technologies has recently been published [115]. In order to design the most effective process, the detailed molecular mechanisms and site actions of the involved processes should be known. The basic molecular knowledge of novel non-thermal technologies is still unclear for most techniques, and in depth research is necessary [9]. The following approaches have been tested until now:

IRE and Temperature Combination

Most aspects of temperature impact on IRE efficiency are discussed in the temperature section above. The temperature has a clear impact on the IRE efficiency in all process stages: inlet, outlet and post-treatment storage. For instance, IRE-thermal treatment at a moderate electric field strength (5 kV/cm) and at mild temperatures (30-50°C) results in damage of a high percentage of *E. coli* cells. In addition, damage efficiency was noticeably improved by addition of nonionic surfactant Triton X-100 [68]. The outlet and post treatment product temperature is important because the ability of the cell membrane to reseal. Saulis et al. proposed that post treatment incubation of cells at low temperatures (4-10°C) should stabilize the pores, increasing the total process efficiency [9]. It has been reported that IRE treatment sensitizes cells to heat, probably as a consequence of the alternation of the chaperone expression in *L. monocytogenes* [93]. IRE treatment (at a field strength of 27.5 kV for 144 μsec) altered the expression of the molecular chaperones GroEL, GroES, and DnaJ and showed substantial sensitivity to subsequent heat treatment (55°C for 10 minutes). The decrease in counts for Scott A and OSY-8578 strains was 6.1 and 2.8 log CFU/ml, respectively [93]. Several additional studies validated the sequenced treatment approach [116] [117]; however, the industrial application of this process was contested due to the fact that IRE sensitizing effect to heat was only observed previously in Gram positive bacteria with electric field intensities similar to those used for combined IRE-temperature treatment, but with longer processing time [9].

Combination of IRE and Microbial Agents

A number of publications reported that IRE in the presence of a nisin/ lysozyme addition resulted in a synergistic effect on the inactivation of certain vegetative microorganisms [115] [118] [119] [120] [121] [122] [123] [124] [125]. In the experiments with novel developed antimicrobial agent AMINA® T 100 it was shown that a low concentration of the active substance (0.025% and 0.00625%) combined with a low intensity IRE treatment (at 40 kJ/kg) showed 5.97 and 4.42 log cycle reduction of *L. innocua* [126], a result unattainable by individual application of each technique. A 4.5-log cycle reduction of *Salmonella enterica* in apple juice was achieved with IRE treatment for 1 msec in combination with 60 μg/ml of enterocin AS-48 and a treatment temperature of 40°C [127]. Synergistic and additive killing effects against *S. Enteritidis* and *E. coli* O157:H7 were reported in fruit juices by combining IRE treatment with natural antimicrobials, citric acid, or cinnamon bark oil [128]. The mechanism underlying this synergy is not yet fully understood [115], but it can be suggested that the additional stress of IRE facilitates the antimicrobial agent's access to the cytoplasmic membrane. Both treatments may inflict damage, resulting in more or larger pores, or pores with greater stability [122] [124]. The synergistic bactericidal effects of nisin/lysozyme mixtures on IRE-treated cells were more pronounced than the addition of either nisin or lysozyme alone [125]. Inactivation studies on the effects of PEF on bacterial spores are scarce and results vary [33]. Early studies [21] [22] reported that *Bacillus* species spores were resistant to exponential wave pulses during IRE treatment with strength fields up to 30 kV/cm. Electric fields do not induce spore germination [129]. However, provided such germination is induced by other factors, the resulting vegetative cells do exhibit sensitivity to electric pulses [129] [130]. Simpson et al. confirmed the high resistance of *B. subtilis* spores to IRE, and subsequently studied a hurdle approach with heat-shock, lysozyme, EDTA, and pH treatments. Only a combination of 80°C heat-shock and lysozyme, followed by IRE at 60°C was able to produce a 2-4 log-cycle reduction of spores [131]. Lysozyme probably dissolved the outer coating of spores, thus making cells more sensitive to electric pulses [38].

Combination of IRE and High Pressure CO_2

Supercritical (SC) CO_2 treatment has been successfully used in the inactivation of a wide range of bacteria and yeasts [132] [133]. The inactivation of spores with SC-CO_2 requires long residence times and high temperatures [134], but the combination of IRE and High Pressure CO_2 may lead to a lower temperature spore inactivation technique. It was found that *E. coli* and *S. aureus* exposed to 10 pulses at 25 KV/cm, and then to SC-CO_2 at 200 bars and 34°C for 10 min were completely inactivated [134]. *B. cereus* spores were found to be more resistant: partial inactivation of the magnitude of a 3 log-cycle reduction was detected after a pretreatment of 20 pulses at 25 KV/cm and SC-CO_2 treatment at 200 bars for 24 h at 40°C [134]. It has been previously shown that the SC-CO_2 process essentially involves the diffusion of CO_2 into the cell and the extraction of vital biomaterials from the cell [132]. It was proposed that the synergistic effect of IRE and SC-CO_2 results from IRE damage to the cell membrane, thus allowing easier CO_2 penetration of the cell [134].

IRE and High Hydrostatic Pressure (HHP)

The possibility of germinating *Bacillus* spores using HHP, following by IRE treatment in order to inactivate the germinated cells, was investigated [135]. It was found that germination of more than 5 log-cycles of spores was initiated by pressurization, and while the germinated cells did become sensitive to a subsequent heat treatment, they were not sensitized to IRE application below 40°C. One suggestion for increased spore inactivation by these combined processes involves adding an intermediate holding step to allow germinated spores to outgrow into vegetative cells [135]. The results of combined HHP first and IRE treatment afterwards, as reported by Heinz and Knorr [136], indicate that sublethal pressure produces a stabilizing effect on vegetative cells against IRE, probably because the high pressure induces a more rigid gel structure of the membrane [9] [137]. Synergistic lethal effects were observed when HHP followed IRE [136]. For instance, a 2 log-cycle reduction of *B.subtilis* vegetative cells was determined for a 10-minute treatment at 200 MPa followed by 24.7 kV/cm 300μsec IRE treatment, as compared to IRE treatment alone [136].

Ultrasound and IRE Combination

Higher-power ultrasound has the ability to inactivate microbes [138]. The cavitation phenomenon and associated shear disruption, localized heating and free radical formation were found to be the contributory causes of sterilization. It was also shown that brief exposure to ultrasound caused a thinning of cell walls attributed to the detachment of the cytoplasm membrane from the cell wall [139]. The results concerning ultrasounds and IRE in combination vary in the literature. Inactivation of *B. subtilis* spores using up to 2000 Hz sonication combined with IRE at 30–40 kV/cm was reported by Jin et al. [140]. Xin et al. reported that the combined approach leaded to the total increased efficiency with some types of bacteria, while total efficiency was reduced with others [141] in oil-field re-injection water. Both orders of combinations, IRE–ultrasound and ultrasound–IRE, resulted in less than the sum of the two treatments alone during inactivation of *Salmonella enteritidis* in whole liquid egg [142]. The log-cycle reductions for the IRE–ultrasound and ultrasound–IRE treatments were 2.30 and 2.25 respectively, while a single treatment caused a 2.4 log-cycle reduction [142]. Similar antagonistic response of *Salmonela* and *L. monocytogines* was reported by Alvarez and Heinz [143].

Ultraviolet Irradiation and IRE Combination

The FDA has approved UV irradiation as a standalone method suitable for fruit juice preservation, provided that turbulent flow occurs throughout the process [144]. The flow requirements would not be necessary where UV exposure is not the sole hurdle [145]. The application a combination of UV irradiation and IRE to freshly squeezed apple juice resulted in a similar total microbial reduction compared to severe heat treatment (94°C) [145]. At the same time, the quality attributes measured in juice processed by this combined approach were similar to those observed in juice treated by the milder heat process (72°C) and consistently superior when compared to the juice subjected to severe heat treatment (94°C). The combination of the two treatments resulted in a greater microbial inactivation (6.2 and 7.1 log-cycles for IRE-UV and UV-IRE, respectively) than each treatment applied separately (5.4 and 2.2 log-cycles for IRE and UV, respectively) [145].

Examples of Product Treatment

Water

According to the WHO, about 2.6 billion people- half of the developing world- lack even a simple 'improved' latrine, and 1.1 billion people have no access to any type of improved source of drinking water or basic sanitation [146]. As a direct consequence, 1.6 million people die every year from diarrheal diseases (including cholera); 90% of these are children under 5, mostly in developing countries. 160 million people are infected with schistosomiasis, causing tens of thousands of deaths yearly. 500 million people are at risk of trachoma, from which 146 million are threatened by blindness and 6 million are visually impaired. Intestinal helminths (ascariasis, trichuriasis and hookworm infection) are plaguing the developing world as a consequence of inadequate drinking water, sanitation and hygiene, with 133 million people suffering from high intensity intestinal helminth infections. There are approximately 1.5 million cases of clinical hepatitis A reported every year in developing countries [146]. Current worldwide implemented technologies such as boiling or chlorinating water fall significantly short of completely solving the problem[147]. A 4 log reduction for *E. coli* in water was reported at field strengths of 110 kV/cm and a 200nsec pulse width, with a total energy consumption of 40 J/cm^3 [44]. Vernhes et al. reported successful inactivation of *N. lovaniensis* amoebae in fresh water [148]. In log-phase growing cells, cell inactivation reached 100% at a pulse amplitude of 2 kV/cm, which affected only 80% of the stationary phase cells [148].

Beer and Wine

In spite of the fact that alcoholic beverages tend to be more resistant to spoilage microorganisms due to their high alcohol content and acidity, they are still susceptible to problems associated with spoilage [149]. Current treatment for beer involves thermal pasteurization, but problems are often encountered during the filling. An additional challenge is that of heat-resistant microorganisms. Wine is currently treated by either heat or the addition of sulfur dioxide; however, the former affects taste and many people are sensitive to the latter [149]. Heinz Doevenspeck, considered to be a father of pulsed electric field technology in food processing, issued a patent in 1971 for a device that reduces the number of microorganisms in beer by implementing IRE at 25°C [150]. *Lactobacillus plantarum* in sample beer was inactivated only above critical values of 13 kV/ cm and 64 kJ/ kg in study performed by Ulmer et al. [151]. The presence of nisin or hop extract during IRE treatment resulted in an additional reduction of cell viability by 1.5 log-cycles [151]. IRE was demonstrated to be effective in inactivating the selected spoilage bacteria and yeasts in beer under the following treatment parameters: 10.5 mL/s flow rate, 22 kV/cm electric field strength, 10 μs pulses with stainless steel electrodes. This experiment resulted in 0.5, 4.1, 4.3, 4.7,

5.8, and 4.8 log cycle reductions of *Saccaromyces uvarum, Rhodotorula rubra, Lactobacillus plantarum, Pediococcus damnosus*, and *Bacillus subtilis,* respectively [152]. However, a migration of electrode components was observed in the same study. IRE treatment using stainless steel material produces significant migration of Fe, Cr, Zn, and Mn in the IRE-treated beer samples, and causes a significant loss in flavor acceptance and mouth feeling of the products [152]. Complete inactivation occurred when 200 2μsec length pulses were given at 26 kV/cm for Guinness beer and 60 such pulses for Sauvignon Blanc wine [149].

Grape Juice

Two pulses of 2μsec length with an electric field strength of 35 kV/cm reduced the population of vegetative cells of *Zygosaccharomyces bailii* by 5 log-cycles and ascospores by 3.5 log-cycles at 20°C [153]. The extent of inactivation of *N.fischeri* ascospores by IRE was negligible even after 20 pulses, each with an electric field strength of 40 kV/cm [154]. After IRE treatment of 8 pulses at 35 kV/cm at 20°C, the population of *B. fulva* conidiospores in grape juice decreased by 5 log cycles [154]. Further work investigated the influence of a hurdle approach for the treatment. In red grape juice, up to a 5.9 log-cycle reduction of the total microbial count was achieved by applying 20 pulses of 65 kV/cm at 50°C with a 2 hour incubation of the juice containing 1:3 lyso:chrisin (1:3 ratio of lysozyme and nisin; 0.4 g/100ml) [155]. A reduction of 6.2 log-cycles was achieved for the total microbial count at 51°C when 20 pulses at 80 kV/cm were applied with a nisin addition (400 U/ml). A reduction of 4.4 log cycles was observed when 20 pulses of 65 kV/cm were applied at 50°C to white grape juice containing lyso:chrisin (0.4 g/1000ml). The vitamin C content of the red grape juice was unaffected by IRE treatment [155].

Apple Juice

Apple juice from concentrate treatment with IRE at 50 kV/cm, 10 pulses, pulse width of 2 μsec and maximum processing temperature of 45 °C had a shelf-life of 28 days compared to a shelf-life of 21 days of fresh-squeezed apple juice[131]. There were no physical or chemical changes in ascorbic acid or sugars in the IRE-treated apple juice, and a sensory panel found no significant differences between untreated and electric field treated juices[131]. Another study reported that IRE extended the shelf-life at 22 – 25 °C of fresh apple juice and apple juice from concentrate to more than 56 days or 32 days, respectively[156]. There was no apparent change in its physicochemical and sensory properties[156]. A pilot plant scale treatment was performed at 85 Liter/hour product flow under IRE at 35 kV/cm, pulse width of 1.92 μsec and total treatment time 94 μsec. The treatment extended the shelf life of the IRE processed apple juice and cider samples to more than 67 days at 4 °C, approximately 67 days at 22 °C, and 14 days at 37 °C , in comparison with control samples with shelf lives of approximately 14 days at 4 °C and less than 3 days at 22 and 37 °C[157]. Natural food color and vitamin C for both juice and cider were not affected by the pilot plant scale treatment. A comparison of conventional high temperature-short time (HTST) pasteurization with IRE treated apple juice showed that both techniques reduced *E.coli* content in juice by 5 log cycles. Although microbial inactivation is normally considered the standard for fruit juice pasteurization [158], when considering the pH [158] [159], color [158] [159] , total acidity[159], phenolics content [159]and volatile compounds[159], IRE produced results superior to HTST. The thermal treatment showed more significant changes in these properties.

Orange Juice

Sitzmann reported on the effectiveness of the ELSTERIL continuous process developed by the food engineers at Krupp Maachinentechnik GmbH in Hamburg, in association with the University of Hamburg, Germany[160], for orange juice processing. They reported the reduction of the native microbial flora of freshly squeezed orange juice by 3-log cycles without significantly affecting its quality, using an applied electric field of 15 kV/cm. Evaluation the shelf-life of reconstituted orange juice treated with an integrated IRE pilot plant system was performed in 1997 by Zhang et al.[161]. The IRE system consisted of a series of co-field chambers. Temperatures were maintained near ambient with cooling devices between chambers. The total aerobic counts were reduced by 3- to 4-log cycles under 32 kV/cm. When stored at 4 °C, both heat- and IRE-treated juices had a shelf-life of more than 5 months. Vitamin C losses were lower and color was generally better preserved in IRE-treated juices compared to the heat-treated ones for up to 90 days (storage temperature of 4 °C or 22 °C) or 15 days (storage temperature of 37 °C) after processing[161]. Later, effects of commercial-scale IRE processing on the microbial stability, ascorbic acid, flavor compounds, color, Brix, pH, and sensory properties of orange juice were studied and compared with those of thermally processed juice [162]. Freshly squeezed orange juice was thermally processed at 90 °C for 90 s or processed by IRE at 40 kV/cm for 97 ms. Both thermally processed and IRE-processed juices showed microbial shelf life at 4 °C for 196 days. IRE-processed juice retained more ascorbic acid, flavor, and color than thermally processed juice. Sensory evaluation of texture, flavor, and overall acceptability were ranked highest for control (fresh) juice, followed by IRE-processed juice and then by thermally processed juice [162]. Additional studies on influence of IRE on color [163] [164], flavor[164], vitamin C retention [164], pH and Bx and compare with thermal treatments were performed [163] [165] [166] . Several works showed the inactivation of particular spoilage microorganisms in orange juice. *Salmonella typhimurium* counts were reduced by 5.9 log cycles in freshly squeezed orange juice (without pulp) treated with 50 pulses at 90 kV/cm and 55 °C[125]. When IRE treatment was carried out in the presence of nisin (100 U/ml of orange juice), lysozyme (2,400 U/ml), or a mixture of nisin (27.5 U/ml) and lysozyme (690 U/ml), cell viability loss was increased by an additional 0.04 to 2.75 log cycles. The combination of nisin and lysozyme had a more pronounced bactericidal effect than either did alone. An additional *Salmonella* count reduction of at least 1.37 log cycles was achieved when the two antimicrobial agents were used in combination[125]. A maximum inactivation of a 5.1-log reduction was achieved after exposure of *S. cerevisiae* to IRE for 1 msec total treatment time with 4-μs pulse width at 35 kV/cm in bipolar mode [167]. *L. brevis* was inactivated to a maximum of 5.8 log reductions when inoculated orange juice was processed at 35 kV/cm for 1000 μs using 4 μs pulse width[166]. *Leuconostoc mesenteroides, E.coli,* and *Listeria innocua* were reported to be inactivated in 100L/hour flow rate by as much as 5 log cycles at 30 kV/cm and 50 °C [168]. A 2-log cycle reduction for the naturally grown microorganisms in juice was achieved after application of 120 pulses/mL at a IRE level of 46 kV/cm in another study [169].

Tomato Juice

The effects of commercial scale IRE processing on the quality of tomato juice were studied and compared with those of thermal processing [170] Tomato juice was thermally processed at 92 °C for 90s or IRE processed at 40 kV/cm for 57 msec. Both thermally and IRE processed juices showed microbial shelf life at 4 °C for 112 days. IRE processed juice retained more ascorbic acid than thermally processed juice at 4 °C for 42 days. No significant differences were observed in the concentration of lycopene, Bx, pH, or viscosity between thermally and IRE processed juices during the storage [170]. Sensory evaluations indicated that flavor and overall acceptability of IRE processed were preferred to those of thermally processed juice [170]. Nutritious tomato juice with relative lycopene content of 131.8%, vitamin C content of 90.2%, and antioxidant capacity retention of 89.4% was attained with IRE treatments of a 1 μsec pulse duration applied at 250 Hz in bipolar mode[171]. At field strength of 80 kV/cm, 20 pulses at 50 °C in the presence of nisin (100 U/mL) produced about 4.4 log reductions in microbial counts of naturally occurring microorganisms[172].

Milk

A challenge test and shelf-life study with homogenized milk inoculated with *Salmonella dublin* and treated with 40 pulses at a field strength of 36.7 kV/cm over a 25-min time period was performed [173] . *Salmonella dublin* was not detected after IRE treatment or after storage at 7-9 ° C for 8 days. The naturally occurring milk bacterial population increased to 10^7 cfu/ml in the untreated milk, whereas the treated milk contained approximately $4x10^2$ cfu/ml at the end of this time period [173]. Further studies indicated less flavor degradation and no chemical or physical changes in milk quality attributes for cheese making[174]. The shelf life of raw skim milk (0.2% milk fat) was studied [175]. Milk was treated with IRE using 30 exponentially decaying pulses at 40 kV/cm and a treatment time of 2 μsec. The shelf-life of the milk was 2 weeks stored at 4 °C; however, treatment of raw skim milk at 80 °C for 6 seconds followed by PEF treatment of 30 pulses at 30 kV/cm, using a pulse width of 2 μsec increased the shelf-life up to 22 days, with a total aerobic plate count of 3.6-log cfu/ml. The processing temperature did not exceed 28 °C during IRE treatment of the raw skim milk[175]. Qin et al. reported that milk (2% milk fat) subjected to 2 steps of 7 pulses each and 1 step of 6 pulses with an electric field strength of 40 kV/cm achieved a shelf-life of 2 weeks at refrigeration temperature[176]. There was no apparent change in the milk's physical and chemical properties and no significant differences in sensory attributes between heat pasteurized and IRE treated milk were observed in this study[176] . The microbial population of *L. innocua* was reduced by a 2.5-log cycle after IRE treatments of skim milk at 30, 40 or 50 kV/cm[177]. The same IRE intensities and subsequent exposure to 10 IU nisin/ml achieved 2-, 2.7- or 3.4-log reduction cycles of *L. innocua,* respectively[177]. *L. monocytogenes* was reduced by 1- to 3-log cycles at 25 °C and 4-log cycles at 50 °C, with no significant differences among whole, 2% and skim milks [178]. A 3-log cycle reduction in the population of *Staphylococcus aureus* in the skim milk samples was obtained by IRE with a 450 msec treatment time and 35kV/cm electric field strength[116]. In the same study, it was shown that subsequent cooling causes the inactivation of the previously damaged bacterial cells, due to resealing inhibition[116]. IRE-thermal treatment at 35 kV/cm and 60 °C was as reported as effective as thermal pasteurization for inactivation of natural microflora in raw skim milk, including total plate count, Pseudomonas and Enterobacteriaceae [179]. At 60 °C, the log reductions for total microbial population, Pseudomonas and Enterobacteriaceae were 2.3, 2.4 and 1.6, respectively without IRE treatment. When IRE was applied, the field intensities of 25– 35 kV/cm had small effect on total plate count while achieving an additional 1–3.5 log reduction on Pseudomonas, and 0.6 log reduction on Enterobacteriaceae. The relatively insignificant effect of IRE on total plate count was explained by the presence of more resistant gram-positive or spore-forming bacteria in raw milk. Low temperatures significantly reduced the effectiveness of IRE for the milk [179]. Additional research demonstrated that at least a 5-log reduction in Pseudomonas species, the major spoilage bacteria in milk, can be achieved by a combination of IRE treatment at 31 kV/cm and mild heat treatment (55 °C) [180].Significant reduction in the inactivation efficiency of Pseudomonas was achieved at 15 or 40 °C compared with 50 or 55 °C. The total microbial shelf life of the IRE-treated milk was reported to be 13 and 11 days for inoculation levels of 103 and 105 CFU/ mL, respectively [180]. A hurdle approach implementing ultrasound followed by IRE led to 6.8 log cycle reductions of *L. innocua* in low fat milk[181]. However, the results of this approach are controversial as described below. It was observed that a pure culture of *E.coli* is more resistant to IRE than the indigenous microbes in the raw milk[182]. The same study showed that the milk fat content had a significant negative effect on the extent of log-reduction for indigenous microbes, when 2% and 18% of fat content in milk levels were compared [182].

Inactivation of *Lactobacillus plantarum* inoculated into an orange juice–milk based beverage was conducted as a model study for implementation of IRE in mixed and complicated products [183]. A maximum inactivation of 2.12 log cycles was achieved at a 40kV/cm field strength,130 μsec total treatment time at

35 °C , which was increased to 2.6 log cycles at 55 °C [183]. The authors observed a reduction of effectiveness in the mixed product treatment in comparison with studies where pure juices or milk were treated [184]. This observation makes sense due to the significant dependence of effectiveness on the product characterization, as discussed earlier.

Reconstructed infant formula milk was treated against the an occasional contaminant *Enterobacter sakazakii,* which is considered thermally resistant clinical isolate [37] . A 1.2 log cycle reduction was achieved when 144 pulses were applied at 40 kV/cm, using a 2.5 μsec pulse width [37].

Yakju (Rice Wine)

Yakju is a Korean traditional rice wine fermented by a wild starter, and has a relatively short shelf life (about 3 weeks) mainly due to the souring caused by residual lactic acid bacteria[185]. When the Yakju is heat treated to extend its shelf life, the quality is remarkably lowered by heat-induced off-flavor. Several studies aim to increase the shelf life of Yakju without affecting its quality. It was shown that IRE treatment stabilized pH and acidity at 4 and 30 °C, and the shelf life of the product improved up to 6 weeks while still keeping the high quality of the product [186] [185].

Egg Processing

Egg-associated salmonellosis is an important public health challenge. The bacterium causing this disease has been associated with the consumption of food products containing raw or inadequately cooked Grade A whole shelled eggs, homemade ice cream, eggnog, and mayonnaise[187]. Denaturation of whole egg, indicative of a change in viscosity, occurred in the temperature range of 56 to 66 °C[188]; howevr at present, liquid whole egg is pasteurized at 60 °C for 3 min [189] to obtain stable and rnicrobiologically safe products.

Some of the earliest studies in egg products were conducted by Dunn and Pearlman. They used a static parallel electrode treatment chamber with 2-cm gap, applying 25 exponentially decaying pulses with peak voltages of around 36 kV [173]. Tests were carried out on liquid eggs, on heat-pasteurized liquid egg products, and on egg products with potassium sorbate and citric acid added as preservatives. Comparisons were made with regular heat-pasteurized egg products with and without the addition of food preservatives when the eggs were stored at low (4 °C) and high (10 °C) refrigeration temperatures. The study demonstrated the importance of the hurdle approach in shelf-life extension. Its effectiveness was even more evident during storage at low temperatures, where egg products with a final count around 500 cfu/ml stored at 10 °C and 4 °C maintained a low count for 4 and 10 days, respectively, versus a few hours for the heat pasteurized samples [173]. Other studies on liquid whole eggs treated with IRE showed that IRE treatment decreased the viscosity but increased the color (in terms of b -carotene concentration) of liquid whole eggs compared to fresh eggs [176] [190]. At a 1.2

mL/s flow rate, 200 pps frequency, 2.12 μsec pulse duration, 25 kV/cm electric field strength, and 250 μsec total treatment time, a 1 log cycle reduction of *Salmonella enteritidis* population in liquid whole egg was observed. The IRE treated product was exposed for post IRE thermal treatment at 55 °C for 3.5 min. The hurdle approach led to 4.3 log cycle reduction of bacteria in the product without significant change in viscosity, electrical conductivity, color, pH, and Bx. Treated samples presented significantly longer shelf-life at 4 °C compared with the control and heat treated samples [191]. Additional studies show that the following results can be achieved by implementing IRE: 3.5 log cycle reductions in *Salmonella enteritidis*, 5.5 log reductions of *Listeria innocua* by means of a synergistic effect of IRE and nisin in liquid whole egg, and 5.6 log reductions of *Escherichia coli* in beaten fresh liquid egg by IRE treatment applied continuously or discontinuously in five steps [192]. The shelf life of IRE-treated fresh liquid egg was extended to 4 weeks in refrigeration. IRE treatment did not cause notable changes in proteins in a solution of ovalbumin and dialyzed fresh egg white; however, structural changes and functional modifications were observed in fresh egg [192]. In addition ,the texture and microstructure of gels were affected by the application of IRE and optimization for the process is needed[192]. A detailed review concerning implementation of IRE with eggs and eggs product was recently published [192].

Green Tea

Tea is one of the most widely consumed beverages in the world, especially in oriental countries[193] . The loss of sensory qualities (color, flavor and taste) and the particular sensitivity of valuable green tea polyphenols to heat processing[194] has limited the development of green tea beverage in the market , thus leaving this highly demanded beverage still in the "home made" phase. 5.6 and 4.9 log cycle reductions of *E. coli* and *S. aureus* by IRE with a field strength of 38.4 kV/cm for 160 and 200 μsec total treatment time was achieved without significant effect on color, tea polyphenols, and free amino acid contents[193]. Following this, a shelf life evaluation was performed. The color stability, green tea polyphenols and free amino acid contents of the green tea beverages did not change significantly during the 4 °C storage, achieving 6 months of shelf life [193].

Green Pea Soup

Pea soup was exposed to 2 steps of 16 pulses at 35 kV/cm to prevent an increase in temperature beyond 55 ° C during treatment [71] . The shelf-life of the IRE-treated pea soup stored at refrigeration temperature exceeded 4 weeks, while 22 or 32 °C temperatures were found inappropriate to store the product. There were no apparent changes in the physical and chemical properties or sensory attributes of the pea soup directly after IRE processing or during the 4 weeks of storage at refrigeration temperatures [71].

Sour Cherry Juice, Peach and Apricot Nectars

Germination tube elongation and spore germination rate of fungi *Penicillum expansum* were completely inhibited by IRE, using 30 kV/cm exponential decay pulses with 3μsec length and 218 μsec total treatment time at a starting treatment temperature of 11 °C [195].

Future Research Needs

Although the research in the field of IRE implementation started almost fifty years ago and its results show that the technique exhibits tremendous potential for microorganism deactivation with only minimal product damage, its commercial implementation is still limited. Many claim that this is due to the lack of enthusiasm from the traditional food industry side to adopt new technologies. However, a look at the essence of the process shows that important basic questions are still in the early research stage. The molecular mechanism of IRE is still unknown, and this results in a significant limit on the prediction ability of the process. Development of broadly accepted validation methods to ensure microbiological effectiveness of the treatment is still not finished. Methods for identification of pathogens most resistant to IRE in specific products is also still needed. Additional variations of IRE, reducing electrode dissolving such as high frequency or non-metal contact IRE should be investigated and validated. Packed food treatment should be tested. A platform for integration of IRE in the existing factories and processes should be developed at the lowest possible cost. It should be noted that Genesis Juice LTD , a world first commercial implementer of IRE technology in juices, shifted to high hydrostatic pasteurization due to the current limit of IRE to treat foaming beverages. A solution to this problem should be found since IRE propose a flow process which is not possible in HHP systems, thus allowing cheaper and more convenient process for large quantities.

References

[1] Diamond, J.: Guns, Germs, and Steel: The Fates of Human Societies (1997)
[2] Flandrin, J.-L., Montanari, M., Sonnenfeld, A.: Food: A Culinary History. Penguin, New York (2000)
[3] Ponting, C.: A Green History of the World: The Environment and the Collapse of Great Civilizations. St. Martin's, New York (1992)
[4] Ray, B.: Fundamental Food Microbiology. CRC Press, Boca Raton (2004)
[5] FAO/WHO, The role of food safety in health and development. Report of Joint FAO/WHO expert omittee on Food Safety. World Health Organisation. Techn. Report Ser. (1984)
[6] WHO, Food safety and foodborne illness, N°237 (2007)
[7] WHO, Practical actions to promote food safety. Final report: FAO/WHO regional conference on food safety Africa (2005)

[8] Rahman, M.S.: Handbook of Food Preservation. CRC Press, Boca Raton (2007)
[9] Lelieveld, H.L.M., Notermans, S., de Haan, S.W.H.: Food Preservation by pulsed electric fields. From Research to Application. CRC, Cambridge (2007)
[10] Stewarta, C.M., Tompkin, R.B., Cole, M.C.: Food safety: new concepts for the new millennium. Innovative Food Science & Emerging Technologies 3, 105–112 (2002)
[11] Shin, S.Y., Kliebenshtein, J., Hayes, D., Shogren, J.F.: Customers willingness to pay for safer food products. Journal of Food Safety 13, 51–59 (1992)
[12] Baltzer, K.: Customers willingness to pay for food quality - the case of eggs. Acta Agriculturae Scandinavica, Section C-Economy 1, 78–90 (2004)
[13] Sivapalasingam, S., Friedman, C.R., Cohen, L., Tauxe, R.V.: Fresh produce: A growing cause of outbreaks of foodborne illness in the United States, 1973 through 1997. J. Food Trot 67, 2342–2353 (2004)
[14] Mertens, B., Knorr, D.: Development of nonthermal processes for food preservation. Food Technol. 5, 124–133 (1992)
[15] Sizer, C.E., Balasubramaniam, V.M.: New invention processes for minimally processed juices. Food Technol. 53, 64–67 (1999)
[16] Rubinsky, B.: Irreversible electroporation in medicine. Technology in Cancer Research and Treatment 6, 255–260 (2007)
[17] Neumann, E., Schaefer-Ridder, M., Wang, Y., Hofschneider, P.H.: Gene transfer into mouse lyoma cells by electroporation in high electric fields. EMBO J. 1, 841–845 (1982)
[18] Fuller, G.W.: Report on the investigations into the purification of the Ohio river water at Louisville Kentucky. D. Van Nostrand Company, New York (1898)
[19] Beattie, J.M., Lewis, F.C.: The electric current (Appart from the Heat Generated) A Bacteriological Agent in the Sterilisation of milk and other fluids. Journal of Hygiene 24, 123–137 (1925)
[20] Moses, B.D.: Electrical Pasterization of milk. Agricultural Engineering 19, 525–526 (1938)
[21] Sale, A.J., Hamilton, W.A.: Effect of high electric field on micro-organisms. I.Killing of bacteria and yiest. Biochimica Biophysica Acta 148, 781–788 (1967a)
[22] Sale, A.J., Hamilton, W.A.: Effect of high electric field on micro-organisms. II.Mechanism of action of the lethal effect. Biochimica Biophysica Acta 148, 788–800 (1967b)
[23] Sale, A.J., Hamilton, W.A.: Effects of high electric fields on microorganisms. III. Lysis of erythrocytes and protoplasts. Biochimica Biophysica Acta 163, 37–43 (1968)
[24] Doevenspeck, H.: Influencing cells and cell walls by electrostatic impulses. Fleischwirtschaft 13, 968–987 (1961)
[25] Zimmerman, U., Riemann, F., Pilwat, G.: Dielectric breakdown of cell membranes. Biophys. J. 14, 881–899 (1974)
[26] Kinosita, K., Tsong, T.Y.: Voltage-induced pore formation and hemolyses of human erythrocytes. Biochim. Biophys. Acta 471, 227–242 (1977b)
[27] Abidor, I.G., Arakelyan, L.V., Chernomordik, L.V., Chimadzev, Y.A., Pastushenko, V.P., Tarasevich, M.R.: Electrochemic breakdown of bilayer lipid membranes. The main experimental facts and their qualitative discussion. Bioelectrochem. Bioenergetics 6, 37–52 (1979)
[28] Weaver, J.C., Chimadzev, Y.A.: Theory of electroporation: A review. Bioelectrochemical Bioengener. 41, 135–160 (1996)

[29] Somolinos, M., Mañas, P., Condón, S., Pagán, R., García, D.: Recovery of Saccharomyces cerevisiae sublethally injured cells after Pulsed Electric Fields. International Journal of Food Microbiology 125, 352–356 (2008)
[30] Saulis, G., Venslauskas, S., Naktinis, J.: Kinetics of pore resealing in cell membrane after electroporation. Bioelectrochemistry Bioenerg. 26, 1–13 (1991)
[31] Saulis, G.: Pore disappearence in a cell after electroporation.Theoretical simulation and comparison with experiments. Biophys. J. 73, 1299–1309 (1997)
[32] Somolinos, M., García, D., Mañas, P., Condón, S., Pagán, R.: Effect of environmental factors and cell physiological state on Pulsed Electric Fields resistance and repair capacity of various strains of Escherichia coli. International Journal of Food Microbiology 124, 260–267 (2008)
[33] FDA, Kinetics of Microbial Inactivation for Alternative Food Processing Technologies, US, Food and Drug Administration Center for Food Safety and Applied Nutrition IFT/FDA Contract No. 223-98-2333 (2000)
[34] Hülsheger, H., Potel, J., Niemann, E.G.: Killing of bacteria with electric pulses of high field strength. Radiat. Environment Biophys. 20, 53–65 (1981)
[35] Hülsheger, H., Pottel, J., Niemann, E.G.: Electric field effects on bacteria and yeast cells. Radiat. Environ. Biophys. 22, 149–162 (1983)
[36] Peleg, M.: A model of microbial survaval after exposure to pulse electric field. J. Sci. Food Agric. 67, 93–99 (1995)
[37] Pina-Perez, M.C., Aliaga, D.R., Bernat, C.F., Enguidanos, M.R., Lopez, A.M.: Inactivation of Enterobacter sakazakii by pulsed electric field in buffered peptone water and infant formula milk. International Dairy Journal 17, 1441–1449 (2007)
[38] Barbosa-Canovas, G.V., Pothakamury, U.R., Palou, E., Swanson, B.G.: Nonthermal Preservation of Foods. Marcel Decker, New York (1998)
[39] Barbosa-Cánovas, G.V., Gongora-Nieto, M.M., Pothakamury, U.R., Swanson, B.G.: Preservation of foods with pulsed electric fields. Academic Press Ltd., London (1999)
[40] Bushnell, A.H., Clark, R.W., Dunn, J.E., Lloyd, S.W.: Process for reducing levels of microorganisms in pumpable food products using a high pulsed voltage system 5, 514, 391, US Patent 5, 514, 391 (1996)
[41] Pizzichemi, M.: Application of pulsed electric fields to food treatments. In: 10th Topical seminar on Innovative particle and Radiation Detectors, vol. 172, pp. 314–316 (2007)
[42] Zhang, Q., Barbosa-Cánovas, G.V., Swanson, B.G.: Engineering Aspects of Pulsed Electric Field Pasteurization. Journal of Food Engineering 25, 261–281 (1995)
[43] Satoshi, U., Makoto, H., Furmiyoshi, T.: Efficient sterelisation of bacteria by pulse electric field in micro-gap. J. of Electrostatics 66, 427–431 (2008)
[44] Narsetti, R., Curry, R.D., McDonald, K.F., Clevenger, T.E., Nichols, L.M.: Microbial Inactivation in Water Using Pulsed Electric Fields and Magnetic Pulse Compressor Technology. IEEE Transactions on Plasma Science 34, 1386–1393 (2006)
[45] Mohan, N., Underland, T.M., Robbins, W.P.: Power electronics converters, Application and design, 2nd edn. John Wiley & Sons, New York (1995)
[46] Roodenburg, B.: Electrochemistry in pulsed electric field treatment chambers. In: Lelieveld, H.L.M., Netermans, S., de Haan, S.W.H. (eds.) Food preservation by pulsed electric fields. From research to application. Woodhead publishing limited, Cambridge (2007)

[47] Caplot, M., Cote, G.: Les technologies nécessaires aux machines de traitment par Champs Electriques Pulsés. In: Formation Technologique en Agro- Alimentaire, La conservation de demain, Bordeaux Pessac, France (1999)
[48] Morren, J., Roodenburg, B., de Haan, S.W.H.: Electrochemical reactions and electrode corrosion in pulsed electric field (PEF) treatment chambers. Innov. Food Sci. Emerg. Technol. 4, 285–295 (2003)
[49] Gongora-Nieto, M.M.: Food preservation by pulsed electric fields: evaluation of critical processing parameters and process optimization. Ph.D. Dissertation, Washington State University, Pullman, WA (2000)
[50] Qin, B.L.: Continuous flowelectrical treatment of flowable food products. US Patent 566, 203 (1997)
[51] Toepfl, S., Heinz, V., Knorr, D.: High intensity pulsed electric fields applied for food preservation. Chemical Engineering and Processing 46, 537–546 (2007)
[52] Davalos, R.D., Mir, L.M., Rubinsky, B.: Tissue ablation with Irreversible Electroporation. Annals of Biomedical Engineering 33, 223–231 (2005)
[53] Zimmerman, U.: The effect of high intensity electric field pulses on eucaryotic cell membranes: fundamentals and application. CRC Press, Boca Raton (1996)
[54] Heinz, V., Alvarez, I., Angersbach, A., Knorr, D.: Preservation of liquid foods by high intensity pulsed electric fields-basic concepts for process design. Trends in Food Science and Technology 12, 103–111 (2002)
[55] Neumann, E.: Gene delivery by membrane elecroporation. In: Lynch, P.T., Davet, M.R. (eds.) Electrical manipuation of cells, pp. 157–184. Chapman and Hall, New York (1996)
[56] Chalise, P.R., Perni, S., Shama, G., Novac, B.M., Smith, I.R., Komg, M.G.: Lethality mechanisms in Escherichia coli induced by intense sub-microsecond electrical pulses. Appl. Phys. Lett. 89, 153902.1–153902.3 (2006)
[57] Hair, P.S., Schoenbach, K.H., Buescher, E.S.: Sub-microsecond intense pulsed electric field applications to cells show specificity of effects. Bioelectrochemistry 61, 65–72 (2003)
[58] Miller, L., Leor, J., Rubinsky, B.: Cancer Cells Ablation with Irreversible Electroporation. Technol. Cancer Res. Treat., 699–705 (2005)
[59] Aly, R.E., Joshi, R.P., Stark, R.H., Svhoenbach, K.H., Bebe, S.J.: Repair time of bacteria after pulsed electric field application. In: Pulsed Power Plasma Science. IEEE Conference Record Abstracts, p. 310 (2001)
[60] De Hann, S.W.H., Willcock, P.R.: Comparison of the energy performance of pulse generation circuits for PEF. Innov. Food Sci. Emerg.Techol., 349–356 (2002)
[61] Keith, D., Chen, G.: Electroporation How different length and shaped electrical pulses affect the permeability of cells. In: IEEE, Annual Report Conference on Electrical Insulation and Dielectric Phenomena (2006)
[62] Töpfl, S.: Pulsed Electric Fields (PEF) for Permeabilization of Cell Membranes in Food- and Bioprocessing – Applications, Process and Equipment Design and Cost Analysis. In: Prozesswissenschaften, Doktor der Ingenieurwissenschaften, p. 194. der Technischen Universität Berlin, Berlin (2006)
[63] Stanley, D.W.: Biological membrane deterioration and associated quality losses in food tissues. In: Critical Reviews in Food Science and Nutrition. F. M. Clydesdale. CRC Press, New York (1991)
[64] Dunn, J.E., Pearlman, J.S.: Methods and apparatus for extending the shelf life of fluid food products, US 4695472 (1987)

[65] Jayaram, S., Castle, G.S.P., Magaritis, A.: The effect of high field DC pulse and liquid medium conductivity on survivability of L. Brevis. Applied Microbiology and Biotechnology 40, 117–122 (1993)

[66] Zhang, B., Qin, L., Barbosa-Cánovas, G.V., Swanson, B.G.: Inactivation of Escherichia coli for food pasteurization by high-strength pulsed electric fields. Journal of Food Processing and Preservation 19, 103–118 (1995)

[67] Pothakamury, U.R., Vega, H., Zhang, Q., Barbosa-Cánovas, G.V., Swanson, B.G.: Effect of growth stage and processing temperature on the inactivation of Escherichia coli by pulsed electric fields. Journal of Food Processing 59, 1167–1171 (1996)

[68] El Zakhem, H., Lanoisellé, J.-L., Lebovka, N.I., Nonus, M., Vorobiev, E.: Influence of temperature and surfactant on Escherichia coli inactivation in aqueous suspensions treated by moderate pulsed electric fields/. International Journal of Food Microbiology 120, 259–265 (2007)

[69] Jayaram, S., Castle, G.S.P., Magaritis, A.: Kinetics of sterlisation of Lacatobacillas brevis cells by the application of high voltage pulses. Biotechnol. and Bioeng. 40, 1412–1420 (1992)

[70] Bazhal, M.I., Ngadi, M.O., Raghavan, G.S.V., Smith, J.P.: Inactivation of Escherichia coli O157:H7 in liquid whole egg using combined pulsed electric field and thermal treatments. LWT—Food Science and Technology 39, 419–425 (2006)

[71] Vega-Mercado, H., Martin-Belloso, O., Chang, F.-J., Barbosa-Cánovas, G.V., Swanson, B.G.: Inactivation of Escherichia coli and Bacillus subtilis suspended in pea soup using pulsed electric fields. J. Food Process Preserv. 20, 501–510 (1996)

[72] Heinz, V., Toepfl, S., Knorr, D.: Impact of temperature on lethality and energy efficiency of apple juice pasteurization by pulsed electric fields treatment. Inno. Food Sci. Emerging Technol. 4, 167–175 (2003)

[73] Toefl, S., Mathys, A., Heinz, V., Knorr, D.: Review: Potential of High Hydrostatic Pressure and Pulsed Electric Fields for Energy Efficient and Environmentally Friendly Food Processing. Food Reviews International 22, 405–423 (2006)

[74] Sensoy, I., Zhang, Q.H., Sastry, S.: Inactivation of Salmonella dublin by pulsed electric field. J. of Food Sci. 20, 367–381 (1997)

[75] Fox, M.B., Esveld, D.C., Mastwijk, H.C., Boom, R.M.: Inactivation of L. Plantaum in a PEF microreactor.The effect of puls width and temperature on the inactivation. Inno. Food Sci. Emerging Technol. (2007)

[76] Fiala, A., Wouters, P.C., van den Bosch, H.F.M., Creyghton, Y.L.M.: Coupled Electrical-fluid Model of Pulsed Electric Field Treatment in a Model Food System. J. Innove. Food Sci. Emerg.Technologies 2, 229–238 (2001)

[77] Lindgren, M., Aronsson, K., Galt, S., Ohlssona, T.: Simulation of the temperature increase in pulsed electric field (PEF) continuous flow treatment chambers. Innovative Food Science and Emerging Technologies 3, 233–245 (2002)

[78] Kinosita, K., Tsong, T.Y.: Formation and resealing of pores of controlled sizes in human erythrocyte membrane. Nature 268, 438–441 (1977a)

[79] Aronsson, K., Lindgren, M., Johansson, B.R., Ronner, U.: Inactivation of microorganisms using pulsed electric fields: the influence of process parameters on Escherichia coli, Listeria innocua, Leuconostoc mesenteroides and Saccharomyces cerevisiae. Innovative Food Science & Emerging Technologies 2, 41–54 (2001)

[80] Gaskova, D., Sigler, K., Janderova, B., Plasek, J.: Effect of high-voltage electric pulses on yeast cells: Factors influencing the killing efficiency. Bioelectrochem. Bioenergetics 39, 195–202 (1996)

[81] Molinari, P., Pilosof, A.M.R., Jagus, R.J.: Effect of growth phase and inoculum size on the inactivation of Saccharomyces cerevisiae in fruit juice, by pulsed electric fields. Food Research International 37, 793–798 (2004)
[82] Bergey, D.H., Holt, J.G.: Bergey's Manual of Determinative Bacteriology. Lippincott William & Wilkins (1994)
[83] Garcia, D., Gomez, N., Raso, J., Pagan, R.: Bacterial resistance after pulsed electric fields depending on the treatment medium pH. Innovative Food Science and Emerging Technologies 6, 388–395 (2005)
[84] Wouters, P.C., Alvarez, I., Raso, J.: Critical factors determining inactivation kinetics by pulsed electric field food processing. Trends in Food Science and Technology, 112–121 (2001)
[85] De Jong, P., Van Heesch, E.J.M.: Review: Effect of pulse electric field on the quality of food products. Milchwissenschaft 53, 4–8 (1998)
[86] Fernandez-Molina, J.J., Bermudez-Aguerre, D., Altunakar, B., Swanson, B.G., Barbarosa-Canovas, G.V.: Inactivation of Listeria inocua and Pseudomonas flourescens by pulsed electric fields in skim milk: Energy requirements. J. of food processing engineering 29, 561–573 (2006)
[87] Zhang, Q., Monsalve-Gonzalez, A., Qin, B.L., Barbosa-Canovas, G.V., Swanson, B.G.: Inactivation of Saccharomyces cerevisiae by square wave and exponential decay electric field. Journal of Food Process Engineering 17, 469–478 (1994)
[88] Griffiths, S., Smith, S., MacGregor, S.J., Anderson, J.G., van der Walle, C., Beveridge, J.R., Grant, M.H.: Pulsed electric field treatment as a potential method for microbial inactivation in scaffold materials for tissue engineering: the inactivation of bacteria in collagen gel. Journal of Applied Microbiology 105, 963–969 (2008)
[89] Jacob, H.E., Forster, W., Berg, H.: Microbial implications of electric field effects. II. Inactivation of yeast cells and repair of their cell envelope. Z. Allg. Microbial. 21, 225–232 (1981)
[90] Wouters, P.C., Dutreux, N., Smelt, J.P.P.M., Lelieveld, H.L.M.: Effects of Pulsed Electric Fields on Inactivation Kinetics of Listeria innocua. Applied and Environmental Microbiology 65, 5364–5371 (1999)
[91] Hohmann, S., Mager, W.H.: Molecular biology intelligence unit. Yeast stress responses. Landes Bioscience, Austin, TX, USA (1997)
[92] Robey, M., Benito, A., Hutson, R.H., Pascual, C., Park, S.F., Mackey, B.M.: Variation in resistance to high hydrostatic pressure and rpoS heterogeneity in natural isolates of Escherichia coli O157:H7. Applied Environmental Microbiology 67, 4901–4907 (2001)
[93] Lado, B.H., Bomser, J.A., Dunne, P.C., Yousef, A.E.: Pulsed Electric Field Alters Molecular Chaperone Expression and Sensitizes Listeria monocytogenes to Heat. Applied and environmental microbiology 70, 2289–2295 (2004)
[94] Dutreux, N., Notermans, S., Wijtzes, T., Gongora-Nieto, M.M., Barbosa-Canovas, G.V., Swanson, B.G.: Pulsed electric fields inactivation of attached and free-living Escherichia coli and Listeria innocua under several conditions. International Journal of Food Microbiology 54, 91–98 (2000)
[95] Tsong, T.Y.: Review on electroporation of cell membranes and some related phenomena. Biochem. Bioenergy 24, 171 (1990)

[96] Dunne, C.P., Dunn, J., Clark, W., Ott, T., Bushnell, A.H.: Application of high energy electric field pulses to preservation of foods for combat rations. Science and Technology for Force XXI. Department of the Army. Norfolk, Virginia, June 24-27, 7 (1996)

[97] Álvarez, I., Pagán, R., Condón, S., Raso, J.: Environmental factors influencing the inactivation of Listeria monocytogenes by pulsed electric fields. Letters in Applied Microbiology 20, 691–700 (2002)

[98] Liu, X., Yousef, A.E., Chism, G.W.: Inactivation of E. coli0157:H7 by the combination of organic acids and pulsed electric field. Journal of Food Safety 16, 287–299 (1997)

[99] Vega-Mercado, H., Pothakamury, U.R., Chang, F.-J., Barbosa-Cánovas, G.V., Swanson, B.G.: Inactivation of Escherichia coli by combining pH, ionic strength and pulsed electric fields hurdles. Food Res. Int. 29, 117–121 (1996)

[100] García, D., Gómez, N., Mañas, P., Condón, S., Raso, J., Pagán, R.: Occurrence of sublethal injury after pulsed electric fields depending on the microorganism, the treatment medium pH and the intensity of the treatment investigated. Journal of Applied Microbiology 99, 94–104 (2005)

[101] Gomez, N., Garcıa, D.D., Alvarez, I., Raso, J.S., Condon, S.: A model describing the kinetics of inactivation of Lactobacillus plantarum in a buffer system of different pH and in orange and apple juice. Journal of Food Engineering 70, 7–14 (2005)

[102] Aronsson, K., Ronner, U.: Influence of pH, water activity and temperature on the inactivation of Escherichia coli and Saccharomyces cerevisiae by pulsed electric fields. Innovative Food Science & Emerging Technologies 2, 105–112 (2001)

[103] Min, S., Reina, L., Zhang, Q.H.: Water activity and the inactivation of Enterobacter cloacae inoculated in chocolate liquor and a model system by pulsed electric field treatment. Journal of food processing and preservation 26, 323–337 (2002)

[104] Keith, W., Harris, L., Hudson, L., Griffiths, M.: Pulsed electric field fields as a processing alternative for microbial reduction in spice. J. Food Res. Int. 30, 185–191 (1997)

[105] Keith, W., Harris, L., Hudson, L., Griffiths, M.: Reduction of bacterial levels in flour by pulsed electric fields. J. Food Processing Eng. 21, 263–269 (1998)

[106] Barbosa-Canovas, G.V., Maria, S.T., Cano, M.P.: Novel Food Processing Technologies. CRC, Boca Raton (2004)

[107] Neidhardt, F.C., Ingraham, J.L., Schaechter, M.: The effects of temperature, pressure, and pH. In: Physiology of the Bacterial Cell. A Molecular Approach, pp. 226–246. Sinauer, Sunderland (1990)

[108] Palou, E., Lopez_Malo, A., Barbosa-Canovas, G.V., Welti-Chanes, J., Swanson, B.G.: High hydrostatic pressure as a hurdle for Zygosaccharomyces bailii inactivation. Journal of Food Science 62, 855–857 (1997)

[109] Jay, J.M.: Modern Food Microbiology, 4th edn. Chapman & Hall, New York (1992)

[110] Raso, J., Pagán, R.R., Condón, S.: Nonthermal technologies in combination with other preservation factors. In: Barbosa-Cánovas, G.V., Tapia, M.S., Cano, M.O. (eds.) Novel food processing technologies, pp. 453–476. Marcel Decker/CRC Press, Boca Raton (2005)

[111] Leistner, L.: Basic aspects of food preservation by hurdle technology. International Journal of Food Microbiology 55, 181–186 (2000)

[112] Leistner, L.: Food preservation by combined methods. Food Research International 25, 151–158 (1992)

[113] Leistner, L.: Further developments in the utilization of hurdle technology for food preservation. Journal of Food Engineering 22, 421–432 (1994)
[114] Yousef, A.E.: Efficacy and limitations of non-thermal preservation technologies. IFT Annual Meeting Book of Abstracts, Session 9-1 (2001)
[115] Ross, A.I.V., Griffiths, M.W., Mittal, G.S., Deeth, H.C.: Combining nonthermal technologies to control foodborne microorganisms. International Journal of Food Microbiology 89, 125–138 (2003)
[116] Evrendilek, G.A., Zhang, Q.H., Richter, E.R.: Application of Pulsed Electric Fields to Skim Milk inoculated with Staphylococcus aureus. Biosystems Engineering 87, 137–144 (2004)
[117] Góngora-Nieto, M.M., Sepulveda, D.R., San-Martin, M.F., Barbarosa-Canovas, G.V.: Influence of treatment temperature on the inactivation of Listeria innocua by pulsed electric fields. Lebensmittel-Wissenschaft und Technologie 38, 167–172 (2005)
[118] Calderón-Miranda, M.L., Barbosa-Cánovas, G.V., Swanson, B.G.: Inactivation of Listeria innocua in liquid whole egg by pulsed electric fields and nisin. Int. J. Food Microbiol. 51, 7–17 (1999)
[119] Calderón-Miranda, M.L., Barbosa-Cánovas, G.V., Swanson, B.G.: Inactivation of Listeria innocua in skim milk by pulsed electric fields and nisin. Int. J. Food Microbiol. 51, 19–30 (1999)
[120] Dutreux, N., Notermans, S., Góngora-Nieto, M.M., Barbosa-Cánovas, G.V., Swanson, B.G.: Effects of combined exposure of Micrococcus luteus to nisin and pulsed electric fields. Int. J. Food Microbiol. 60, 147–152 (2000)
[121] Lu, J., Mittal, G.S., Griffiths, M.W.: Reduction in levels of Escherichia coli O157:H7 in apple cider by pulsed electric fields. Journal of Food Protection 64, 964–969 (2001)
[122] Pol, I.E., Maswijk, H.C., Bartels, P.V., Smid, E.J.: Pulsed-electric field treatment enhances the bactericidal action of nisin against Bacillus cereus. Appl. Environ. Microbiol. 66, 428–430 (2000)
[123] Jagus, R., Terebiznik, M.R., Cerrutti, P., Pilosof, A.M.R., Huergo, M.S.: Combined effects of pulsed electric field technology and nisin on E. coli inactivation. IFT Annual Meeting Book of Abstracts, Session 83A-6 (1999)
[124] Smith, K., Mittal, G.S., Griffiths, M.W.: Pasteurization of milk using pulsed electric field and antimicrobials. Journal of Food Science 67, 2304–2308 (2002)
[125] Liang, Z., Mittal, G.S., Griffiths, M.W.: Inactivation of Salmonella Typhimurium in orange juice containing antimicrobial agents by pulsed electric field. Journal of Food Protection 65, 1081–1087 (2002)
[126] Buranasompob, A.: Kinetics of the inactivation of microorganisms by water insoluble polymers with antimicrobial activity. In: Institut für Lebensmitteltechnologie Lebensmittelbiotechnologie und prozesstechnik. Doktor der Ingenieurwissenschaften, p. 135. der Technischen Universität Berlin, Berlin (2005)
[127] Martínez, V., Sobrino López, A., Ben Omar, N., Abriouel, H., Lucas López, R., Valdivia, E., Martín Belloso, O., Gálvez, A.: Enhanced bactericidal effect of enterocin AS-48 in combination with high-intensity pulsed-electric field treatment against Salmonella enterica in apple juice. International Journal of Food Microbiology 128, 244–249 (2008)

[128] Mosqueda-Melgar, J., Raybaudi-Massilia, R.M.R., Martín-Bellos, O.: Non-thermal pasteurization of fruit juices by combining high-intensity pulsed electric fields with natural antimicrobials. Innovative Food Science and Emerging Technologies 9, 328–340 (2008)

[129] Knorr, D., Geulen, M., Grahl, T., Sitzmann, W.: Food application of high voltage field pulses. Trends in Food Science and Technology 5, 71–75 (1994)

[130] Marquez, V.O., Mittal, G.S., Griffiths, M.W.: Destruction and inhibition of bacterial spores by high voltage pulsed electric field. Journal of Food Science 62, 399–401, 409 (1997)

[131] Simpson, M.V., Barbosa-Cánovas, G.V., Swanson, B.G.: The Combined inhibitory effect of lysozyme and high voltage pulsed electric fields on the growth of Bacillus subtilis spores. IFT Annual Meeting: Book of Abstracts 267 (1995)

[132] Spilimbergo, S., Elvassore, N., Bertucco, A.: Microbial inactivation by high-pressure. J. Supercrit. Fluids 22, 55–63 (2002)

[133] Dillow, A., Dehghani, F., Hrkach, J., Foster, N., Langer, R.: Bacterial inactivation by using near- and supercritical carbon dioxide. Proc. Natl. Acad. Sci. USA 96, 10344–10348 (1999)

[134] Spilimbergo, S., Dehghani, F., Bertucco, A., Foster, N.R.: Inactivation of bacteria and spores by pulse electric field and high pressure CO2 at low temperature. Biotechnology and Bioengineering 82, 118–125 (2003)

[135] Pagan, R., Esplugas, S., Gongora-Nieto, M.M., Barbosa-Canovas, G.V., Swanson, B.G.: Inactivation of Bacillus subtilis spores using high intensity pulsed electric fields in combination with other food conservation technologies. Food Science and Technology International 4, 33–44 (1998)

[136] Heinz, V., Knorr, D.: Effect of pH, ethanol addition and high hydrsatic pressure on the inactivation of Bacillus subtilis by pulsed electric field. Inno. Food Sci. Emerging Technol. 1, 151–159 (2000)

[137] Herberhold, H., Fahsel, S., Ulmer, H.M., Gänzle, M.G., Winter, R., Vogel, R.F.: Effects of Pressure-Induced Membrane Phase Transitions on Inactivation of HorA, an ATP-Dependent Multidrug Resistance Transporter, in Lactobacillus plantarum. Applied and Environmental Microbiology 68, 1088–1095 (2002)

[138] Piyasena, P., Mohareb, E., McKellar, R.C.: Inactivation of microbes using ultrasound: a review. International Journal of Food Microbiology 87, 207–216 (2003)

[139] Mason, T.J., Joyce, E., Phull, S.S., Lorimer, J.P.: Potential uses of ultrasound in the biological decontamination of water. Ultras. Sonoch. 10, 319–323 (2003)

[140] Jin, Z.T., Su, Y., Tuhela, L., Singh, B., Zhang, Q.H.: Inactivation of Bacillus subtilis using high voltage pulsed electric fields and ultrasonication. IFT Annual Meeting Book of Abstracts, Session 83A-6 (1998)

[141] Xin, Q., Zhang, X., Li, Z., Lei, L.: Sterilization of oil-field re-injection water using combination treatment of pulsed electric field and ultrasound. Ultrasonics Sonochemistry 16, 1–3 (2009)

[142] Huang, E., Mittal, G.S., Griffiths, M.W.: Inactivation of Salmonella enteritidis in Liquid Whole Egg using Combination Treatments of Pulsed Electric Field, High Pressure and Ultrasound. Biosystems Engineering 94, 403–413 (2006)

[143] Alvarez, I., Heiz, V.: Hurdle technology and the preservation of food by pulsed electric fields. In: Lelieveld, H.L.M., Notermans, S., de Haan, S.W.H. (eds.) Food preservation by pulsed electric fields. From research to applications. Wood publishing limted, Cambridge England (2007)

[144] FDA, Irradiation in the production, processing and handling of food. Federal Register -Rules and Regulations 65(230) (2000)
[145] Noci, F., Riener, J., Walkling-Ribeiro, M., Cronin, D.A., Morgan, D.J., Lyng, J.G.: Ultraviolet irradiation and pulsed electric fields (PEF) in a hurdle strategy for the preservation of fresh apple Juice. Journal of Food Engineering 85, 141–146 (2008)
[146] WHO, Health through safe drinking water and basic sanitation (2008), `http://www.who.int/water_sanitation_health/mdg1/en/index.html`
[147] Gadgil, A.J., Derby, E.A.: Providing Safe Drinking Water to 1.2 Billion Unserved People. Presented at 96th Annual AWMA conference, San Diego, CA (2003)
[148] Vernhesa, M.C., Benichoua, A., Perninb, P., Cabanesc, P.A., Teissi, J.: Elimination of free-living amoebae in fresh water with pulsed electric fields. Water Research 36, 3429–3438 (2002)
[149] Beveridge, J.R., Wall, K., Scott, J., Macgregor, S.J., Anderson, J.G., Rowan, H.J.: Pulsed Electric Field Inactivation of Spoilage Microorganisms in Alcoholic Beverages. Proceeding of the IEEE 92, 1138–1143 (2004)
[150] Doevenspeck, H.: Method for treating beer. USA3625843 (1971)
[151] Ulmer, H.M., Helnz, V., Gänzle, M.G., Knorr, D., Vogel, R.F.: Effects of pulsed electric fields on inactivation and metabolic activity of Lactobacillus plantarum in model beer. J. Appl. Microbiol. 93, 326–335 (2002)
[152] Evrendilek, G.A., Li, S., Dantzer, W.R., Zhang, Q.H.: Pulsed Electric Field Processing of Beer: Microbial, Sensory, and Quality Analyses. Journal of Food Science 69, 228–232 (2004)
[153] Raso, J., Calderon, M.L., Gongora, M., Barbosa-Canovas, G.V., Swanson, B.G.: Inactivation of Zygosaccharomyces bailli in fruit juices by heat, high hydrostatic pressure and pulsed electric fields. Journal of Food Science 63, 1042–1044 (1998)
[154] Raso, J., Calderon, M.L., Gongora, M., Barbosa-Canovas, G.V., Swanson, B.G.: Inactivation of mold ascospores and conidiospores suspended in fruit juices by pulsed electric fields. Journal of Food Science and Technology 31, 668–672 (1998)
[155] Wu, Y., Mittal, G.S., Griffiths, M.W.: Effect of Pulsed Electric Field on the Inactivation of Microorganisms in Grape Juices with and without Antimicrobials. Biosystems Engineering 901, 1–7 (2005)
[156] Vega-Mercado, H., Martin-Belloso, O., Qin, B.-L., Chang, F.-J., Gongora-Nieto, M.M., Barbosa-Cánovas, G.V., Swanson, B.G.: Non-thermal food preservation: pulsed electric fields. Trends Food Sci. Technol. 8, 151–157 (1997)
[157] Evrendilek, G.A., Jin, Z.T., Ruhlman, K.T., Qiu, X., Zhang, Q.H., Richter, E.R.: Microbial safety and shelf-life of apple juice and cider processed by bench and pilot scale PEF systems. Innovative Food Science & Emerging Technologies 1, 77–86 (2000)
[158] Charles-Rodriguez, A.V., Nevaíez-Moorillon, G.V., Zhang, Q.H., Ortega-Rivas, E.: Comparison of thermal processing an pulsed electric fields treatment in pasteurization of apple juice. Food and Bioproducts Processing 87, 93–97 (2007)
[159] Agular Rosa, S.F., Ballinas-Casarubias, M.L., Nevarez-Moorillon, G.V., Martin-Beloso, O., Ortega-Rivas, E.: Thermal and electric fields pasteurization of apple juice: effects of physicochemical properties and flavour compounds. Presented at International Symposium of CIGR Secton IV on Future of Food Engineering N2, Warsaw, Pologne (2007)

[160] Sitzmann, V.: High voltage pulse techniques for food preservation. In: Gould, G.W. (ed.) New methods for food preservation, pp. 236–252. Blackie Academic and Professional, London (1995)

[161] Zhang, Q.H., Qiu, X., Sharma, S.K.: Recent development in pulsed electric field processing. In: National Food Processors Association. New Technologies Yearbook, Washington, DC, pp. 31–42 (1997)

[162] Min, S., Jin, Z.T., Min, S.K., Yeom, H., Zhang, Q.H.: Commercial-Scale Pulsed Electric Field Processing of Orange Juice. Journal of Food Science 68, 1265–1271 (2006)

[163] Cortésa, C., Estevea, M.J., Frígola, A.: Color of orange juice treated by High Intensity Pulsed Electric Fields during refrigerated storage and comparison with pasteurized juice. Food Control 19, 151–158 (2008)

[164] Ayhan, Z., Yeom, H.W., Zhang, Q.H., Min, D.B.: Flavor, color, and vitamin C retention of pulsed electric field processed orange juice in different packaging materials. J. Agric. Food Chem. 49, 669–674 (2001)

[165] Yeom, H.W., Streaker, C.B., Zhang, Q.H., Min, D.B.: Effects of pulsed electric fields on the quality of orange juice and comparison with heat pasteurization. J. Agric. Food Chem. 48 (2000)

[166] Escolà-Hernández, J., Elez-Martínez, P., Soliva-Fortuny, R.C., Martín-Bellos, O.: Inactivation of Lactobacillus brevis in orange juice by high-intensity pulsed electric fields. Food Microbiology 22, 311–319 (2005)

[167] Elez-Martínez, P., Escolà-Hernández, J., Soliva-Fortuny, R.C., Martín-Belloso, O.: Inactivation of Saccharomyces cerevisiae Suspended in Orange Juice Using High-Intensity Pulsed Electric Fields. Journal of Food Protection 67, 2596–2602 (2004)

[168] Mc Donald, C.J., Lloyd, S.W., Vitale, M.A., Petersson, K., Innings, F.: Effects of Pulsed Electric Fields on Microorganisms in Orange Juice Using Electric field Strengths of 30 and 50 kV/cm. Journal of Food Science 65, 984–989 (2000)

[169] El-Hag, A.H., Jayaram, S.H., Griffiths, M.W.: Inactivation of Naturally Grown Microorganisms in Orange Juice Using Pulsed Electric Fields. IEEE transactions on plasma science 34, 1412–1415 (2006)

[170] Min, S.M., Jin, Z.T., Zhang, Q.H.: Commercial scale pulsed electric field processing of tomato juice. J. Agric. Food Chem. 51, 3338–3344 (2003)

[171] Odriozola-Serrano, I., Aguiló-Aguayo, I., Soliva-Fortuny, R., Gimeno-Añó, V., Martín-Belloso, O.: Lycopene, Vitamin C, and Antioxidant Capacity of Tomato Juice as Affected by High-Intensity Pulsed Electric Fields Critical Parameters. J. Agric. Food Chem. 55, 9036–9042 (2007)

[172] Nguyen, P., Mittal, G.S.: Inactivation of naturally occurring microorganisms in tomato juice using pulsed electric field (PEF) with and without antimicrobials. Chemical engineering and processing 46, 360–365 (2007)

[173] Dunn, J.E., Pearlman, J.S.: Methods and apparatus for extending the shelf-life of fluid food products. I. Maxwell Laboratories, Ed. U. S. Patent 4,695,472 (1987)

[174] Dunn, J.: Pulsed light and pulsed electric field for foods and eggs. Poul Sci. 75, 1133–1136 (1996)

[175] Fernandez-Molina, J.J., Barkstrom, E., Torstensson, P., Barbosa-Cánovas, G.V., Swanson, B.G.: Shelf-life extension of raw skim milk by combining heat and pulsed electric fields. Food Res. Int. (1999)

[176] Qin, B., Pothakamury, U.R., Vega, H., Martin, O., Barbosa-Cánovas, G.V., Swanson, B.G.: Food pasteurization using high intensity pulsed electric fields. J. Food Technol. 49, 55–60 (1995)

[177] Calderon-Miranda, M.L.: Inactivation of listeria inocua by pulsed electric fields and nisin. Washington State University, Pullman (1998)
[178] Reina, L.D., Jin, Z.T., Yousef, A.E., Zhang, Q.H.: Inactivation of Listeria monocytogenes in milk by pulsed electric field. J. Food Protect. 61, 1203–1206 (1998)
[179] Shamsi, K., Versteeg, C., Sherkat, F., Wan, J.: Alkaline phosphatase and microbial inactivation by pulsed electric field in bovine milk. Innovative Food Science and Emerging Technologies 9, 217–223 (2008)
[180] Craven, H.M., Swiergon, P., Midgely, J., Ng, S., Versteeg, C., Coventry, M.J., Wan, J.: Evaluation of pulsed electric field and minimal heat treatments for inactivation of pseudomonads and enhancement of milk shelf-life. Innovative Food Science and Emerging Technologies 9, 211–216 (2008)
[181] Noci, F., Walkling-Ribeiro, M., Cronin, D.A., Morgan, D.J., Lyng, J.G.: Effect of thermosonication, pulsed electric field and their combination on inactivation of Listeria innocua in milk. International Dairy Journal, 1–6 (2008)
[182] Otunola, A., El-Hag, A., Jayaram, S., Anderson, W.A.: Effectiveness of Pulsed Electric Fields in Controlling Microbial Growth in Milk. International Journal of Food Engineering 4, Article 1 (2008), doi:10.2202/1556-3758.1494
[183] Sampedro, F., Rodrigo, D., Martineza, A., Barbarosa-Canovas, G.V., Rodrigo, M.: Review: Application of pulsed electric fields in egg and egg derivatives. Food Science and Technology International 12, 397–405 (2006)
[184] Sampedro, F., Rivas, A., Rodrigo, D., Martınez, A., Rodrigo, M.: Pulsed electric fields inactivation of Lactobacillus plantarum in an orange–juice milk based beverage: Effect of process parameters. Journal of Food Engineering 80, 931–938 (2007)
[185] Chullkyoon, M., Sangki, L.: Sterilization of Yakju (Rice Wine) on a Serial Multiple Electrode Pulsed Electric Field Treatment System. Korean J. of Food Science and Technology 32, 356–362 (2000)
[186] Su, Y.K., Yong, S.P., Chulkyoon, M.: Sterilisation of yakju (rice wine) using a batch-type high voltage pulsed field system. Korean J. of Food Science and Technology 31, 1247–1253 (1999)
[187] Poppe, C.: Epidemiology of Salmonella enterica serovar Enteritidis. In Salmonella enterica Serovar Enteritidis. In: Gast, R.K., Saeed, A.M., Potter, M.E., Wall, P.G. (eds.) Humans and Animals, Epidemiology, Pathogenesis, and Control, pp. 3–19. Iowa State University Press, Ames (1999)
[188] Cunningham, F.E.: Egg product pasteurization. In: Cotterill, O.J., Stadelman, W.J. (eds.) Egg Science and Technology, pp. 289–323. AVI Publishing Co., Westport (1986)
[189] Foegeding, P.M., Leasor, S.: Heat resistance and growth of Listeria monocytogenes in liquid whole egg. J. Food Prot. 53, 9–14 (1990)
[190] Ma, L., Chang, F.J., Barbosa-Cánovas, G.V.: Inactivation of E. coli in liquid whole eggs using pulsed electric fields technologies. New frontiers in food engineering. In: Proceedings of the Fifth Conference of Food Engineering. American Institute of Chemical Engineers, pp. 216–221 (1997)
[191] Hermawan, N., Evrendilek, G.A., Dantzer, W.R., Zhang, Q.H., Richter, E.R.: Pulsed electric field treatment of liquid whole egg inoculated with salmonella enteritidis. Journal of Food Safety 24, 71–85 (2007)

[192] Sampedrdo, F., Rodrigo, D., Martineza, A., Barbarosa-Canovas, G.V., Rodrigo, M.: Review: Application of pulsed electric fields in egg and egg derivatives. Food Science and Technology International 12, 397–405 (2006)

[193] Zhao, W., Yang, R., Lu, R., Wang, M., Qian, P., Yang, W.: Effect of PEF on microbial inactivation and physical–chemical properties of green tea extracts. LWT 41, 425–431 (2008)

[194] Wang, L.F., Kim, D.M., Chang, L.Y.: Effects of heat processing and storage on flavanols and sensory qualities of green tea beverage. Journal of Agriculture and Food Chemistry 48, 4227–4232 (2000)

[195] Evrendilek, G.A., Tok, F.M., Soylu, E.M., Soylu, S.: Inactivation of Penicillium expansum in sour cherry juice, peach and apricot nectars by pulsed electric fields. Food Microbiology 25, 662–667 (2008)

Author Index

Author Index

MIX
Papier aus verantwortungsvollen Quellen
Paper from responsible sources
FSC® C105338

If you have any concerns about our products,
you can contact us on
ProductSafety@springernature.com

In case Publisher is established outside the EU,
the EU authorized representative is:
Springer Nature Customer Service Center GmbH
Europaplatz 3, 69115 Heidelberg, Germany

Printed by Libri Plureos GmbH
in Hamburg, Germany